OVARIAN FOLLICULAR DEVELOPMENT AND FUNCTION

Ovarian Follicular Development and Function

Editors

A. Rees Midgley, M.D.
Professor
Department of Pathology
Reproductive Endocrinology Program
University of Michigan
Ann Arbor, Michigan

William A. Sadler, Ph.D.
Chief
Population and Reproduction Branch
Center for Population Research
National Institute for Child Health and Human Development
National Institutes of Health
Bethesda, Maryland

Raven Press ▪ New York

Raven Press, 1140 Avenue of the Americas, New York, New York 10036

Made in the United States of America

International Standard Book Number 0-89004- 186-5
Library of Congress Catalog Card Number 77-17750

Preface

This meeting was possible only because the Center for Population Research of the National Institute for Child Health and Human Development chose to provide from highly limited discretionary funds sufficient support to permit the attendance of many of the participants from the United States and Canada. We hope this book will justify the wisdom of that decision and will encourage them to continue to support similar workshops in the future. Grateful appreciation is also extended to Mr. Antonio Scarpa of Serono Ltd. who made it possible for several European scientists to attend. In addition, the editors wish to acknowledge the special efforts of Dr. Robert B. Jaffe in arranging for the meeting to be held in Santa Cruz and the efforts of other members of the Organizing Committee (Dr. Georgianna Jagiello, Dr. Bruno Lunenfeld, Dr. A.V. Nalbandov, Dr. Griff T. Ross, Dr. Kenneth J. Ryan, and Dr. Robert J. Ryan) whose planning assistance and chairing of sessions contributed to all phases of the workshop. Finally, we wish to give special acknowledgment to Ms. Susan Bareis. To a large extent, her organizational efforts were responsible for making this workshop so successful.

Acknowledgments

Few areas in biology have received greater attention in recent years than that of ovarian follicular development and function. Buried within the complex of interactions that result in the selection for ovulation of but a very few oocytes from the thousands available are the potentialities for fertility regulation at many levels. The problems of hormonally and nonhormonally controlled progressive functional and morphologic differentiation of thecal cells, granulosa cells and oocytes; exogenous and endogenous hormonal regulation of receptors for protein and steroid hormones; eventual death through atresia or rescue by ovulation and luteinization are complex but the solutions will be highly relevant to our understanding not only of reproductive biology but of eucaryotic regulatory biology in general. These problems have relevancy for biologists in many disciplines.

Largely as a consequence of advances in technology, work on these problems has been progressing rapidly in many unrelated disciplines. In an attempt to bring together representatives of these disciplines, to define important problems ripe for experimental investigation, and to achieve collectively a better understanding of the latest developments in the field, a workshop was held at the University of California at Santa Cruz. The workshop consisted of eight formal presentations, each followed by a number of brief contributions and an active discussion. To facilitate discussion written versions of all presentations and contributions were circulated to the participants in advance of the meeting. For this publication the presentors, working with typed transcripts, summarized the discussion following their presentation. Thus this book consists of eight sections, each prepared by a different presentor and each consisting of the presentor's text, contributions largely submitted in advance of the workshop, and a brief summary reflecting the discussions following each presentation. If the readers of this book realize even a small amount of the excitement and understanding achieved by participants at the workshop, this publication will have fulfilled its purpose.

Contents

OOGENESIS

Contributors

Barry D. Albertson
Thorndike Memorial Laboratory
Section of Endocrinology and Metabolism
Boston City Hospital
Boston University School of Medicine
Boston, Massachusetts 02118

A. Amsterdam
Department of Hormone Research
Weizmann Institute of Science
Rehovot, Israel

Everett Anderson
Department of Anatomy and Laboratory of Human Reproduction and Reproductive Biology
Harvard Medical School
Boston, Massachusetts 02115

Winston A. Anderson
Department of Zoology and Howard University Cancer Research Center
Howard University
Washington, D.C.

David T. Armstrong
Departments of Physiology and Obstetrics and Gynaecology
University of Western Ontario
London, Canada

T.G. Baker
Hormone Laboratory
Department of Obstetrics and Gynaecology
University of Edinburgh
Edinburgh EH3 9ER, Scotland

Lee Beck
Department of Pediatrics
University of Alabama Medical Center
University Station
Birmingham, Alabama 35294

William H. Beers
The Rockefeller University
New York, New York 10021

John D. Biggers
Laboratory of Human Reproduction and Reproductive Biology and
Department of Physiology
Harvard Medical School
Boston, Massachusetts 02115

Lutz Birnbaumer
Department of Cell Biology
Baylor College of Medicine
Houston, Texas 77030

J. Blankstein
Institute of Endocrinology
Chaim Sheba Medical Center
Tel Hashomer
and
Department of Life Sciences
Bar Ilan University
Ramat Gan, Israel

G. Eric Bradford
Animal Science Department
University of California at Davis
Davis, California

Anne Grete Byskov
The Finsen Laboratory
The Finsen Institute
Copenhagen, Denmark

S.C. Chang
Department of Molecular Medicine
Mayo Medical School
Rochester, Minnesota 55901

Cornelia P. Channing
Department of Physiology
University of Maryland School of Medicine
Baltimore, Maryland 21201

James R. Clark
Department of Animal Science
Texas Tech University
Lubbock, Texas 79409

Martin R. Clark
Departments of Obstetrics and Gynaecology and
The Endocrine Laboratory
University of Miami School of Medicine
Miami, Florida

W.A. Condon
University of Pennsylvania
New Bolton Center
Section of Reproductive Studies
Kennett Square, Pennsylvania 19348

P.B. Coulson
Department of Zoology
Division of Life Sciences
University of Tennessee
Knoxville, Tennessee 37916

Frances E. DeLisle
Department of Research
Sinai Hospital of Detroit
Detroit, Michigan

Donald J. Dierschke
Wisconsin Regional Primate Research Center
University of Wisconsin
Madison, Wisconsin 53706

Michael J. Dimino
Department of Research
Sinai Hospital of Detroit
Detroit, Michigan

Jennifer H. Dorrington
Banting and Best Department of Medical Research
University of Toronto
Toronto, Canada

James R. Downing
Department of Research
Sinai Hospital of Detroit
Detroit, Michigan

B.G. England
Department of Pathology
University of Michigan
Ann Arbor, Michigan 48104

Lawrence L. Espey
Trinity University
and
The University of Texas Health Science Center
San Antonio, Texas 78284

William H. Fletcher
Department of Anatomy
Duke University Medical Center
Durham, North Carolina 27710

Joanne E. Fortune
Departments of Obstetrics and Gynaecology and Physiology
University of Western Ontario
London, Canada

V.K. Ganjam
University of Pennsylvania
New Bolton Center
Section of Reproductive Studies
Kennett Square, Pennsylvania 19348

Irving I. Geschwind
Animal Science Department
University of California at Davis
Davis, California

Alan K. Goff
Departments of Physiology and Obstetrics and Gynaecology
University of Western Ontario
London, Canada

Arnold L. Goodman
Section on Endocrinology
Reproduction Research Branch
National Institue of Child Health and Human Development
National Institues of Health
Bethesda, Maryland 20014

Gilbert S. Greenwald
Departments of Obstetrics and Gynaecology and Anatomy
University of Kansas Medical Center
Kansas City, Kansas 66103

Howard S. Grob
Adelphi University
Garden City, New York

Seong S. Han
Reproductive Endocrinology Program
Departments of Anatomy and Oral Biology
University of Michigan
Ann Arbor, Michigan 48109

E.R. Hauser
Department of Meat and Animal Science
University of Wisconsin
Madison, Wisconsin 53706

Anne N. Hirschfield
Reproductive Endocrinology Program
University of Michigan
Ann Arbor, Michigan

Gary D. Hodgen
Section on Endocrinology
Reproduction Research Branch
National Institute of Child Health and Human Development
National Institutes of Health
Bethesda, Maryland 20014

Mary Hunzicker-Dunn
Department of Biochemistry
Northwestern University Medical School
Chicago, Illinois 60611

Georgiana M. Jagiello
Columbia University
College of Physicans and Surgeons
New York, New York 10032

Richard A. Jungmann
Department of Biochemistry
Northwestern University Medical School
Chicago, Illinois 60611

Yuan-Hsu Kang
Department of Zoology
and
Howard University Cancer Research Center
Howard University
Washington, D.C.

R.M. Kenney
University of Pennsylvania
New Bolton Center
Section of Reproductive Studies
Kennett Square, Pennsylvania 19348

Eric Knox
Department of Pediatrics
University of Alabama Medical Center
University Station
Birmingham, Alabama 35294

Y. Koch
Department of Hormone Research
The Weizmann Institute of Science
Rehovot, Israel

Z. Kraiem
Institute of Endocrinology
Chaim Shelba Medical Center
Tel-Hashomer, Israel

S. Lamprecht
Department of Hormone Research
The Weizmann Institute of Science
Rehovot, Israel

Florence Ledwitz-Rigby
Department of Biological Sciences
Northern Illinois University
DeKalb, Illinois 60115

William J. LeMaire
Department of Obstetrics and Gynaecology
and
The Endocrine Laboratory
University of Miami
School of Medicine
Miami, Florida 33152

H.R. Lindner
Department of Hormone Research
The Weizmann Institute of Science
Rehovot, Israel

Ada M. Lindsey
Department of Physiology
University of Maryland School of Medicine
Baltimore, Maryland 21201

Suzanne Lowry
Departments of Obstetrics and Gynaecology
University of Alabama Medical Center
University Station
Birmingham, Alabama 35294

B. Lunenfeld
Institute of Endocrinology
Chaim Shelba Medical Center
Tel-Hashomer, Israel

John M. Marsh
Departments of Obstetrics and Gynaecology
and
The Endocrine Laboratory
University of Miami School of Medicine
Miami, Florida

Eileen A. Merz
Department of Meat and Animal Science
University of Wisconsin
Madison, Wisconsin 53706

Thomas M. Mills
Department of Endocrinology
Medical College of Georgia
Augusta, Georgia 30902

Patricia L. Morris
New York University
New York, New York

A. Nimrod
Department of Hormone Research
Weizmann Institute of Science
Rehovot, Israel

Wilbert E. Nixon
Section on Endocrinology
Reproduction Research Branch
National Institute of Child Health and Human Development
National Institutes of Health
Bethesda, Maryland 20014

John J. Peluso
Reproductive Physiology Laboratories
and
C.S. Mott Center for Human Growth and Development
Wayne State University School of Medicine
Detroit, Michigan 48201

Hannah Peters
The Finsen Laboratory
The Finsen Institute
Copenhagen, Denmark

R. Douglas Powers
Laboratory of Human Reproduction and Reproductive Biology
and
Department of Physiology
Harvard Medical School
Boston, Massachusetts 02115

Mrinalini C. Rao
Reproductive Endocrinology Program
Department of Pathology
University of Michigan
Ann Arbor, Michigan 48109

Jon M.R. Rawson
Trinity University
and
The University of Texas Health Science Center
San Antonio, Texas 78284

JoAnne S. Richards
Reproductive Endocrinology Program
Department of Pathology
University of Michigan
Ann Arbor, Michigan 48109

R.J. Ryan
Department of Molecular Medicine
Mayo Medical School
Rochester, Minnesota 55901

Y. Salomon
Department of Hormone Research
The Weizmann Institute of Science
Rehovot, Israel

James R. Schreiber
Reproduction Research Branch
National Institute of Child Health and Human Development
Bethesda, Maryland 20014

Ralph A. Sorensen
Department of Biological Sciences
DePaul University
Chicago, Illinois 60614

Jimmy L. Spearow
Animal Science Department
University of California at Davis
Davis, California

Sidney Strickland
The Rockefeller University
New York, New York

K.H. Thanki
Department of Physiology
School of Medicine
University of Maryland at Baltimore
Baltimore, Maryland 21201

Judith L. Vaitukaitis
Thorndike Memorial Laboratory
Section of Endocrinology and Metabolism
Boston City Hospital
Boston University School of Medicine
Boston, Massachusetts 02118

R. Weissenberg
Institute of Endocrinology
Chaim Sheba Medical Center
Tel Hashomer, Israel
and
Department of Life Sciences
Bar Ilan University
Ramat Gan, Israel

Richard F. Wilkinson, Jr.
Laboratory of Human Reproduction and Reproductive Biology
and
Department of Anatomy
Harvard Medical School
Boston, Massachusetts 02115

Richard C. Wolf
Wisconsin Regional Primate Research Center
University of Wisconsin
Madison, Wisconsin 53706

Debra J. Wolgemuth-Jarashow
Columbia University College of Physicans and Surgeons
New York, New York 10032

E.V. Younglai
Department of Obstetrics and Gynaecology
Reproductive Biology Program
McMaster University Medical Center
Hamilton, Ontario, Canada L8S 4J9

Ovarian Follicular Development and Function,
edited by A. R. Midgley and W. A. Sadler.
Raven Press, New York © 1979.

Some Aspects of Early Follicular Development

Hannah Peters

The Finsen Laboratory, The Finsen Institute
Copenhagen, Denmark

Follicular development represents one of the essential functions of the ovary. It depends on the interplay of several control mechanisms, which lie partly outside the ovary--mainly in the pituitary and the central nervous system--and partly within the ovary and the follicle itself.

Follicular development has many facets, including (1) follicular organization, which establishes the pool of nonproliferating small follicles and (2) follicular growth, which begins with the follicle's emergence from the nongrowing state and at times results in the cyclic event of ovulation, but most often ends in atresia.

FOLLICULAR ORGANIZATION

The Establishment of the Nonproliferating Follicular Pool

The majority of the germ cells are found enclosed in small nongrowing follicles in the ovary and form a most important (and perhaps a most neglected) portion of the total follicle stock. How is the stock of nonproliferating follicles established? In most mammals at birth the oocytes are lying close together in cell nests and are the most conspicuous within the ovary at that time, although they share the space within the organ with a fine and intricate network of cords and tubules, i.e., the intraovarian rete. Byskov and Lintern-Moore (1973) have shown that the oocytes actually lie within this rete system, and that cells from the rete become attached to the surface of the oocyte, thus establishing the first granulosa cells. In the early stage of follicular formation, there are open connections between the forming follicles and the rete tubules, but at a certain point the basement membrane surrounding the oocyte and the cells that have become attached to it closes, and the follicle becomes an independent unit. These units, consisting of a small oocyte, the cells attached to it, and the basement membrane surrounding them, become the pool of nonproliferating follicles, from which the follicle emerges when it starts to grow.

The rete ovarii within which the oocytes originally lie represents the rest of mesonephric tubules, which in the male develops into the epididymus. The epididymus is known to have the ability to transform steroid precursors such as acetate and cholesterol into testosterone as well as to change testosterone into various metabolites (Fawcett et al., 1969; Rivarola and Podesta, 1972). Therefore, it is understandable that the stroma of the ovary, part of which consists of mesonephric tubule derivatives, i.e., the rete, might later become implicated in androgen metabolism.

Influence of the Small Follicular Pool on Growth Initiation

Very little is known about the nonproliferating small follicles and their influence on ovarian physiology. One of the few known facts is that their total number has an influence on the number of follicles that start to grow. The larger their number, the larger the number of follicles that begin to develop (Krohn, 1967; Krarup et al., 1969). With advancing age, the number of small follicles is reduced (Jones and Krohn, 1961) and so is the number of follicles that start to grow per unit time (Pedersen, 1972). That the pool size, i.e., the number of small, nonproliferating follicles, is apparently the main controlling factor in the immature animal can be shown by artificially reducing their number in infancy. This can be accomplished by giving a single small dose (20 R) of radiation or a single oral dose of 9:10 dimethyl 1:2 benzanthracene (DMBA) to young mice. The immediate reduction of the number of small follicles is followed by a secondary reduction in the number of developing follicles (Peters, 1969a; Krarup et al., 1969). How this control is mediated is at present unknown, and whether the mass of small follicles acts via the feedback mechanism or directly within the ovary cannot be determined at this time.

However, that the (small) follicular pool size is not the only controlling factor is likely if we consider the number of follicles that start to grow in the mouse ovary per 24 hr at different ages (Pedersen,

1972). There is a rapid reduction in the number of follicles beginning to develop between 1 and 3 weeks (Table 1). The very high starting rate during the immediate neonatal period falls to half 3 weeks later and remains approximately constant in the mature cycling animal. This suggests that in addition to the

TABLE 1. *Number of follicles starting to grow in the mouse ovary per 24-hr period at different ages (Pedersen, 1972)*

Weeks					Months				
1	2	3	4	5	3	3*	9	12	16
41	31	21	20	17	19	14	10	10	3

*Pregnant

pool size another factor restraining growth initiation becomes active. This factor is apparently present in the follicular fluid of degenerating follicles. Fluid-containing large follicles are not present during the first week of life and growth initiation is high. However, if fluid from large follicles is injected into neonatal mice, growth initiation during the first week is markedly reduced (Peters et al., 1973a). The factor involved in growth initiation restraint has not yet been identified.

FOLLICULAR GROWTH

When a follicle starts to grow, its oocyte begins to enlarge and granulosa cells begin to multiply. Whether a stimulus is necessary for the individual follicle to start to develop or the removal of an inhibitor is involved is not clear. It is doubtful that gonadotropins or estrogens are involved. Exogenous gonadotropin or the injection of estradiol benzoate (EB) failed to increase the number of follicles in the beginning stages (Table 2; Peters et al., 1973b). Injection of 1 IU PMSG three times a day or 0.25 μg EB once daily for 4 days between days 3 and 6 did not alter the number of follicles starting to grow. Nor does deprivation of endogenous gonadotropin by antigonadotropin prevent follicular growth initiation (Eshkol et al. 1970).

Follicular growth initiation is continuous throughout the animal's life: it is established during infancy, when ovulation has not yet started (Peters, 1969b); during

TABLE 2. *Number of follicles with more than 20 cells on largest cross section on day 6*

0.25 μg EB daily: days 3-6	0.1 ml Sesame oil	3 IU PMSG daily: days 3-6	0.1 ml NaCl
110 ± 9	96 ± 10	121 ± 14	114 ± 13

maturity when ovulation occurs follicular growth continues through all stages of the cycle (Pedersen, 1970b); and during pregnancy, when ovulation is held in abeyance, follicles continue to develop (Greenwald et al., 1967; Schwartz and Talley, 1968; Govan, 1968, 1970; Pedersen and Peters, 1971). Follicular growth process (Peters et al., 1975), on which the cyclic surges of FSH and LH are superimposed leading to ovulation (Richards and Midgley, 1976).

Early Stages of Follicular Development

Early stages of follicular growth depend on gonadotropin in the sense that in its absence the ordered growth of the follicle is disturbed. Studies of infant mice showed that deprivation of endogenous gonadotropin by injection of antigonadotropin leads to an abnormal development of the granulosa layer and a lack of thecal development, which is corrected by exogenous replacement of the gonadotropin (Eshkol et al., 1970; Eshkol and Lunenfeld, 1971; Hardy et al., 1974; Purandare et al., 1974). Similar gonadotropin experiments in rats failed to alter granulosa and theca cell development but led to a deficient interstitial cell development in the antiserum-treated infant rats (Schwartz et al., 1974). However, antigonadotropic studies designed to localize specific intraovarian binding sites for FSH and LH in the immature rat ovary could show that ^{125}I-FSH bound to follicles at many stages of development including those with one and two layers of granulosa cells, i.e., follicles that are in early developmental stages (Richards and Midgley, 1976). This strongly suggests that early growing follicles in the rat also have the equipment necessary to react to FSH.

X

Theca Development

Early follicular development includes oocyte growth with active RNA synthesis within the oocyte (Oakberg, 1968; Moore et al., 1974), the formation of the zona pellucida, and the multiplication of granulosa cells. It further includes the differentiation of the theca cells. Although little is known about the formation of this layer, Midgley et al. (1974) suggested that the cells might differentiate in response to a theca cell organizer produced by the granulosa cells. Cells that can be defined as theca cells appear early during follicular growth. Follicles with a growing oocyte and one layer of granulosa cells (Type 3b) show cells that lie in close proximity to the basement membrane, which delineates the granulosa layer (Fig. 1). These are the earliest cells populating the theca layer. In early preantral stages the theca layer is often well defined (Fig. 2). At later stages, as the antrum forms, theca cells increase in size and number (Figs. 3 and 4). In some species the theca layer develops early and is almost as wide as the granulosa layer (Fig. 5). Perhaps one of the reasons why the definition of the theca layer has been difficult and uncertain is that this layer is most sensitive to and dependent on a fine interplay of hormones to succeed in its full development. Therefore, not all follicles in comparable stages of development within the same ovary

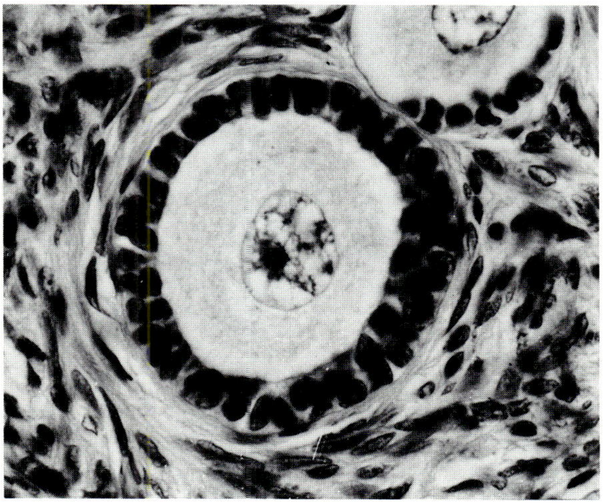

FIG. 1. An early growing follicle. A theca layer begins to form. Ovary of a 4-year-old child. ×600.

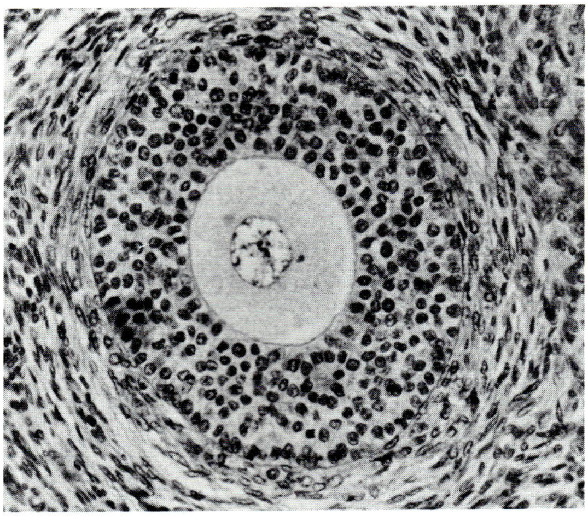

FIG. 2. A preantral follicle with many cells forming a theca layer. Ovary of a 7-year-old child. ×245.

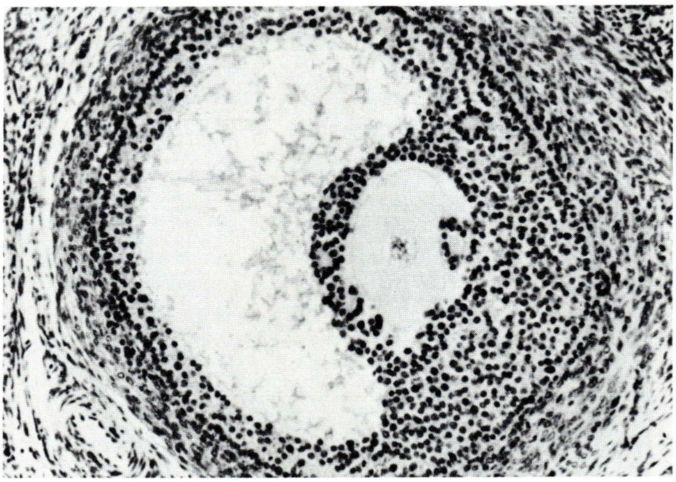

FIG. 3. In early antral follicles theca cells increase in number and size. Ovary of a 4-year-old child. ×155.

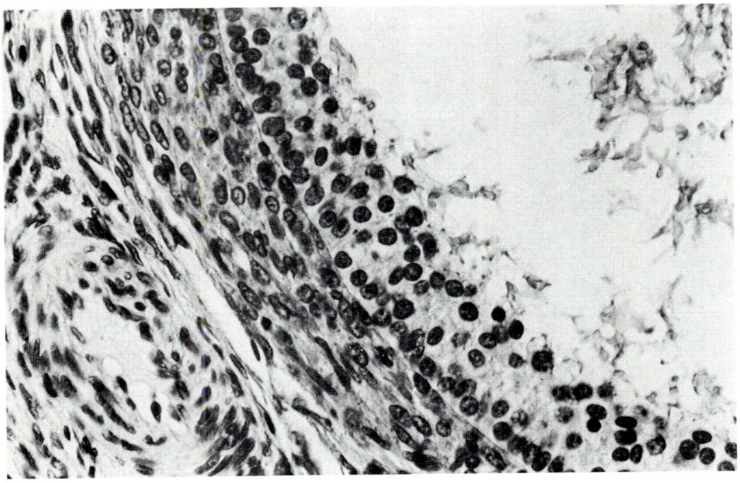

FIG. 4. Ovary of a four-year-old child. The theca layer is wide and well defined. X 420

show similar thecal layer development. Most characteristic are early growing follicles in which the theca layer is totally absent. These are invariably follicles that do not develop any further. Thus, it is likely that a competent theca layer is already necessary in early follicular growth.

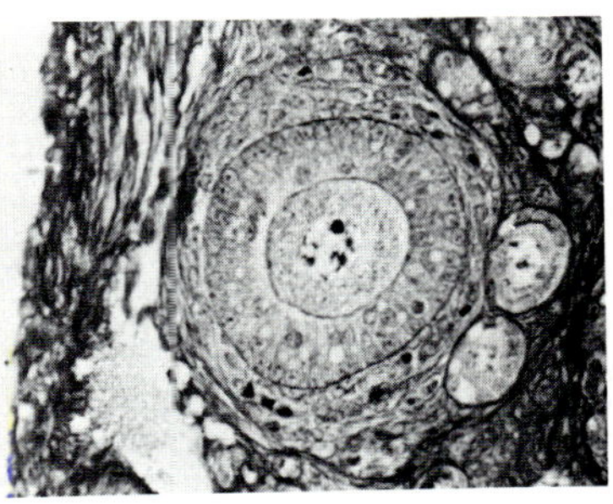

FIG. 5. In same species, the theca layer is as wide as the granulosa layer in early developing follicles. Djungarian hamster (Phodopus sungorus cambelli), 32 days old. ×245.

Continued Follicular Development

Gonadotropin is necessary not only for the ordered development during the early stages of follicular evolution but most critically to support continued follicular growth. Deprivation of gonadotropins by hypophysectomy leads to atresia of medium and large follicles (Ingram, 1953). Simultaneous injection of gonadotropin prevents atresia and maintains follicular growth (Welschen, 1973). It has been suggested that FSH stimulates renewed growth and full development of large follicles (Schwartz, 1969; Greenwald, 1974) that are held in reserve (Greenwald, 1973). However, a pool of resting large follicles does not seem to reside in the ovary: follicular growth is sequential and continuous until atresia or ovulation occurs (Peters and Levy, 1966; Pedersen, 1972). It has recently been suggested that the action of FSH is not to restart growth of resting follicles but rather to prevent granulosa cell degeneration and to support continued follicular growth. It was shown that injection of PMSG into 21-day-old mice, in whom large follicle atresia is at a maximum, did not increase the total number of large follicles, but changed the balance between healthy and atretic follicles (Peters et al., 1975). Recent autoradiographic studies gave further information on the effect of PMSG on the growing follicles (S. Hoffmann, unpublished). Twenty-one-day-old mice were injected either with 5 IU PMSG or with 0.1 ml NaCl and killed 12 hr later. One hour before death they received 50 μCi ^{3}H-thymidine. Autoradiograms were prepared of the serially sectioned ovaries, follicle counts (Types 5a and 6) were done, and a distinction was made between healthy (no pyknotic cells among the granulosa cells) and atretic follicles (one or more pyknotic granulosa cells on the largest cross section of the follicle). The labelling index, i.e., the percent of labelled granulosa cells, was determined. A change in the labelling index would indicate a change in the growth rate of the follicle (Pedersen, 1970a). The mitotic index, i.e., percent of mitosis among the granulosa cells, was noted. Among the atretic follicles, the pyknotic index, i.e., percent of pyknotic cells among the granulosa cells, was determined.

Twelve hours after PMSG injection the total number of large follicles was comparable in the experimental and control groups (Table 3), but considerably fewer atretic follicles were seen 12 hr after PMSG injection. Only about half the follicles were healthy in the control animals, while about 80% remained healthy after

TABLE 3. *Effect of 5 IU PMSG on large follicles in 21-day-old mice*

Animals	No.	Total number large follicles (Type 5a and 6) Mean ± SEM	Atretic follicles (%) Mean ± SEM	Labelling index granulosa cells (%) Mean ± SEM	Mitotic index granulosa cells (%) Mean ± SEM	Pyknotic index atretic follicles (%) Mean ± SEM
0.1 ml NaCl plus 12 hr (control)	4	67 ± 13.7	49.9 ± 9.1	16.9 ± 0.7	1.7 ± 0.1	8.9 ± 0.6
5 IU PMSG plus 12 hr	5	73.0 ± 14.2	16.7[a] ± 4	23.1[a] ± 0.7	2.9[a] ± 0.3	3.1[a] ± 0.4

[a]Significantly different from its control ($p < 0.001$).

injection of gonadotropin. In those follicles that were atretic, the pyknotic index had fallen from 9% in the controls to 3%, and the mitotic index had increased. The labelling index rose after PMSG injection, suggesting that more granulosa cells had entered the cell cycle. It thus appears that PMSG acted on the granulosa cells by preventing them from becoming atretic and by inducing more cells to proliferate. The atresia-preventing action of FSH on granulosa cells has also been shown in tissue culture (McNatty et al., 1975). Human granulosa cells remained viable for 10 days when grown in serum devoid of LH or prolactin. But they degenerated when the serum was devoid of FSH. Daily addition of FSH to the serum devoid of FSH prevented the granulosa cell atresia.

CONCLUSIONS

Many facets of early follicular development remain unclear. We wanted to stress the importance of follicular formation that occurs early in life and establishes the pool of nonproliferating small follicles remaining dormant until follicular growth is initiated. The early growing follicles depend on gonadotropins for the normal development of the granulosa and theca layers. The theca layer develops early and is apparently essential for further development. In preantral and early antral development, gonadotropins specifically protect against atresia of granulosa cells and sustain continued sequential follicular development.

ACKNOWLEDGMENT

I wish to thank Ruth Himelstein-Braw and Steen Hoffmann for their valuable contributions. Published and unpublished observations from this laboratory have been supported in part by Euratom contract 120-73-1 BIODK.

REFERENCES

Byskov, A. G., and Lintern-Moore, S. (1973): Follicle formation in the immature mouse ovary: The role of the rete ovarii. *J. Anat.*, 116:207-217.

Eshkol, A., and Lunenfeld, B. (1971): Biological effects of antibodies to antigonadotrophins, hormones and antagonists. *Gynecol. Invest.*, 2:23-56.

Eshkol, A., Lunenfeld, B., and Peters, H. (1970): Ovarian development in infant mice. Dependence on gonadotrophic hormones. In: *Gonadotrophins and Ovarian Development*, edited by W. R. Butt, A. C. Crooke, and M. Ryle, pp. 249-258. Livingstone, London.

Fawcett, D. W., Long, J. A., and Jones, A. L. (1969): The ultrastructure of endocrine glands. *Recent Progr. Horm. Res.*, 25:315-380.

Govan, A. D. T. (1968): The human ovary in early pregnancy. *J. Endocrinol.*, 40:421-428.

Govan, A. D. T. (1970): Ovarian follicular activity in late pregnancy. *J. Endocrinol.*, 48:235-241.

Greenwald, G. S., Keever, J. E., and Grady, K. L. (1967): Ovarian morphology and pituitary FSH and LH concentration in the pregnant and lactating hamster. *Endocrinology*, 80:851-856.

Greenwald, G. S. (1973): Distinction between developing and reserve follicles in the cycling hamster. *Ann. Biol. Anim. Biochim. Biophys.*, 13:199-210.

Greenwald, G. S. (1974): Quantitative aspects of follicular development in the untreated and PMS-treated cyclic hamster. *Anat. Rec.*, 178:139-144.

Hardy, B., Danon, D., Eshkol, A., and Lunenfeld, B. (1974): Ultrastructural changes in the ovaries of infant mice deprived of endogenous gonadotrophins after substitution with FSH. *J. Reprod. Fertil.*, 36:345-352.

Ingram, D. L. (1953): The effect of hypophysectomy on the number of oocytes in the adult albino rat. *J. Endocrinol.*, 9:307-311.

Jones, E. C., and Krohn, P. L. (1961): The relationship between age, number of oocytes and fertility in virgin and multiparous mice. *J. Endocrinol.*, 21:469-495.

Krarup, T., Pedersen, T., and Faber, M. (1969): Regulation of oocyte growth in the mouse ovary. *Nature*, 224:187-188.

Krohn, P. L. (1967): Factors influencing the number of oocytes in the ovary. *Arch. Anat. Micros. Morphol. Exp.*, 56:151-159.

McNatty, K. P., Hunter, W. M., McNeilly, A. S. and Sawers, R. S. (1975): Changes in the concentration of pituitary and steroid hormones in the follicular fluid of graafian follicles throughout the menstrual cycle. *J. Endocrinol.*, 64:555-571.

Midgley, A. R., Jr., Zeleznik, A. J., Richards, J. D., and Han, S. S. (1974): Gonadotropin and estradiol receptors and follicular development. In: *Advances in Fertility Regulation Through Basic Research*, edited by W. A. Sadler and S. Segal, In press.

Moore, G. P. M., Lintern-Moore, S., Peters, H., and Faber, M. (1974): RNA synthesis in the mouse ovary. *J. Cell Biol.*, 60:416-422.

Oakberg, E. F. (1968): Relationship between stage of follicle development and RNA synthesis in the mouse ovary. *Mutat. Res.*, 6:155-165.

Pedersen, T. (1970a): Determination of follicle growth rate in the ovary of the immature mouse. *J. Reprod. Fertil.*, 21:81-93.

Pedersen, T. (1970b): Follicle kinetics in the ovary of the cyclic mouse. *Acta Endocrinol. (Kbh.)*, 64:304-323.

Pedersen, T. (1972): Follicle growth in the mouse ovary. In: *Oogenesis*, edited by J. D. Biggers and A. W. Schuetz, pp. 361-376. University Park Press, Baltimore.

Pedersen, T., and Peters, H. (1971): Follicle growth and cell dynamics in the mouse ovary during pregnancy. *Fertil. Steril.*, 22:42-52.

Peters, H. (1969a): The effect of radiation in early life on the morphology and reproductive function of the mouse ovary. In: *Advances in Reproductive Physiology*, edited by A. McLaren, Vol. 4, pp. 149-185.

Peters, H. (1969b): The development of the ovary from birth to maturity. *Acta Endocrinol. (Kbh.)*, 62:98-116.

Peters, H., Byskov, A. G., and Faber, M. (1973a): Intraovarian regulation of follicle growth in the immature mouse. In: *The Development and Maturation of the Ovary and Its Functions*, edited by H. Peters, pp. 20-23. International Congress Series No. 267, Excerpta Medica, Amsterdam.

Peters, H., Byskov, A. G., Himelstein-Braw, R., and Faber, M. (1975): Follicular growth: The basic event in the mouse and human ovary. *Reprod. Fertil.*, 45:559-566.

Peters, H., Byskov, A. G., Lintern-Moore, S., Faber, M., and Andersen, M. (1973b): The effect of gonadotrophin on follicle growth initiation in the neonatal mouse ovary. *Reprod. Fertil.*, 35:139-141.

Peters, H., and Levy, E. (1966): Cell dynamics of the ovarian cycle. *J. Reprod. Fertil.*, 11:227-236.

Purandare, T. V., Munshi, S. R., and Rao, S. S. (1974): Follicular development in mice treated with antisera to FSH and LH in the neonatal period. In: *Proceedings of the Fifth Asia and Oceania Congress of Endocrinology*, Vol. 1.

Richards, J. S., and Midgley, A. R. (1976): Protein hormone action: A key to understanding ovarian follicular and luteal cell development. *Biol. Reprod.*, 14:82-96.

Rivarola, M. A., and Podesta, E. J. (1972): Metabolism of testosterone-^{14}C by seminiferous tubules of mature rats: Formation of 5α-androstan-3α, 17β-diol-^{14}C. *Endocrinology*, 90:618-623.

Schwartz, N. B. (1969): A model for the regulation of ovulation in the rat. *Recent Progr. Res.*, 25:1-54.

Schwartz, N. B., Andersen, C. H., Neguin, L. G., and Ely, C. A. (1974): Follicular maturation. In: *The Control of the Onset of Puberty*, edited by M. M. Grumbach, G. D. Grave, and F. E. Mayer, pp. 367-385. John Wiley & Sons, New York.

Schwartz, N. B., and Talley, W. L. (1968): Daily measurement of pituitary LH content during pregnancy in the rat: Do cyclic changes persist? *J. Reprod. Fertil.*, 15:39-45.

Welschen, R. (1973): Amounts of gonadotrophins required for normal follicle growth in hypophysectomized adult rats. *Acta Endocrinol. (Kbh.)*, 72:137-155.

Ovarian Follicular Development and Function,
edited by A. R. Midgley and W. A. Sadler.
Raven Press, New York © 1979.

Comments on Some Aspects of Early Follicular Development

Gilbert S. Greenwald

Department of Physiology, University of Kansas Medical Center, Kansas City, Kansas 66103

Dr. Peters' interesting presentation raises several questions and comments:

1. Dr. Byskov's findings on the role of the rete ovarii as the precursor of the granulosa cells and the initiator of the meiotic prophase border on heresy, and if they hold true, they will revolutionize our concepts on the morphology and physiology of gonadal differentiation. A recent report by Ruth Deanesly claims that the rete is not involved in germ cell regulation in the rabbit and guinea pig. This raises the question: How widespread is the phenomenon of rete-oocyte interactions? Do species differences exist or is the relationship more universal?

2. The data in Tables 1 and 2 of Dr. Peters' chapter show that in mice the number of small growing follicles is greater in the first 3 weeks of life than thereafter and is drastically reduced in androgen-treated animals. The results point out the dilemma in trying to decide whether intraovarian or gonadotropin feedback mechanisms are involved. The prepubertal rodent (at least the rat) has elevated levels of FSH, which may be responsible for the initially faster rate of follicular development. Moreover, the rat receiving testosterone on day 5 has significantly lower serum FSH concentrations for at least 10 days. Hence, changes in gonadotropin levels rather than intraovarian mechanisms could account for the variations observed in the number of growing follicles.

3. One of the most fascinating findings from Dr. Peters' laboratory is that injecting 3-day-old mice with fluid aspirated *at random* from large follicles of cow ovaries prevents the growth of about one-third the normal number of growing follicles. Whether the follicular product is steroidal or protein in nature has

not been determined. Dr. Peters has postulated that large degenerating follicles release a product that reduces the rate at which follicles move out of the nonproliferating pool. Although this is an attractive hypothesis, final proof rests on collection of fluid from antral follicles that are *definitely* known to be either atretic or normal.

4. I agree with Dr. Peters that intraovarian mechanisms may be of considerably greater importance in regulating follicular development than they have been credited with. The direct action of estrogen and gonadotropins on the ovary dates back to the work of Williams and Bradbury, but a plausible theory can also be postulated for interactions between androgens or progesterone and gonadotropins. This raises a number of points. Several *in vitro* studies suggest that estrogen secretion is initiated in preantral follicles, but without further evidence I am unwilling to concede at this time that this is necessarily true *in vivo*. If only antral follicles secrete estrogen, then the interesting possibility arises that a cluster of smaller follicles depend for their source of estrogen on follicles that are one to several cycles ahead of them in development. The possible role of the interstitium as a steroidal precursor also cannot be excluded.

Another relationship worth exploring is the influence of corpora lutea on adjacent follicles. The locally high concentrations of progesterone could conceivably affect follicular growth. Similarly, the extreme vascularity of corpora lutea could be a factor influencing adjacent follicles. A related topic of interest is the *topography* of the ovary: the relationship of growing follicles to the previous generation of corpora lutea or the present crop of antral follicles.

5. Dr. Peters pointed out that early growing follicles in which the theca layer is missing are the ones most prone to become atretic. To my knowledge, this is the first definite proof substantiating Hisaw's belief that thecal underdevelopment is largely responsible for atresia. It is tempting to tie this observation in with the results to be presented later in this meeting by Dr. Armstrong and his colleagues.

6. Dr. Peters states that in some of my earlier work I implied that the reserve follicles--recruited for superovulation--represent a "resting" population, held in abeyance until stimulated by FSH. This is a misinterpretation on her part because I agree that once a follicle is launched out of the resting pool it is committed to continuous growth, culminating in either atresia or in ovulation. There is no resting

stage in which growing follicles enter a period of metabolic quiescence.

The reserve follicles--at least in the cyclic hamster--constitute a population of preantral follicles with four to five layers of granulosa cells that normally require 8 to 12 days before developing into antral follicles. Under the influence of FSH preparations, these reserve follicles are prematurely recruited and transformed into antral follicles within 1 day.

7. I agree with Dr. Peters that one of the roles of PMSG is to prevent atresia and to support continued growth of these "rescued" follicles (Table 4). However, it seems unlikely that this is the sole role of PMSG. With higher doses of PMSG, superovulation of 60 ova can be obtained in prepubertal mice and rats and cyclic hamsters. These numbers cannot be accounted for solely by reduced atresia but mostly represent the additional *recruitment* of smaller preantral follicles by accelerated growth. Thus, the effects of PMSG are dose dependent; smaller doses prevent some follicles from undergoing atresia, and larger doses, in addition, increase the pool of follicles developing into antral stages.

8. In a number of studies, Dr. Peters and her colleagues have shown the value of quantitative studies of follicular development, based on serially sectioned ovaries. This approach, dating back to the 1920s and 1930s, still has considerable merit. This is especially true of the human ovary where we are appallingly ignorant of the follicular changes occurring during the menstrual cycle. The only thorough study of follicular development in the primate is Dr. Koering's investigation of the rhesus ovary, which was limited to changes in the population of antral follicles. While recognizing the problems inherent in collecting human ovaries at timed stages of the cycle, a cooperative study involving investigators from several countries should be able to collect such material. This information is indispensable if the basic research to be discussed at this meeting is to have any relevancy to new techniques of fertility control.

Ovarian Follicular Development and Function,
edited by A. R. Midgley and W. A. Sadler.
Raven Press, New York © 1979.

The Role of FSH in the Recruitment of Large Follicles

Anne N. Hirshfield

Reproductive Endocrinology Program, University of Michigan, Ann Arbor, Michigan

Ovarian follicular development appears to be a continuous process. Once growth has been initiated, it can be terminated only by ovulation and luteinization or by atresia(1) In order to study the mechanisms whereby those follicles that will ovulate are chosen from the total pool of growing follicles, we examined the course of follicular development in the cycling rat using histological and morphometric techniques. Our findings suggest that there is a critical period during follicular development (corresponding to a mean follicular diameter of 350 to 400 μm) during which most follicles become atretic. We present data indicating that the gonadotropic surge of one estrous cycle is instrumental in "rescuing" a few follicles and enabling them to complete the final stages of development in preparation for ovulation during the subsequent cycle. These data had led us to propose the hypothesis that FSH is the hormone responsible for recruiting a cohort of follicles into the final stages of growth and development. This hypothesis is supported by three lines of evidence: (1) In normal cycling rats, there appeared to be a discontinuous movement of follicles into the largest sized class. This movement was associated temporally with the proestrous surge of FSH. (2) Administration of exogenous FSH 1 day ahead of the usual time caused an influx of follicles into the largest size class, 1 day ahead of the usual time. (3) Abolishment of the gonadotropic surge on proestrus eliminated the influx of follicles into the largest size class normally seen on the following day.

For purposes of brevity, follicles greater than 390 μm in diameter will be called "pregraafian" follicles.

SIZE DISTRIBUTION OF PREGRAAFIAN FOLLICLES THROUGHOUT THE NORMAL ESTROUS CYCLE

Rats displaying at least three consecutive 5-day estrous cycles were killed at noon on each day of the cycle. Morphometric analysis of follicles was carried out on complete 5-μm serial sections of three ovaries from three different rats killed on each day of the cycle. Figure 1 shows the size distribution of pregraafian follicles in these ovaries. Only small pregraafian follicles were seen on estrus and metestrus. No significant influx of small pregraafian follicles was seen on subsequent days, suggesting that those small pregraafian follicles seen on estrus and metestrus are the same follicles that will ultimately ovulate in response to the LH surge on the following proestrus. FSH appeared to be the gonadotropin most likely responsible for the influx of follicles because the FSH surge is prolonged and lasts over 24 hr, while the LH surge is of very brief duration. The influx of follicles appears to extend over 2 days.

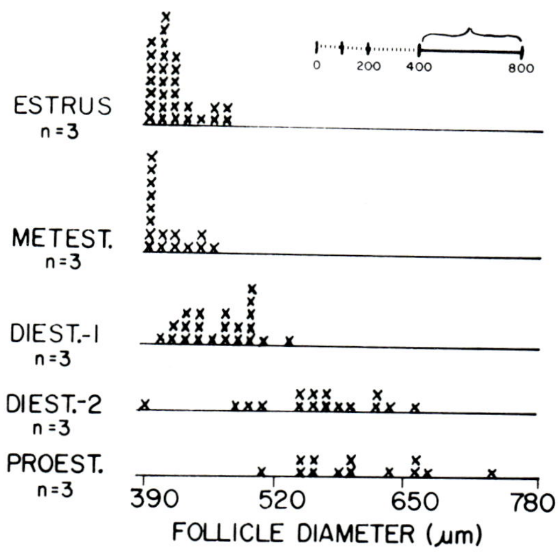

FIG. 1. Size distribution of largest follicles throughout the estrous cycle.

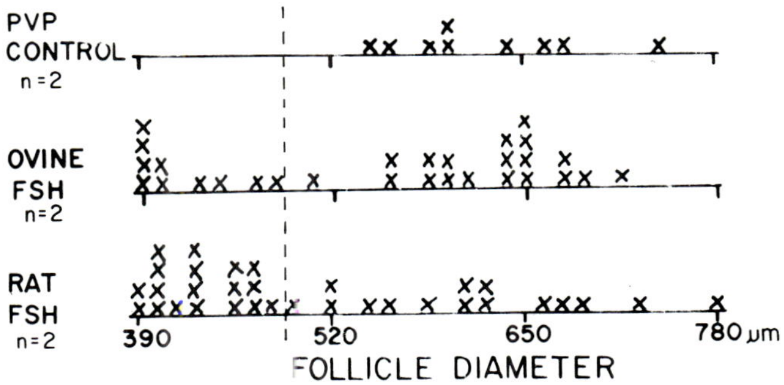

FIG. 2. Effect of exogenous FSH on number and size of largest follicles on proestrus.

THE EFFECT OF EXOGENOUS FSH ON THE SIZE DISTRIBUTION OF PREGRAAFIAN FOLLICLES

Ovine FSH or rat FSH was administered in polyvinylpyrrolidone (PVP; a delaying vehicle)(2) ondiestrus-2. PVP was used in an attempt to provide serum concentrations of FSH which, over time, would approximate those of the FSH surge. Rats were killed 1 day later, on proestrus. Figure 2 shows that the ovaries of rats given FSH in PVP contained numerous small pregraafian follicles, not normally seen on proestrus, as well as large pregraafian follicles, presumably those that were destined to ovulate in response to the LH surge later that afternoon. Ovaries of rats given PVP alone contained only large pregraafian follicles.

THE EFFECT OF ELIMINATION OF THE FSH/LH SURGE ON THE SIZE DISTRIBUTION OF PREGRAAFIAN FOLLICLES

At noon, on the day of proestrus, rats were given pentobarbital (35 mg/kg) or phenobarbital (75 mg/kg) in order to block the proestrous surge of gonadotropins(3) Rats were killed at noon on the following day ("estrus"). Figure 3 shows the size distribution of pregraafian follicles in these rats. Barbiturate blockade of the proestrous surge of gonadotropins prevented the influx of small pregraafian follicles normally seen on the following day. Ovaries from untreated control rats killed on estrus contained numerous small pregraafian follicles. Ovaries from barbiturate-blocked rats resembled those of control rats killed on proestrus.

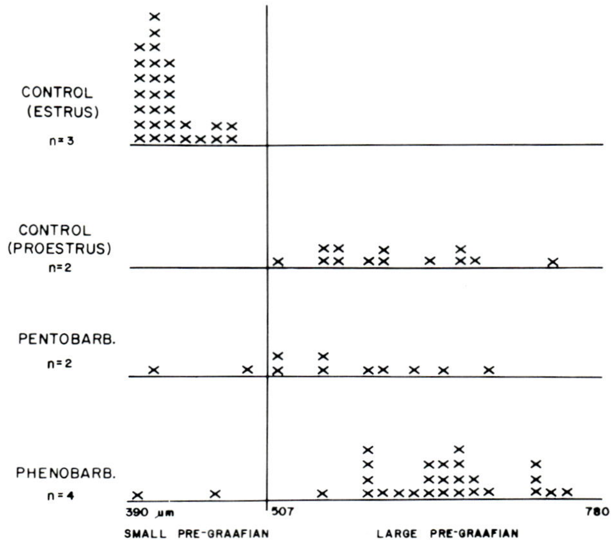

FIG. 3. Effect of barbiturate blockade on the size distribution of pregraafian follicles.

These results suggest that the surge of gonadotropins of one cycle recruits follicles for the next cycle. We believe that FSH is the gonadotropin responsible for this recruitment.

REFERENCES

1. Pederson, T. (1970): Follicle kinetics in the ovary of the cyclic mouse. Acta. Endocrinol. 64:304.
2. Morishige, W.K., and Rothchild I. (1974): Temporal aspects of the regulation of corpus luteum function by luteinizing hormone, prolactin, and placental luteotrophin during the first half of pregnancy in the rat. Endocrinology 95:260.
3. Blake, C.A. (1974): Differentiation between "critical period," the "activation period," and the "potential activation period" for neurohumoral stimulation of LH release in proestrous rats. Endocrinology 95:572.

Ovarian Follicular Development and Function,
edited by A. R. Midgley and W. A. Sadler.
Raven Press, New York © 1979.

Steroidal Modification of the Effects of Atropine Sulfate Treatment on the Ovaries of Hypophysectomized Mice

Patricia L. Morris and *Howard S. Grob

New York University, New York, New York, and *Adelphi University, Garden City, New York

A series of pharmacologic experiments was designed to investigate the possibility of an intraovarian neural component in the control of follicular development. Morris and Grob (1975) reported that daily subcutaneous administration of atropine sulfate and atropine methylbromide, tertiary and quartenary parasympatholytic agents respectively, to intact mice for a duration of 2 weeks, resulted in a significant increase in the mean follicle mitotic index (X_M) of all medium and large follicles observed. This effect was present and statistically significant at all stages of the estrous cycle (determined by vaginal smears prior to autopsy).

To eliminate the possibility of a drug effect on hypothalamic/hypophyseal centers, mice were surgically hypophysectomized 10 days prior to the initiation of atropine sulfate treatment. All follicle compartments examined (Types 4, 5a, and 5b) in tertiary atropinized, hypophysectomized mice showed a significant increase ($p < 0.005$) in X_M values compared to the saline-injected, hypophysectomized controls.

Morris and Grob (1975) also observed that the relative increase in mean mitotic index showed an intracycle variation, with the maximum mitogenic effect at early proestrus and proestrus, and minimum stimulation at diestrus. Since there are cyclic variations in gonadotropins and steroidal hormones throughout the estrous cycle, the following experiments were undertaken to explore the possibility of hormonal modification of the drug effect in the hypophysectomized mouse:

Protocol

Animal group	Treatment
A	Saline and sesame oil (vehicle control)
B	ATR SO_4 + oil
C	Saline + (E)
D	ATR SO_4 + (E)
E	Saline + (P)
F	ATR SO_4 + (P)
G	Saline and sequential (E)/(P)
H	ATR + sequential (E)/(P)

Drugs

Atropine sulfate (ATR SO_4)
β-Estradiol-3-benzoate (E)
Δ^4-Pregnene-3, 20-dione progesterone (P)

Regimen

Duration:	14 days
Route:	Subcutaneous
Time:	Between 1400 and 1600

Drug		Daily dose
ATR SO_4		5 mg/kg
(E)		1 μg/animal
(P)		1 μg/animal
SEQUENTIAL	3 days	1 μg/animal (E)
	2 days	1 μg/animal (P) and 0.5 μg/animal (E)

Virgin female CD-1 mice were hypophysectomized by the parapharyngeal approach at 42 days of age. Ten days postoperatively, drug regimens were instituted. The animals were killed by ether 24 hr after the last injection of the regimen. Both ovaries were removed and fixed in Bouin's solution. The ovaries were dehydrated through an alcohol series and embedded in paraffin wax. One ovary of each pair was randomly chosen for serial sectioning at 6 μm. The tissue was stained with hematoxylin and eosin. The present report summarizes the light microscopic findings.

Our data demonstrate that atropinization of hypophysectomized mice results in the appearance of nests and cords of large, lightly staining cells in the interstitium. These cells have oval nuclei, which share the same staining characteristics of granulosa cell nuclei. The vacuolated cells exhibit a low nuclear/cytoplasmic ratio. The nuclei appear both centric and

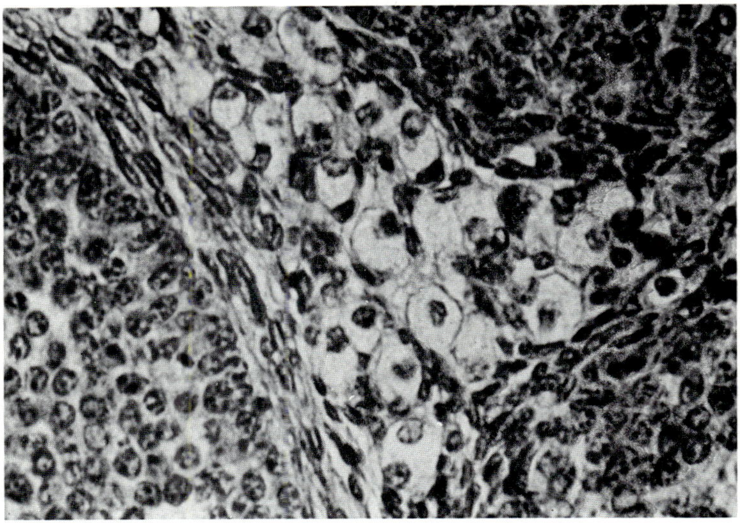

FIG. 1. Ovary of an atropinized, estrogen-treated hypophysectomized mouse with a small nest of large cells bordered by the boundaries of three follicles. Hematoxylin and eosin. ×280.

eccentric, but the apparent differences may be due to the plane of section. The cells are not eosinophilic; they do not appear luteinized or pyknotic. The thecal layers of both atretic and normal-appearing medium-sized follicles appear to be boundaries for the small nests and cords (Fig. 1). These nests and cords do not appear in control ovaries.

The addition of estradiol to the atropine regimen results not only in an increased frequency of the appearances of these cells (Fig. 2), but also in the presence of large corpora containing these cells (Fig. 3). These corpora appear different from those observed in atropinized, progesterone-treated mice. The ovaries of the latter group display very large corpora with pyknotic cores.

In view of the electron microscopic demonstration of bouton contact within the cytoplasm of interstitial cells (Unsicker, 1970) and the pattern of cholinesterase distribution in the ovary, we speculate that nerves may regulate interstitial cell function.

We propose the atropinized, hypophysectomized mouse as a system in which to explore the physiologic role of interstitial tissue in the ovary. Furthermore, our data indicate that intraovarian events may be subject to the influence of the peripheral nervous system.

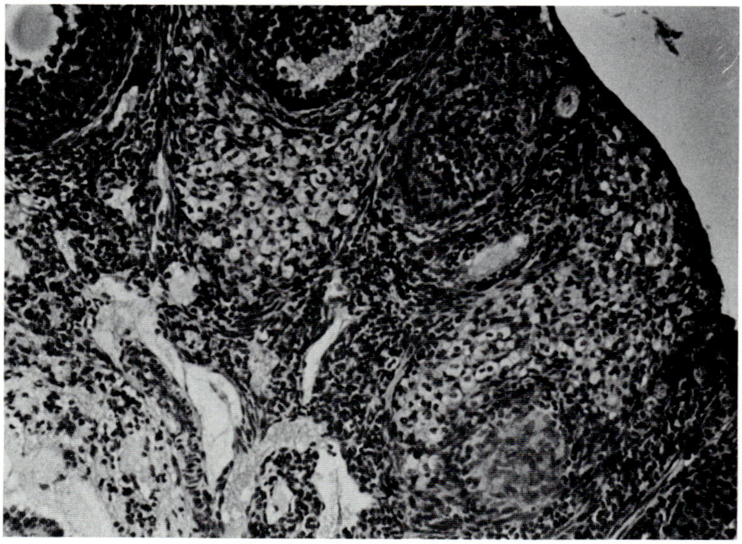

FIG. 2. Ovary of an atropinized, estrogen-treated mouse. Note areas of large interstitial cells. Hematoxylin and eosin. ×70.

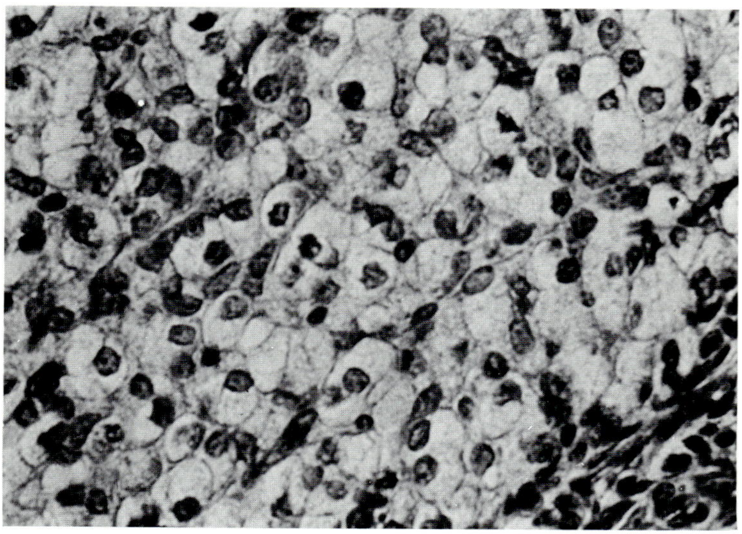

FIG. 3. Ovary of an atropinized, estrogen-treated hypophysectomized mouse. Cells from a large corpora. Hematoxylin and eosin. ×330 .

REFERENCES

Morris, P. L., and Grob, H. S. (1975): Effect of tertiary and quartenary atropine compounds on the ovarian follicles of intact and hypophysectomized mice. *Fed. Proc.*, 34:383 (abstract).

Unsicker, K. (1970): Zur innervation der interstitiellen Druse im Ovar der Maus. *Z. Zellforsch.*, 109:495-516.

Ovarian Follicular Development and Function,
edited by A. R. Midgley and W. A. Sadler.
Raven Press, New York © 1979.

Regulation of Folliculogenesis in the Cycling Rhesus Monkey

Arnold L. Goodman, Wilbert E. Nixon, and Gary D. Hodgen

Section on Endocrinology, Reproduction Research Branch, National Institute of Child Health and Human Development, National Institutes of Health, Bethesda, Maryland 20014

It is generally accepted that a follicle destined to ovulate during each cycle originates from a pool of nonproliferating primordial follicles that had been formed prenatally. There is evidence suggesting that ovulating follicles are the survivors of cohorts of growing follicles recruited from the nonproliferating pool; that is, these follicles escaped atresia during their postnatal development to reach maturity. Among mammalian genera, the number of ova ovulated during each cycle is species related, but in all species the number appears to be fewer than that observed growing in any cycle. In monotocous species such as the higher primates, some kind of selection process is especially evident: for each cycle typically just a single ovum is ovulated by only one of the two ovaries. In this and other respects, the regulation of folliculogenesis in primates may differ qualitatively, as well as quantitatively, from that in the oft-studied rodents, in which both ovaries contribute about equally to the total number of ova shed.

The present *in vivo* study was designed to examine certain aspects of the regulation of follicular growth in the adult rhesus monkey by ablating the ovary's cyclic structures: the preovulatory follicle and functioning corpus luteum (CL), to impose a sudden discontinuity on the ovarian cycle. In addition, certain endocrine and gross morphological events accompanying the resumption of cyclic ovarian activity were also observed. These included patterns of circulating LH, FSH, estradiol, estrone, and progesterone before and after ablation, as measured by radioimmunoassay of daily serum samples; the time and side of the next ovulation, as observed at a second laparotomy performed 2 to 3 weeks after the ablations; and the pattern of vaginal bleeding, which was checked daily.

The effects of follicle cautery are shown in Fig. 1. Destruction of the largest visible follicle on cycle days 8 to 12 (day 1 = onset of menses) abolished the expected midcycle surge of LH in seven of eight monkeys. In the eighth cautery had been performed on the day of the surge. By one-way analysis of variance, no differences were found in average LH ($F_{10,77} = 1.25$) or FSH ($F_{10,77} = 0.32$, $p > 0.25$) levels in the interval from 4 days before until 6 days after cautery. In contrast, follicle ablation immediately terminated the presumed incipient estradiol surge apparent in half the group. This observation is further circumstantial evidence that, in the monkey, the preovulatory follicle is the primary source of midcycle increment in serum estradiol. In the remaining four monkeys, serum estrogens were slightly reduced after cautery. In all animals, serum progesterone remained low after cautery.

The next LH surge occurred an average of 12.4 ± 0.9 days after cautery (mean ± SE; range, 10 to 17 days) and was accompanied by a surge of FSH secretion. Surges of estrogen release preceded by 1 day or accompanied the gonadotropin surges. The subsequent pattern of serum progesterone is indicative of ovulation and corpus luteum function, which lasted about 14.5 ± 2.0 days (range, 6 to 22 days). Ovulation was confirmed at laparatomy; in six of eight, the ovulation was contralateral to follicle ablation. The ovarian cycle apparently resumed shortly after ablation, since the interval from follicle cautery to the onset of the next menses approximated the length of a normal menstrual cycle--about 27 days.

The effects of luteectomy are depicted in Fig. 2. Luteectomy was performed on days 15 to 17, about 5 days after the midcycle LH surge. The completeness of the ablation is indicated by the prompt fall in circulating progesterone levels, and the premature onset of menses 3 to 4 days after luteectomy. However, peripheral levels of LH, FSH, and estrogens were, in general, maintained after CL removal. Thereafter, surges of LH and FSH secretion coincident with a rise in serum estrogens occurred 12.8 ± 0.9 days after luteectomy. These surges were followed by a pattern of elevated progesterone levels indicative of ovulation and CL function. Single, contralateral ovulations in each of the monkeys were confirmed at laparotomy.

As was apparent in the follicle cautery group, the magnitude and duration of the postablative hormone patterns, including their temporal sequence, were typical of the normal menstrual cycle, after luteectomy, as

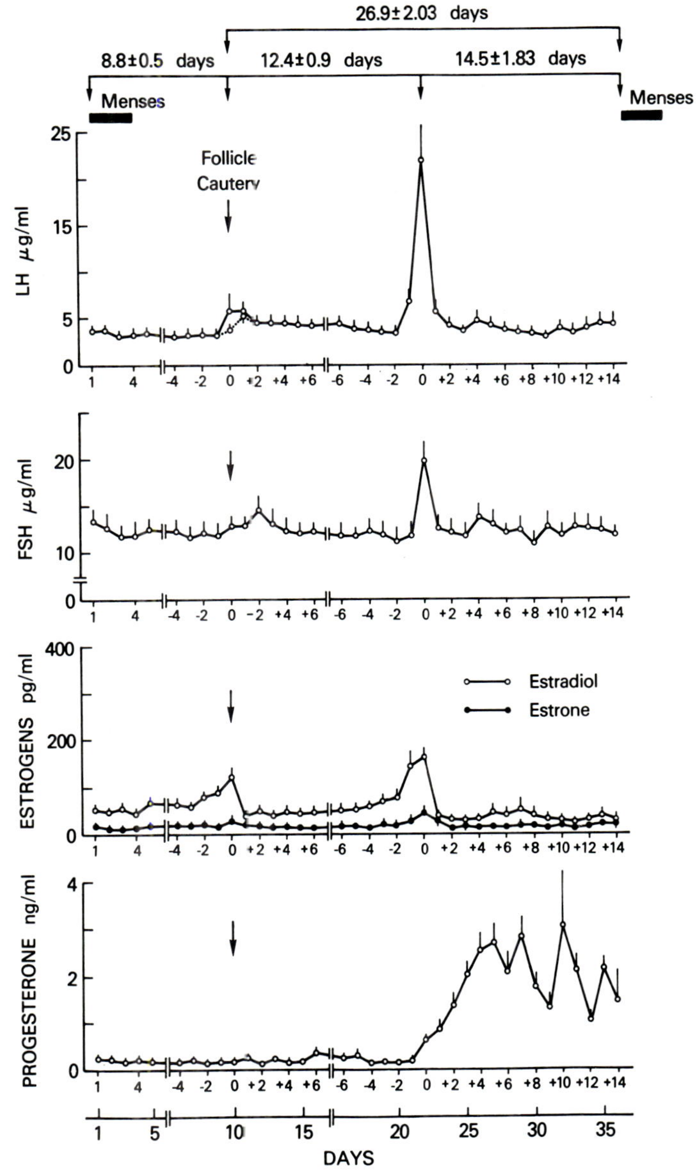

FIG. 1. Patterns of circulating levels of gonadotropins and ovarian steroids before and after follicle cautery in eight ovulating monkeys. Since the interval from the onset of menses to follicle cautery and the interval from cautery to the next LH surge were variable, the data have been synchronized to the onset of menses, the day of follicle cautery, and the day of the next LH surge. A progressive time scale is provided at the bottom of the figure. One monkey had an LH surge at the time of cautery that tended to bias the group mean (*solid line*). When the LH values from this animal are excluded on these 2 days, average serum LH of the remainder was not different from the preceding days. Vertical bars indicate 1 SE.

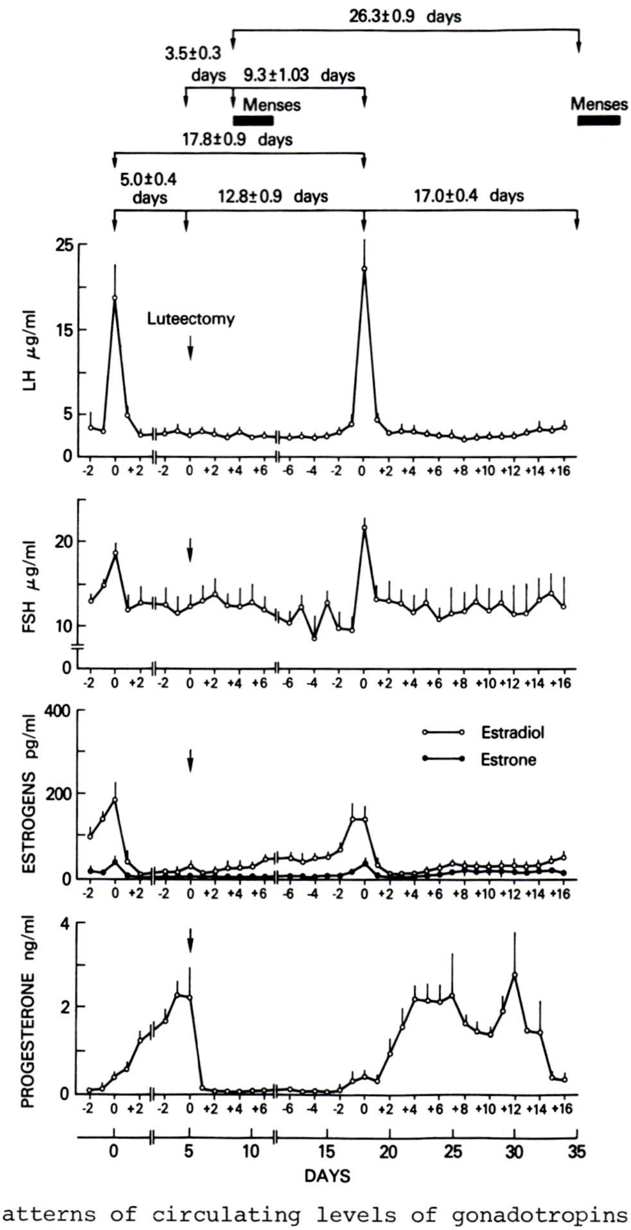

FIG. 2. Patterns of circulating levels of gonadotropins and ovarian steroids before and after luteectomy in four ovulating animals. Since the interval from the antecedent LH surge to luteectomy and the interval from luteectomy to the next LH surge were variable, the data have been synchronized to the day of the first LH surge, the day on which luteectomy was performed, and the day of the next LH surge. A progressive time scale is shown at the bottom of this figure. Vertical bars indicate 1 SE.

well. Again, similar to the follicle ablation group, the ovarian cycle apparently resumed soon after luteectomy, since the interval from the postoperative menses to the next menses was about 27 days. However, luteectomy reduced the interval between LH surges to about 18 days.

In two of two control monkeys, sham follicle cautery of the ovary, at a site distal to the largest follicle, did not interrupt the ovarian cycle, that is, did not prevent ipsilateral ovulation or impair subsequent progesterone secretion.

Since destruction of the largest visible follicle abolished midcycle gonadotropin surges, and blocked ovulation, we conclude that the follicle destined to ovulate had been destroyed by cautery. Such a follicle may have arisen from a cohort of growing follicles, but we do not exclude the possibility of a cohort of one. Our findings imply that the preovulatory follicle apparently became dominant over the others of the presumed cohort; that is, the dominant follicle was "selected" by the middle of the follicular phase. The absence of a ready substitute indicates that submission of any other follicles of the cohort was irreversible.

That the interval from ablation to the next ovulation was about the same after interruption of either the follicular or the luteal phase indicates the endocrine milieu at ablation had no discernible differential effect on the reinitiation of follicular growth. Further, postablative follicular growth was reinitiated in the face of apparently unaltered circulating tonic levels of gonadotropins. This, in turn, suggests that the dominant follicle or active CL influences locally the responsiveness of the follicular apparatus to extraovarian stimulation. However, since the rhesus monkey is monotocous it is not unlikely that there is communication between ovaries. That is, the dominant follicle or active CL on one ovary may influence the responsiveness of the contralateral organ as well.

In conclusion, follicular dynamics during the primate menstrual cycle may differ in several respects from that found in species with an estrous cycle. In the rhesus monkey, in particular, there seems to be a virtually complete suppression of follicular development on both ovaries in the presence of the dominant follicle or active CL. Following ablation of these structures, follicular development resumes anew, seemingly more as a result of a local disinhibition than as a compensatory increase in gonadotropic stimulation.

Ovarian Follicular Development and Function,
edited by A. R. Midgley and W. A. Sadler.
Raven Press, New York © 1979.

A Characterization of the Effects of Selection for Increased Litter Size

Jimmy L. Spearow, Irving I. Geschwind, and G. Eric Bradford

Animal Science Department, University of California at Davis, Davis, California

The following series of experiments was conducted in order to determine the physiological factors that have genetic variation and that are most limiting to natural ovulation rate, i.e., the number of eggs released at a particular estrus. To do this, I compared lines of mice that originated from a common base population but that were randomly bred (line C), the control line, and lines selected for increased litter size (line SI) (Bradford, 1969).

Data collected up to this time indicate (1) that the natural ovulation rate has increased from 11.0 to 16.9 in response to selection for large litter size; (2) the ovaries of line SI mice are heavier than those of line C mice at 4 and at 8 weeks of age (Table 1); (3) the

TABLE 1. *Ovary weight in milligrams (both ovaries)*

Age (weeks)	Line C	Line SI
4	1.48 ± 0.40	2.15 ± 0.55
8 (at proestrus)	2.53 ± 0.91	5.00 ± 0.35

Mean ± SE

gonadotropin-induced superovulation rate of these strains has been characterized by Bindon and Pennyuik (1974), who found that line SI females are equally as responsive as line C females at low doses of PMSG, but are more responsive at high doses of PMSG; and (4) in order to determine the kinetics of follicular growth, I have compared autoradiographs of serial sections of SI and C ovaries labeled for 1 hr *in vivo* with ^{3}H-thymidine, the method of Pedersen (1970). At proestrus, SI females have more small follicles (Type 3a) and medium follicles (Types 3b and 4). The number of medium-large follicles (Type 5a) is similar for the two strains. SI females have significantly more large (Type 5b) follicles at this stage of the cycle but have a nonsignificant increase in the number of large (Types 6 to 8) follicles (Table 2). Analysis of labeling

TABLE 2. *Number of follicles*

Follicle size, type	Line C	Line SI
Small, 3a	540.0 ± 97.8	999.0 ± 57.6[a]
Medium, 3b	264 ± 32	440.8 ± 27[a]
Medium, 4	160.5 ± 22.9	256.9 ± 12.4[a]
Medium-large, 5a	60.0 ± 11.5	73.0 ± 6.1
Large, 5b	47.1 ± 8.4	75.8 ± 8.2[b]
Large, 6 Large, 7	33.2 ± 9.4	41.8 ± 9.2

[a]Indicates highly significant differences at $p < 0.01$.
[b]Indicates significant differences at $p < 0.05$.
Mean ± SE

index (the percent of the granulosa cells labeled with ^3H-thymidine) shows that (1) the small follicles (Type 3a) of line SI grow at the same rate or perhaps slower than those of line C; (2) the medium follicles (Types 3b and 4) of line SI grow slower than those of line C; (3) the medium-large follicles (Type 5a) of line SI grow at the same rate as those of line C; and (4) the large follicles (Type 5b) of line SI grow faster at first than the analogous follicles in line C, then grow at the same rate as line C (Type 6) follicles, and finally grow faster than line C (late Type 6 and Type 7) follicles (Table 3). Line SI mice had more follicles in early atresia than line C mice, but the percentage of the large follicles that are undergoing early atresia is the same in both strains, i.e., 48%.

TABLE 3. *Labeling index*

Follicle size, type	Line C	Line SI
Small, 3a	0.064 ± 0.008	0.049 ± 0.007
Medium, 3b	0.107 ± 0.009	0.081 ± 0.006[a]
Medium, 4	0.095 ± 0.005	0.075 ± 0.006[a]
Medium-large, 5a	0.101 ± 0.008	0.116 ± 0.009
Large, 5b	0.161 ± 0.017	0.232 ± 0.021[a]
Large, 6	0.417 ± 0.022	0.406 ± 0.023
Large, late 6	0.417 ± 0.014	0.493 ± 0.015[a]
Large, 7	0.438 ± 0.032	0.425 ± 0.032

[a]Indicates highly significant differences at $p < 0.01$.
Mean ± SE

This data indicates that selection for litter size has increased ovulation rate by increasing the number of follicles, and the growth rates of specific follicle types, but not by decreasing the incidence of atresia.

REFERENCES

Bradford, G. E. (1969): *Genetics*, 61:905.

Bindon, B., and Pennyuik, P. (1974): J. Reprod. Fertil., 36:221-224.

Pedersen, T. (1970): Acta Endocrinol. 64:304.

Ovarian Follicular Development and Function,
edited by A. R. Midgley and W. A. Sadler.
Raven Press, New York © 1979.

Discussion Summary: Session on Follicular Development

Hannah Peters

Many questions were raised about the role of the rete ovarii in relation to follicular organization and the establishment of the pool of resting, nonproliferating small follicles. Doubts were expressed that the rete-oocyte interaction was a widespread phenomenon and the question was asked whether or not the inducing role of the rete was confined to rodents. Byskov pointed out that many different species have been examined and that in all cases the rete ovarii was found to invade the ovary, that meiosis started in those germ cells that come into contact with the rete system first, and that follicles always form in the central part of the cortex first. In the early stages of follicular development, a connection to the rete system can be demonstrated. The presence of the rete seems necessary for the normal early development of the gonad in both sexes. It is important for sexual development as well as for germ cell differentiation. A question was asked whether the "primary interstitial tissue" of the ovary, which occupies the ovary for about 2 weeks after birth (rodents), could have any relation to the rete. It was stated that the "primary interstitial tissue" is indeed identical with the intraovarian rete cords and tubules.

In discussing the effect of gonadotropin on the follicular population, it was suggested that during the development of the follicle there is a critical period during which most follicles become atretic, and that the gonadotropic surge of the estrus cycle is instrumental in rescuing some follicles that continue to grow and complete the final stage of development in the subsequent cycle. This concept coordinates well with results of experiments in which injection of PMSG rescued some follicles from early atresia by (1) preventing granulosa cells from becoming pyknotic, (2) eliminating those pyknotic cells that already are present in the granulosa cell layer via phagocytosis, and (3) stimulating an increased number of granulosa cells to enter the cell cycle.

During the final stages of follicular differentiation, a distinct heterogeneity of granulosa cells occurs: cells in pseudostratification lying close to the basement membrane are clearly distinguishable from those close to the antrum. The latter are proliferating cells as shown by ^{3}H-thymidine incorporation in autoradiograms, whereas the cells close to the basement membrane are not proliferating but can be shown to form luteinizing hormone (LH) receptors. It was suggested that some time prior to the LH surge peripheral granulosa cells cease to grow and undergo cellular differentiation, one aspect of which is the acquisition of LH receptors.

It was suggested that an intraovarian control of the number of follicles that start to grow was doubtful, and the argument was voiced that a change in the follicular growth initiation rate after a reduction of the pool of small follicles by testosterone propionate, for example, or in old age might change the gonadotropin release from the pituitary, which could then directly influence the growth initiation rate. However, intraovarian control mechanisms originating from local production of estrogen and progesterone were thought to act on the development of growing follicles.

In considering the differentiation of the theca layer, it was reported that cells with ultrastructural characteristics of smooth muscle have been observed in the theca externa of ovarian follicles. Using the immunofluorescence technique, the presence of two contractile proteins, i.e., actin and myosin, could be shown. With the differentiation of the theca layer in the growing follicle, smooth muscle cells develop that become the target of prostaglandins.

At the close of the discussion it was pointed out that both intra- and extraovarian control mechanisms guide the rate and speed of follicular development as well as atresia.

A plea was made for the necessity of examining human material; as yet we have little knowledge of follicular kinetics in the ovary of women of different ages.

ATRESIA

Ovarian Follicular Development and Function,
edited by A. R. Midgley and W. A. Sadler.
Raven Press, New York © 1979.

Atresia

Anne Grete Byskov

The Finsen Laboratory, The Finsen Institute,
Copenhagen, Denmark

The term "atresia" is used to denote those processes during which the ovarian follicle loses its integrity and the egg is expelled by means other than ovulation.

It is well known that by far the largest number of eggs in the mammalian ovary is eliminated not by the ovulatory process, but during atresia.

At all ages during the life span, follicles are initiated to leave the pool of nongrowing small follicles and become growing follicles, i.e., medium slowly growing and large fast-growing follicles (Pedersen, 1970). Once the follicle leaves the pool of small follicles, it either ovulates or becomes atretic. However, atresia also attacks the nongrowing small follicle. Thus, follicles may become atretic at any stage during their development (Fig. 1) (for review see Ingram, 1962).

The dynamics of the atretic process, when initiated in follicles, are determined by cell growth and cell loss. It is therefore difficult to follow the sequence of transformations a follicle undergoes during atresia, particularly if the follicle consists of a large cell number when atresia begins. Only in follicles in the early stages of atresia, when the rate of cell loss is still low, is it possible by morphological criteria to judge from which "healthy" type of follicle is originated.

The mechanisms that initiate and control atresia are not understood, but it is well known that different hormonal treatments influence the rate at which follicles become atretic. Our limited knowledge of the physiology of the atretic process is presumably related to the fact that at the present time we are not able to recognize the incipient stages of atresia. It is, for example, impossible to distinguish morphologically between "healthy-looking" follicles, which are committed to undergo atresia, and "healthy" ones,

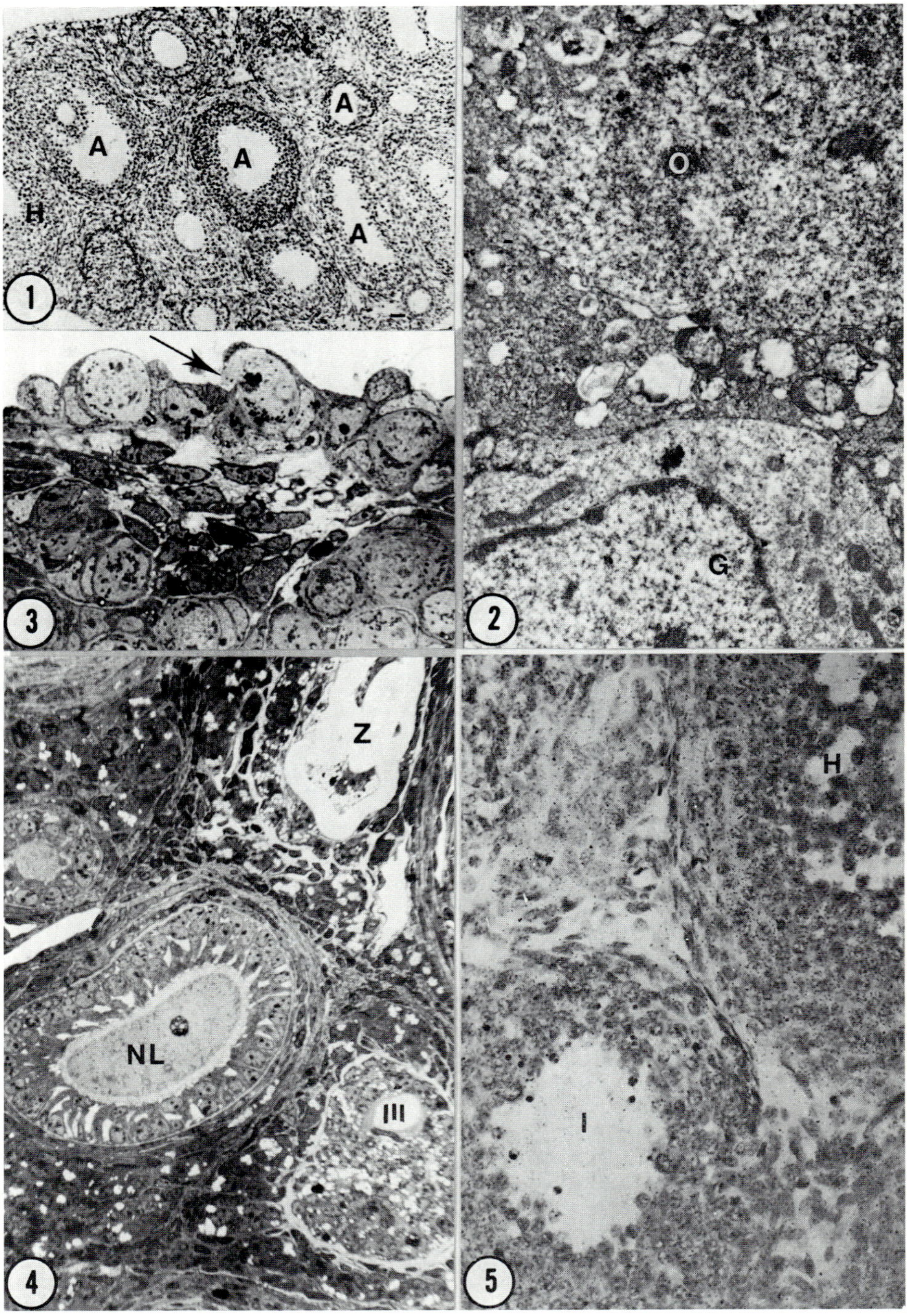
A
A
A
A
A
H
1
O
G
2
3
Z
NL
III
4
H
I
5

which will terminate their growth phase by ovulating the egg. In the mature mouse ovary (Bagg strain) the number of large antral "healthy" follicles, i.e., without pyknotic granulosa cells, that are present around 1½ days before expected ovulation is about 15 in each ovary, i.e., twice the maximal number that will ovulate. Half of these follicles must therefore become atretic before or during ovulation, since the growth is arrested only by atresia or ovulation and no large healthy follicles remain in the ovary immediately after ovulation.

Other follicles that actually show morphological signs of atresia as, for example, presence of pyknotic granulosa cell nuclei may become rescued and proceed through ovulation. This was accomplished in hamsters after unilateral ovariectomy, where compensatory ovulation of the remaining ovary results from a decrease in follicular atresia (Greenwald, 1961). A restoration of large atretic follicles can also be induced by injection of PMSG, thus explaining the excessive number of eggs ovulated in response to PMSG (Peters et al., 1975).

←

FIG. 1. Paraffin section with Feulgen stain of 4-week-old mouse ovary. Follicles of different sizes are atretic (A). With this stain, the atretic follicles show up easily, as the necrotic nuclei will become heavily stained. A, atretic follicle; H, "healthy" follicle. ×180

FIG. 2. Electron micrograph of small atretic follicle. The oocyte (O) is necrotic with blown-up perinuclear space and mitochondria. The granulosa cell (G) seems unaffected. × 6750

FIG. 3. Part of ovarian cortex from a 2-day-old mouse. Oocytes are situated in the surface epithelium, one with a pseudopodium (*arrow*), which has penetrated the epithelium (1-μm plastic sections). × 1260

FIG. 4. Five-week-old mouse ovary with a nonluteinized atretic follicle (NL) with a distorted oocyte, a late Stage III of a luteinized large follicle (III), and remnant of a luteinized follicle with a coiled zona pellucida (Z) (1-μm plastic section). × 605

FIG. 5. Autoradiogram of a 4-week-old mouse ovary prepared 1 hr after injection of ^{3}H-proline. Many grains are present over the healthy follicle (H), whereas the grain number over the atretic follicle of Stage I (I) is much lower. ×605

All follicles do not follow a uniform pattern during atresia; the atretic process varies with the size (i.e., stage of differentiation) of the follicle. Therefore I shall first consider atresia in relation to follicle size. Since very little information about atresia in small follicles is available, I shall only briefly discuss small follicles and mainly focus on the medium and large follicles that undergo atresia. Second, I shall discuss cellular response of large follicles to gonadotropins (PMSG). More attention will be focused on the fate of the granulosa cells during atresia rather than on the other components of the follicle. It is important, however, to emphasize that the normal follicular growth as well as atresia depend on interaction between the theca layer, the granulosa layer, and the oocyte.

ATRESIA IN RELATION TO FOLLICLE SIZE

Atresia of Small Follicles

It has been reported that elimination of small follicles occurs mainly by lysis or phagocytosis of the pyknotic oocyte (Franchi and Mandl, 1962; Gondos, 1969). The granulosa cells, which surround such dying oocytes, often appear to be unaffected (Fig. 2). In early life, small oocytes may also disappear from the ovary by passing through the surface epithelium (Peters, 1969; Byskov and Rasmussen, 1973) (Fig. 3). It has not been possible to stimulate growth in small follicles, whereas elimination of small follicles, for example, can be the result of irradiation or of hormonal treatment.

Atresia of Medium-Sized Follicles

Only a small fraction of the medium-sized follicles become atretic and most of these become atretic without luteinization, i.e., without accumulation of lipids in the granulosa or the theca cells. The atretic process may begin in either cell compartment first. The granulosa cells stop growing and become necrotic, and the theca layer as well as the granulosa layer often become depleted of lipid. The oocyte may shrink and become distorted (Fig. 4). The nucleus may show necrotic changes, but the formation of a spindle or other events simulating the maturation division do not occur. The lack of meiotic maturation in atretic oocytes of similar follicle size was also noted in the human ovary (Himelstein-Braw et al., 1976).

Atresia of Large Follicles

The majority of all large follicles will be trapped by atresia. Atresia in large follicles often leads through different stages of luteinization. Healthy follicles and three consecutive stages of follicular atresia of large follicles were defined in the mouse ovary (Byskov, 1974) and will be used here: healthy follicles, without pyknotic granulosa cells; Stage I atresia, in which mitosis as well as pyknotic nuclei are present within the granulosa layer; Stage II atresia, in which few cells are proliferating with many pyknotic granulosa cells; and Stage III atresia, which signifies a collapsed nongrowing follicle.

Some of the initial changes of the atretic process are related to an increased activity of hydrolytic enzymes, which can be noted both in the theca and granulosa layers and in the oocyte (Lobel et al., 1961). That other metabolic activities change when atresia is initiated is also indicated by a lower incorporation of ^{3}H-proline in atretic follicles than that in healthy ones. In autoradiograms of mouse ovaries obtained 1 hr after injection of ^{3}H-proline, the grain number over granulosa cells of healthy follicles was three times higher than that over granulosa cells from atretic Stage I follicles (Fig. 5) (Byskov, *unpublished data*).

Similar to the luteinization of the preovulatory follicle (Byskov, 1969), the accumulation of lipid droplets in the incipient stages of atresia are first seen in the granulosa cells close to the basement membrane and in the cells of theca interna (Fig. 8). In healthy large follicles very little lipid is present (Fig. 9). Also characteristic of the early stages of atresia is the formation of pyknotic granulosa cell nuclei, concentrated in the center of the follicle, a relative expansion of the follicular cavity and a decrease in the growth rate of the granulosa cells. These changes, parallel with the accumulation of lipids in granulosa and theca cells, gradually lead to Stage II atresia. As atresia progresses, the number of granulosa cells becomes reduced by lysis and phagocytosis. Granulosa cells as well as invading leukocytes and macrophages participate in the removal of granulosa cell debris.

The granulosa cells vary in size due to different degrees of pyknosis (Fig. 7), whereas the size and shape of granulosa cells in healthy follicles are uniform (Fig. 6). In late Stage II atresia, when the follicle is thinned, the follicle collapses. In the mouse it takes about 3 to 4 days, i.e., approximately the legnth of the cycle time, for large follicle to reach the collapses luteinized state,

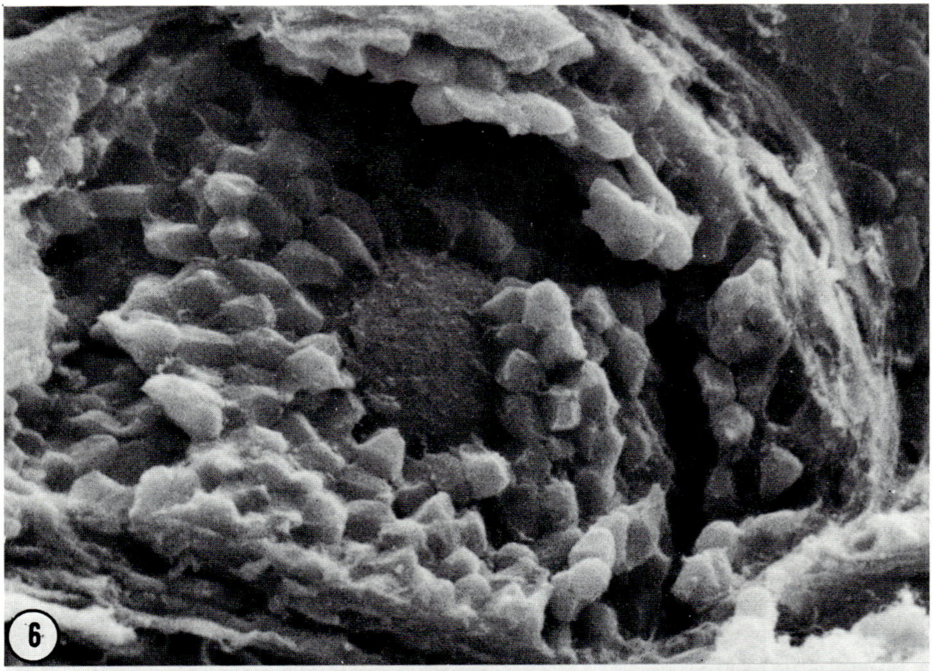
6

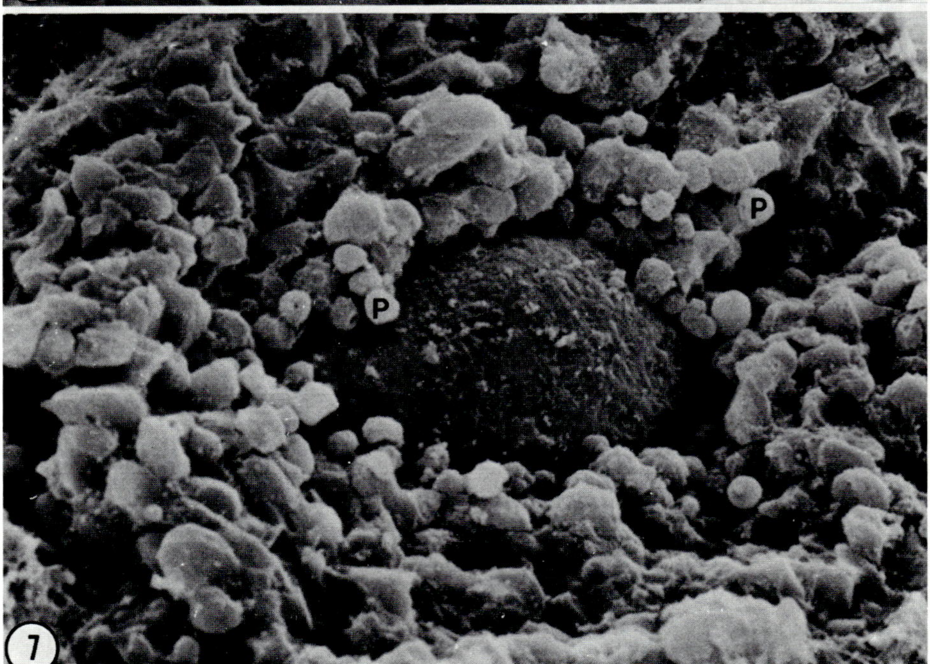
P
P
7

classified as Stage III atresia (Fig. 10) (Byskov, 1974). It is unknown how long the luteinized follicular remnants remain in the ovary.

During the atretic differentiation few theca cells become necrotic. In contrast, the cells of the theca interna frequently hypertrophy and accumulate lipid droplets. As luteinization and degradation of the granulosa layer progresses beyond Stage III atresia, the luteinized theca cells and blood vessels of the theca layer mix with the remnants of the luteinized granulosa layer.

In some follicles the signs of atresia appear first in the oocyte, which may shrink, become necrotic, and lose its connections with the cumulus cells. These events often occur simultaneously with or are followed by maturation division, pseudocleavage, or fragmentation of the oocyte (Fig. 11). Subsequently, phagocytosing cells penetrate the zona pellucida and remove the oocyte. The zonae pellucida are usually the last recognizable components of the follicles (Ingram, 1962). They are frequently seen in the center of the ovary (Fig. 4), between luteinized follicle layers, which may now be termed interstitial tissue.

←

FIG. 6. Scanning electron micrograph of a "healthy" follicle. The shape of the granulosa cells is very uniform. ×1350

FIG. 7. Scanning electron micrograph of an atretic follicle. The sizes of the granulosa cells vary due to the different degree of shrinkage of pyknotic cells (P) mixed with larger, more normally shaped cells. ×1350

Micrographs courtesy of Dr. Richard F. Wilkinson, Jr., Laboratory of Human Reproduction and Reproductive Biology and Department of Anatomy, Harvard Medical School, Boston, Mass.

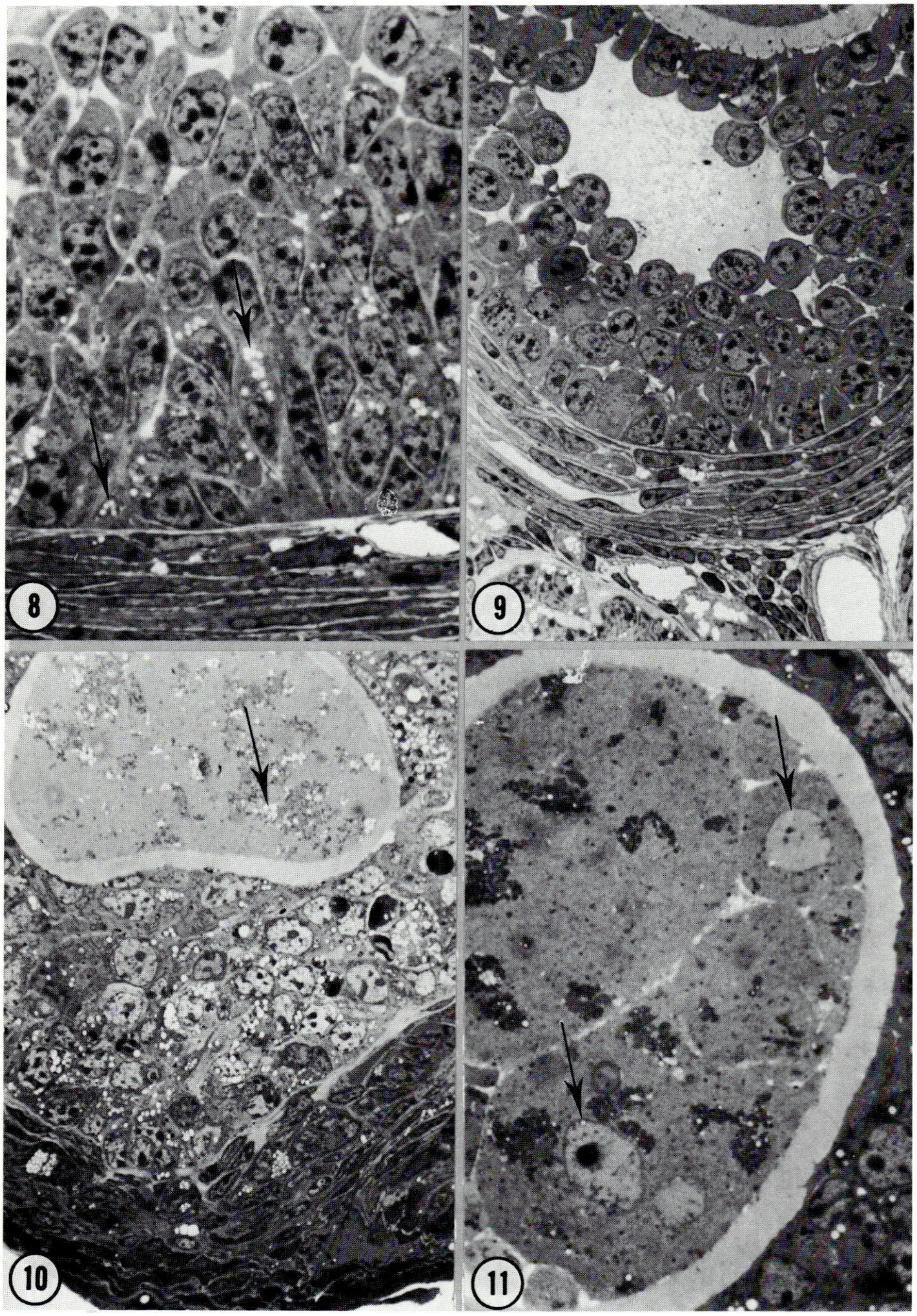
8
9
10
11

CELLULAR RESPONSE OF HEALTHY AND ATRETIC LARGE FOLLICLES TO GONADOTROPINS (PMSG)

By injecting different hormones--gonadotropins, estrogen, progesterone, and testosterone--into intact or hypophysectomized mammals, it is possible to manipulate the pattern of growth and atresia of large follicles (for review see Greenwald, 1974). Small and medium-sized follicles are often much more refractory to these hormones. Neal and Baker (1973), however, reported that PMSG treatment, both *in vivo* and *in vitro*, induced meiotic maturation in both large and medium-sized mouse follicles.

The well-established mitogenic effect of estrogens, in addition to their antiatretic effect (Goldenberg et al., 1972), and the increased incidence of atresia after injection of testosterone (Payne and Runser, 1958) suggest the important role of these hormones in the determination of the fate of follicles. Experiments by Harman et al. (1975) indicate that estrogens plus HCG given to hypophysectomized rats increase follicular atresia. This effect could be inhibited by treatment with antiandrogens (Louvet et al., 1975). Louvet et al. (1975) also feel that the increased growth rate in follicles of hypophysectomized rats seen after exposure to gonadotropins was caused by the initiation of local estrogen secretion. Thus, the antiatretic effect observed 24 hr after injection of PMSG into intact 21-day-old mice (Peters et al., 1975) might in fact be a result of a local stimulation of estrogen production.

FIG. 8. Part of the granulosa and theca layers of a mouse follicle in Stage I atresia. Lipid droplets are accumulating in the cells close to the basement membrane (*arrows*), whereas the theca cells are comparable to "healthy" follicles (see Fig. 9) (1-μm plastic section). ×1408

FIG. 9. Part of a "healthy" mouse follicle. No lipid accumulation is seen in the granulosa cells. Some lipid droplets are present in the theca cells (1-μm plastic section). × 1056

FIG. 10. Luteinized large follicle in Stage III atresia. The granulosa cells as well as some of the theca cells are stuffed with lipid. Lipid droplets are also present in the distorted oocyte (*arrow*) (1-μm plastic section). ×1056

FIG. 11. Fragmented oocyte from an atretic follicle. Nuclear fragments are present in different parts (*arrows*) (1-μm plastic section). × 1672

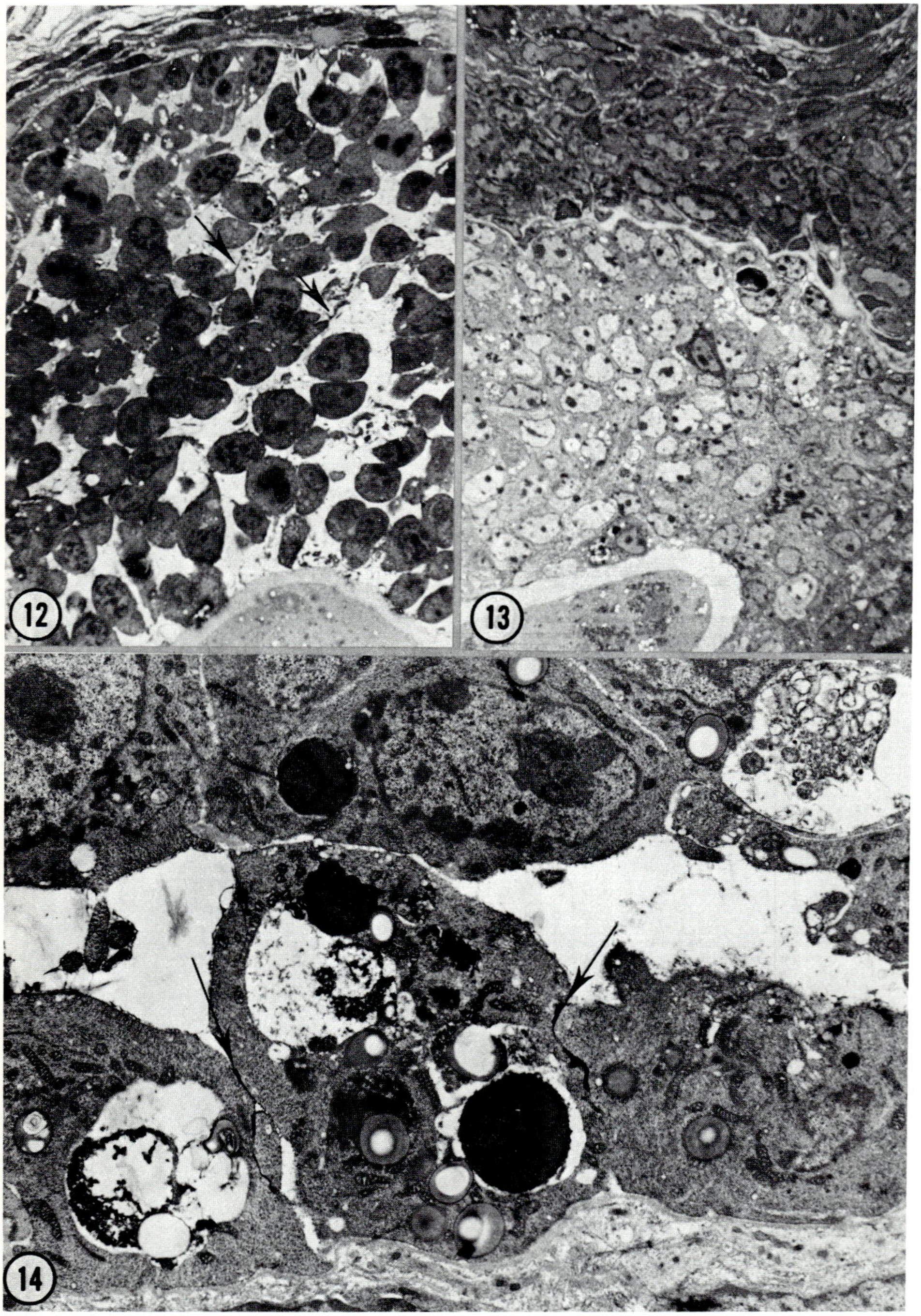
12
13
14

In order to investigate the cellular response of large follicles to the influence of gonadotropic hormones, we studied ovarian follicles of 29-day-old mice, a number of which had received a single injection of 5 IU PMSG 24 hr previously. In one group of untreated and treated animals, the largest follicles present in the ovaries were dissected out in McCoy's culture medium and photographed. During the dissection procedure, it was noticed that the granulosa cells from the PMSG-treated follicles were firmly glued together, whereas granulosa cells of the untreated follicles easily became separated from each other. Each follicle was processed for light or electron microscopy. By comparing different freshly dissected follicles, it was possible to judge whether a follicle was atretic or healthy (Wilkinson and Byskov, 1978). In other control and PMSG-injected animals, the ovaries were immediately fixed for electron microscopy with or without pretreatment with lanthanum to visualize gap junctions (Anderson, 1971). The morphological response of the follicle cells to PMSG treatment 24 hr previously was only studied in healthy large follicles and in follicles in all three stages of atresia.

The PMSG-treated ovaries contained some follicles in Stage I atresia, but very few were in Stage II. The frequency of Stage III seemed to be comparable with the untreated animal.

In the light microscope, many large follicles were judged to be "healthy" by their lack of pyknotic granulosa cells (Fig. 12). However, electron microscopy revealed that most of these "healthy" large follicles were in some stage of atresia but had been "rescued"

←

FIG. 12. "Rescued" follicle after PMSG treatment 24 hr before death. The granulosa layer is loosely arranged. Most cells, which formed microvilli, lie apart from each other. Only a few cells contain lipid droplets (1-μm plastic section). X 1080.

FIG. 13. Follicle in Stage III atresia from a PMSG-treated mouse. The granulosa and theca cells contain little lipid compared to a Stage III follicle in an untreated mouse (see Fig. 10) (1-μm plastic section). ×1080

FIG. 14. Electron micrograph of granulosa cells close to the basement membrane from the "rescued" follicle seen in Fig. 12. The cells contain large vacuoles with cell debris. The intercellular spaces are large and no gap junctions are seen here. The black depositions of lanthanum (*arrows*) between the cells indicate the intercellular space. Lanthanum-treated. × 3870

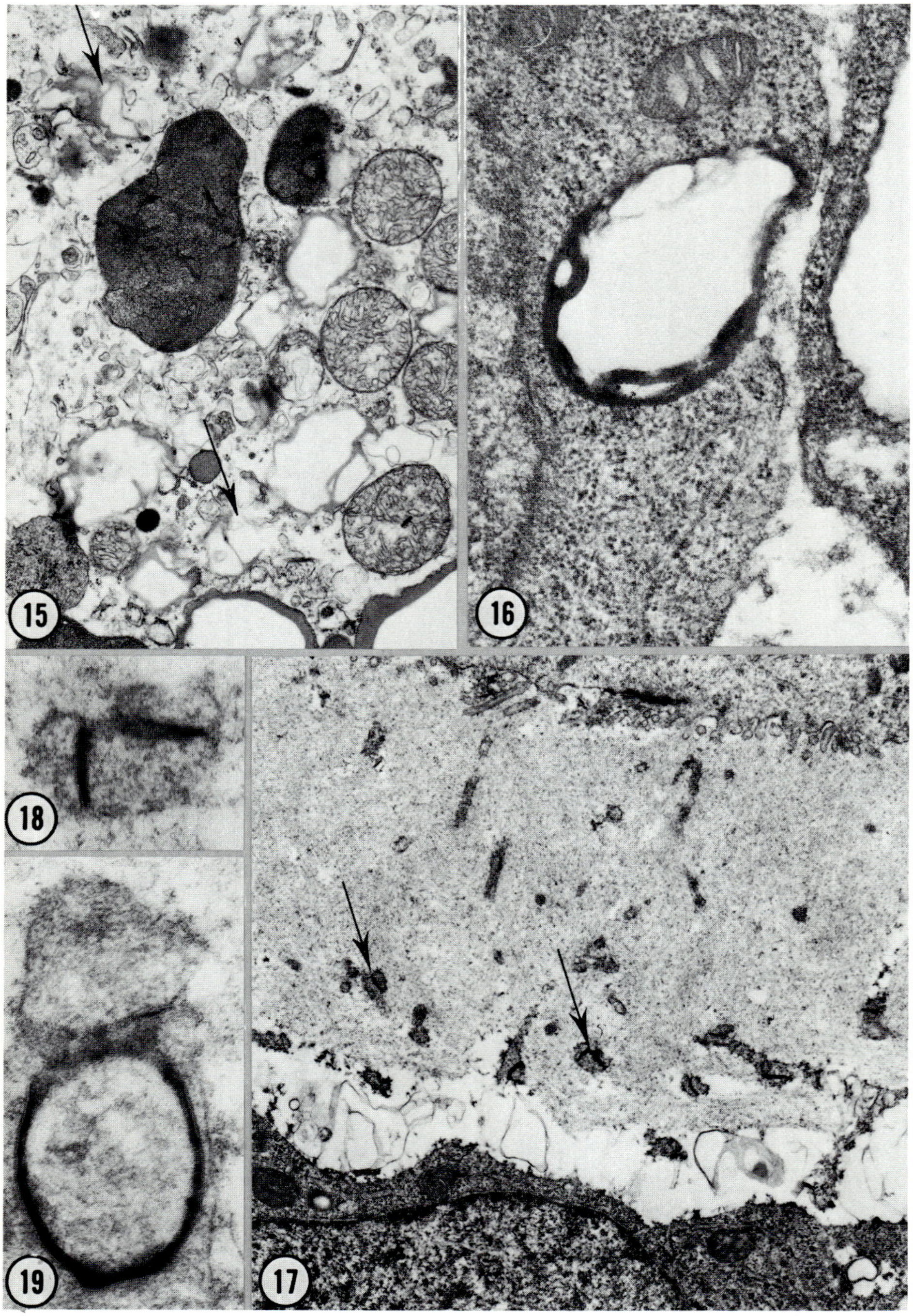
15
16
18
19
17

after injection of PMSG: many healthy-looking granulosa cells were loaded with cell debris, which they apparently had phagocytozed (Fig. 14). Many granulosa cells were in mitosis in follicles in Stage I atresia, but mitotic figures also were seen in Stage II atresia, in which follicles normally are slow- or nongrowing. Thus, some of these atretic follicles seem to become rescued by a self-elimination of dying cells and by a recovered growth ability of the granulosa cells.

A characteristic effect of PMSG treatment 24 hr previously on atretic follicles was a decrease in the cytoplasmic content of lipid droplets compared to controls. In "rescued atretic" follicles the lipid droplets seemed to be dissolved and/or become exocytosed (Fig. 16). If lipids are released into the fluid, the properties of the fluid will probably change. Such changes might be responsible for the striking adherence between the granulosa cells 24 hr after PMSG treatment. The luteolytic effect was particularly evident in follicles in Stage III atresia (Fig. 13; compare with Fig. 10), and affected both the granulosa cells and the theca cells. Lipid droplets seem to become dispersed within the cytoplasm of granulosa cells in Stage III atresia (Fig. 15).

The luteolytic response of the atretic follicle cells seen 24 hr after PMSG treatment is not necessarily a direct effect of PMSG, but rather could be a result of an increased estrogen production as previously discussed.

←

FIG. 15. Electron micrograph of luteinized cell from a follicle in Stage III atresia after PMSG treatment. The lipid droplets seem to be dispersed within the cytoplasm (*arrows*). × 30,600

FIG. 16. Electron micrograph of granulosa cell of "rescued" follicle after PMSG treatment showing exocytosis of a lipid droplet. ×72,900

FIG. 17. Electron micrograph of zona, part of oocyte and granulosa cell of "rescued" follicle after PMSG treatment. The peripheral part of zona contains many microvilli forming gap junctions between each other (*arrows*). Lanthanum-treated. × 13,500

FIG. 18. Electron micrograph of gap junctions from zona of follicle shown in Fig. 17. Lanthanum-treated. × 78,300

FIG. 19. Electron micrograph of gap junctions (from zona of follicle shown in Fig. 17) being internalized within the cytoplasm of the granulosa cell. Lanthanum-treated. ×78,300

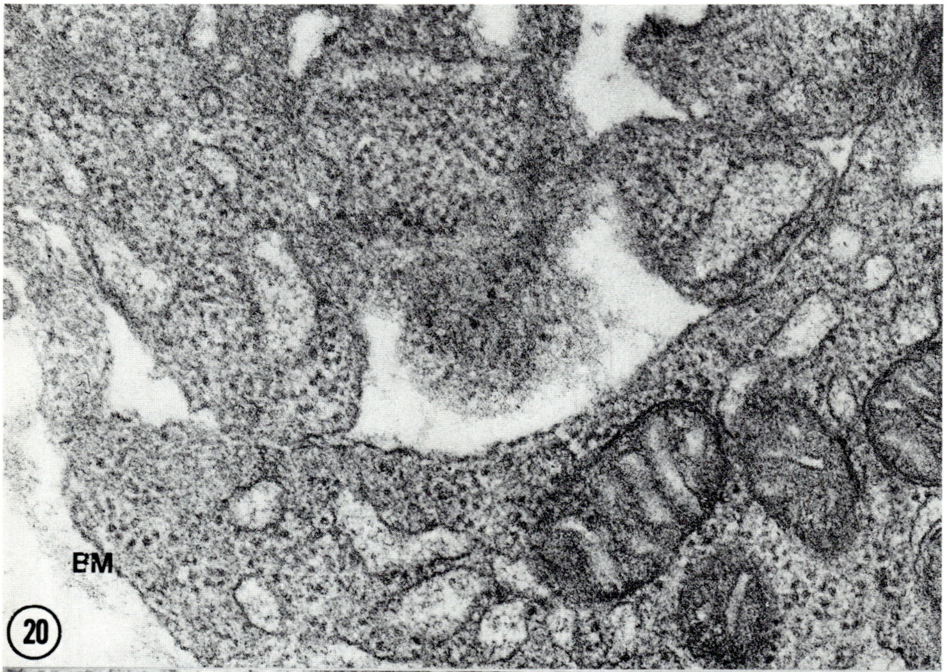
BM
20

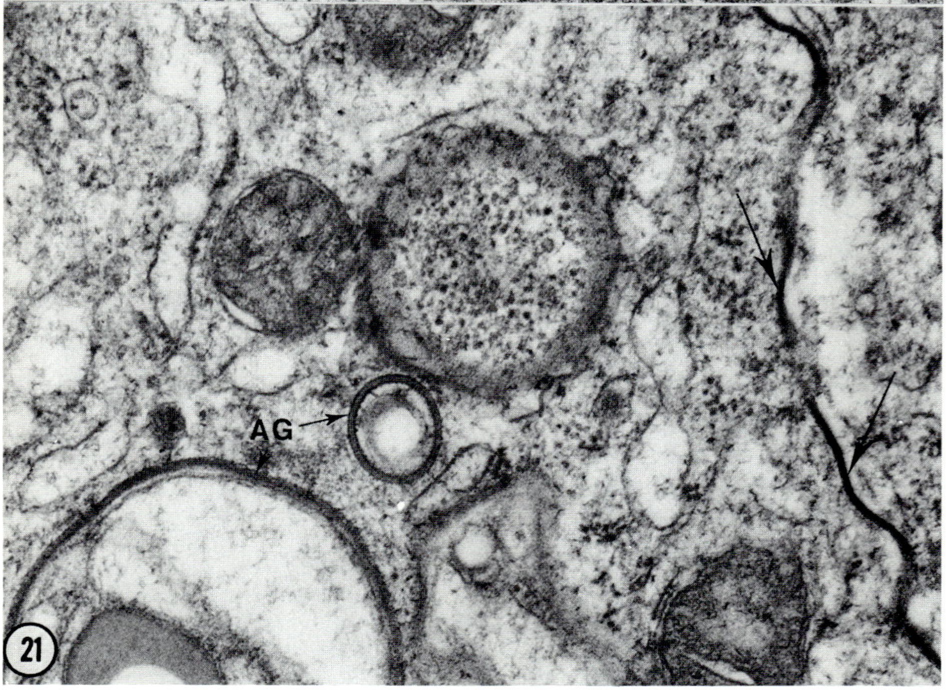
AG
21

The granulosa layers in "healthy," "rescued," and atretic PMSG-treated large follicles were loosely arranged compared to the untreated follicles. The cells were often rounded up but had formed numerous slender microvilli that protruded from their surfaces (Fig. 12). They were well separated from each other even close to the basement membrane (Fig. 12). The number of desmosomes as well as gap junctions seemed to be decreased compared to untreated large follicles (Fig. 21). The sticky appearance of the granulosa layer in follicles of treated animals, noted during the dissecting procedure, is therefore presumably not due to an increased formation of membraneous attachments, but rather due to altered properties of the follicle fluid, which serves as a glue.

Only between cumulus cells were the gaps frequently seen. However, many gap junctions present between microvilli within the zona (Figs. 17 and 18) seemed to be internalized in the microvilli protruding from the cumulus cells (Fig. 19).

In healthy large follicles, numerous gap junctions exist between the granulosa cells (Albertini and Anderson, 1974). These membrane specializations have been proposed to be involved in transport and exchange of ions and molecules, and to serve as electric and metabolic couplers. If the number of these junctions decreases, as seen in large follicles of the PMSG-treated mice, the intrafollicular coordination systems may be disturbed. This would probably influence the differentiation of all follicular compartments. It is possible that when the delicate balance in this highly coordinated system is disturbed, phenomena such as oocyte maturation during atresia or ovulation commence.

In conclusion, follicular atresia in the mature mouse ovary may occur in all follicle types, but the nature and the rate of atresia vary with the size (i.e., stage of differentiation) of the follicle. Medium-sized follicles often become atretic without a concomitant luteinization, whereas luteinization characterizes

←

FIG. 20. Electron micrograph of parts of granulosa cells at the basement membrane (BM) of a large "healthy" follicle after PMSG treatment. The intercellular spaces are wide, leaving only a few small areas where the cells come in close apposition. ×78,300

FIG. 21. Electron micrograph of parts of granulosa cells close to the basement membrane of a large "healthy" follicle of an untreated animal. The cells are closely connected by gap junctions (*arrows*) and annular gaps (AG). Lanthanum-treated. ×72,900

atretic large follicles. The oocyte of medium-sized follicles does not enter the maturation division during atresia, whereas this is the rule in atretic large follicles. The rate of atresia among medium-sized follicles is much lower than among large follicles. It appears that the medium-sized follicles with few hormonal receptors and none or very small gap junctions are rather resistant to the "atretic attack," while the more differentiated large follicles with many hormonal receptors and numerous gap junctions are much more vulnerable.

Finally, it is indicated that large follicles in early stages of atresia may become rescued and proceed through ovulation if a sufficient amount of gonadotropins is available.

REFERENCES

Albertini, D. F., and Anderson, E. (1974): The appearance and structure of intercellular connection during the ontogeny of the rabbit ovarian follicle with particular reference to gap junctions. *J. Cell Biol.*, 63:234-250.

Anderson, E. (1971): Intercellular junctions in the differentiating Graafian follicle of the mouse. *Anat. Rec.*, 169:473a.

Byskov, A. G. S. (1969): Ultrastructural studies on the preovulatory follicle in the mouse ovary. *Z. Zellforsch.*, 100:285-299.

Byskov, A. G. S. (1974): Cell kinetic studies of follicular atresia in the mouse ovary. *J. Reprod. Fertil.*, 37:277-285.

Byskov, A. G. S., and Rasmussen, G. (1973): Ultrastructural studies of the developing follicle. In: *The Development and Maturation of the Ovary and Its Functions*, edited by H. Peters, pp. 55-62. Excerpta Medica, Amsterdam.

Franchi, L. L., and Mandl, A. M. (1962): The ultrastructure of oogonia and oocytes in the foetal and neonatal rat. *Proc. R. Soc. Lond. [Biol.]*, 157:99-114.

Goldenberg, R. L., Vaitukaitis, J. L., and Ross, T. G. (1972): Estrogen and follicle stimulating hormone interactions on follicle growth in rats. *Endocrinology*, 90:1492-1498.

Gondos, B. (1969): The ultrastructure of granulosa cells in the newborn rabbit ovary. *Anat. Rec.*, 165:67-78.

Greenwald, G. S. (1961): Quantitative study of follicular development in the ovary of the intact or unilaterally ovariectomized hamster. *J. Reprod. Fertil.*, 2:351-361.

Greenwald, G. S. (1974): Role of follicle-stimulating hormone and luteinizing hormone in follicular development and ovulation. In: *Handbook of Physiology, Sect. 7: Endocrinology*, edited by S. R. Geiger, pp. 293-323. Waverly Press, Baltimore.

Harman, M. S., Louvet, J.-P., and Ross, G. T. (1975): Interaction of estrogen and gonadotrophins on follicular atresia. *Endocrinology,* 96:1145-1152.

Himelstein-Braw, R., Byskov, A. G., Peters, H., and Faber, M. (1976): Follicular atresia in the infant human ovary. *J. Reprod. Fertil.,* 46:55-59.

Ingram, D. L. (1962): Atresia. In: *The Ovary,* Vol. 1, edited by S. Zuckerman, A. M. Mandl, and P. Eckstein, pp. 247-275. Academic Press, New York.

Lobel, B. L., Rosenbaum, R. M., and Deane, H. W. (1961): Enzymic correlates of physiological regression of follicles and corpora lutea in ovaries of normal rats. *Endocrinology,* 68:232-247.

Louvet, J.-P., Harman, S. M., Schreiber, J. R., and Ross, G. T. (1975): Evidence for a role of androgens in follicular maturation. *Endocrinology,* 97:366-372.

Neal, P., and Baker, T. G. (1973): Response of mouse ovaries *in vivo* and in organ culture to pregnant mare's serum gonadotrophin and human chorionic gonadotrophin. I. Examination of critical time intervals. *J. Reprod. Fertil.,* 37:399-404.

Payne, R. W., and Runser, R. H. (1958): The influence of estrogen and androgen on the ovarian response of hypophysectomized immature rats to gonadotrophins. *Endocrinology,*62:313-321.

Pedersen, T. (1970): Follicle kinetics in the ovary of the cyclic mouse. *Acta Endocrinol. (Kbh.),* 6:304-323.

Peters, H. (1969): The development of the mouse ovary from birth to maturity. *Acta Endocrinol. (Kbh.),* 62:98-116.

Peters, H., Byskov, A. G., Himelstein-Braw, R., and Faber, M. (1975): Follicular growth: The basic event in the mouse and human ovary. *J. Reprod. Fertil.,* 45:559-566.

Wilkinson, R. F., and Byskov, A. G. (1978): *In vivo* criteria that distinguish atretic from healthy murine ovarian follicle. In: *Proceedings of Workshop on Ovarian Follicular Development and Function.* Santa Cruz, Calif. (*in press*).

Ovarian Follicular Development and Function,
edited by A. R. Midgley and W. A. Sadler.
Raven Press, New York © 1979.

Intraovarian Inhibitors of Follicular Function

Cornelia P. Channing

Department of Physiology, University of Maryland
School of Medicine, Baltimore, Maryland 21201

Normally, granulosa cells are kept from luteinizing and the oocyte is kept in the arrested dictyate state of the first meiotic division until shortly prior to ovulation. It is not too unreasonable to propose that there are inhibitors in follicular fluid responsible for keeping the granulosa cells from luteinizing and the oocyte from maturing and extruding the first polar body until the time of ovulation. These inhibitors appear not to be steroidal and may have a basic role in control of follicular atresia either alone or in combination with steroids such as testosterone.

FOLLICULAR FLUID GRANULOSA CELL LUTEINIZATION INHIBITOR

We have demonstrated that addition of 50% follicular fluid obtained from small (1- to 2-mm) porcine follicles to the culture medium can inhibit morphological luteinization and progesterone secretion by monolayer cultures of granulosa cells obtained from large (6- to 12-mm) porcine follicles (Ledwitz-Rigby et al., 1973, 1977). The inhibitory influence was absent in fluid obtained from large follicles. Fifty percent of small follicular fluid in a balance of culture medium 199 could also inhibit the ability of LH to stimulate cyclic AMP accumulation after a 48-hr preincubation with suspensions of granulosa cells obtained from large porcine follicles (Ledwitz-Rigby et al., 1973, 1977). Medium follicular fluid had an intermediate inhibitory effect. Large follicular fluid and serum containing less than 1 ng/ml LH had no inhibitory effect. An assay system for the inhibitory action of small follicular fluid was developed employing a 48-hr preincubation of granulosa cells obtained from large porcine follicles followed by a short 20-min incubation with 1 μg/ml ovine LH (NIH LH-S19) and measurement of cyclic AMP accumulation by the Gilman assay (Channing and

Ledwitz-Rigby, 1975). The inhibitory activity was nondialyzable and was stable at 62 but not at 80°C (Ledwitz-Rigby and Channing, *unpublished observations*). The inhibitory action of small follicular fluid does not appear to be at the level of the LH-HCG receptor, since a 48-hr preincubation of granulosa cells with 50% of small follicular fluid did not alter the ability of the cells to bind iodinated HCG in comparison to preincubation in 50% serum, 50% large follicular fluid, or serum-free culture medium 199D (Ledwitz-Rigby, 1974).

The inhibitory action of the small follicular fluid could be partially overcome by two phosphodiesterase inhibitors, methyl-isobutyl xanthine (MIX) and theophylline (Ledwitz-Rigby, 1974). This observation, as well as our finding that preincubation of granulosa cells in 50% small follicular fluid leads to a stimulation of phosphodiesterase activity above the nondetectable level (Ledwitz-Rigby, 1974), leads us to suspect that the inhibition of luteinization is mediated by a stimulation of phosphodiesterase activity. Stimulation of phosphodiesterase activity could inhibit the increased cyclic AMP levels generated in response to LH and thereby inhibit the further stimulation of actions of cyclic AMP on steroidogenesis. This could keep granulosa cells in small follicles from luteinizing prematurely. That the granulosa cell is kept in a steroidogenically arrested state within the preovulatory monkey follicle *in vivo* as observed by Channing and Coudert (1976) is further support for a follicular fluid luteinization inhibitor. The role of such an inhibitor in the process of follicular atresia may be of importance and is a subject worthy of further investigation.

Further studies on the chemical nature and biological site of inhibitory action of the small follicular fluid luteinization inhibitor are currently being carried out. In the case of medium and large follicles, it is unlikely that the oocyte secretes a luteinization inhibitor as has been originally proposed by Nalbandov and his colleagues (El Fouley et al., 1970; Nekola and Nalbandov, 1971). We have added 20 to 50 cumulus-enclosed porcine oocytes from medium and large follicles to cultures of porcine granulosa cells harvested from medium and large follicles. In no instance, in the presence or absence of 0.5 or 5 μg/ml ovine LH (NIH LH-S18), did the oocytes inhibit granulosa cell luteinization morphologically or in terms of progesterone secretion (Tsafriri and Channing, 1977*a*). Nicosia was also unable to observe oocyte inhibition of morphological luteinization in cultures of rabbit granulosa cells (Nicosia, 1972*a*,*b*).

FOLLICULAR FLUID OOCYTE MATURATION INHIBITOR

The finding that oocytes in medium and large follicles are kept in the arrested dictyate stage of the first meiotic division from shortly before or around the time of birth until immediately prior to ovulation, and only then will resume meiosis up until extrusion of the first polar body stage, prompted us to speculate that there was an inhibitor of oocyte meiosis present in follicular fluid (see Channing and Tsafriri, 1977 for literature review). In 1955 Chang observed that addition of follicular fluid to cultured isolated rabbit oocytes inhibited their spontaneous maturation *in vitro*. We have confirmed and extended this original observation (Tsafriri and Channing, 1975) and demonstrated that addition of 50% porcine follicular fluid from small, medium, or large follicles could inhibit the spontaneous *in vitro* maturation of oocytes isolated from medium and large porcine follicles. Co-culture of porcine oocytes with intact hemisections of follicle wall or with 10^7 granulosa cells, but not with hemisections of plain thecal wall or 10^7 foreskin fibroblasts, also inhibited oocyte maturation (Tsafriri and Channing, 1975). Granulosa cell extract also inhibited oocyte maturation (Tsafriri et al., 1977*b*). These data led us to postulate that an oocyte meiosis inhibitor that originated in granulosa cells was present in follicular fluid. Addition of LH but not prolactin could overcome the inhibitory action of follicular fluid (Tsafriri et al., 1976).

The inhibitory action of porcine follicular fluid was destroyed by trypsin treatment (Tsafriri et al., 1976) and was not extractable with charcoal treatment, which removed all detectable estrogen and progesterone (Channing and Tsafriri, 1977*b*; Tsafriri et al., 1976).

Amicon ultrafiltration of follicular fluid demonstrated that the inhibitor passed through a membrane with a nominal molecular weight (NMW) cutoff of 10,000, but was retained by a membrane of 1,000 NMW cutoff. Chromatography of the UM-2 retentate on a column of Sephadex G-25 demonstrated that the oocyte maturation inhibitor was located in fractions that corresponded to a molecular weight of less than 2,000 (Tsafriri et al., 1976). These data are compatible with the suggestion that the follicular fluid oocyte maturation inhibitor is a polypeptide with a molecular weight of about 2,000. The inhibitory activity of the UM-2 retentate fraction of porcine follicular fluid could be overcome by the addition of 5 μg ovine LH (NIH S19) but not by prolactin (10 μg/ml, NIH-B2) (Tsafriri et al., 1976). The inhibitory action of unfractionated follicular fluid could not be overcome by dibutyryl cyclic AMP (0.04 to 0.4 mM) (Channing and Tsafriri, 1977*b*). The follicular fluid inhibitor of oocyte maturation does not seem to be species specific. It was found that porcine FFl interfered with the maturation of mouse oocytes (Channing and Tsafriri, 1977*b*). Clearly, more studies are needed to further chemically characterize the follicular fluid inhibitor of oocyte maturation as well as to elucidate its mode of action.

OVARIAN INHIBITOR OF LH BINDING

Yang et al. (1976) observed that the high-speed supernatant of homogenates of pseudopregnant and pregnant rat ovaries inhibited LH binding to luteal cell membranes. They further fractionated the inhibitory substance and demonstrated, using Sephadex column chromatography, that the inhibitor(s) was a small molecular weight polypeptide of 3,800 daltons. It was not detectable in immature rat ovary. Further studies are required to determine the precise cellular origin of the inhibitor and its role within the physiological framework of the ovarian function. An attractive speculation is that the follicle that grows and matures secretes less of this inhibitor to enable more LH to bind to it and enable adequate follicular maturation. Channing and Kammerman (1974) have demonstrated that thecal and granulosa cells of large mature but not of small immature porcine follicles can bind iodinated HCG. Alternately, follicles becoming atretic could secrete more inhibitor. This appears to be an area ripe for future investigation.

It is possible that there may be a separate ovarian inhibitor of FSH binding, although it has not been demonstrated. Reichert (1977) demonstrated that an inhibitor of FSH binding to testes tubule is present in testis homogenates. Apparently this inhibitor is specific for FSH and does not inhibit LH binding. Table 1 summarizes some of these intraovarian inhibitors as well as some intra- and extraovarian control mechanisms involved in follicular maturation.

ACKNOWLEDGMENTS

The financial support of the Population Council of New York and of the National Institute of Child Health and Human Development, USPHS grants HD03315 and HD08834, is gratefully acknowledged.

TABLE 1. *Intra- and extraovarian control of follicular maturation*

Intraovarian control of follicular maturation	Extraovarian control of follicular maturation
Stimulators	Stimulators
Estrogen	LH
Testosterone (?)	FSH
Follicular fluid factor(s)	Ovarian growth factor (?)
	Insulin
	Cortisol (physiological levels)
	Thyroxin (physiological levels)
Inhibitors	Inhibitors
Small follicular fluid luteinization inhibitor (Ledwitz-Rigby et al., 1977)	High serum prolactin (?)
LH binding inhibitor (corpus luteum) (Yang et al., 1976)	Adrenal testosterone (?)
Oocyte maturation inhibitor found in follicular fluid (Tsafriri et al., 1976*a*,*b*)	
Testosterone (Louvet et al., 1975)	

REFERENCES

Chang, M. C. (1955): The maturation of rabbit oocytes in culture and their maturation, activation, fertilization and subsequent development in the fallopian tubes. *J. Exptl. Zool.*, 128:378-385.

Channing, C. P., and Coudert, S. P. (1976): Contribution of granulosa cells and follicular fluid to ovarian estrogen secretion in the Rhesus monkey *in vivo*. *Endocrinology*, 98:590-597.

Channing, C. P., and Kammerman, S. (1974): Binding of gonadotropin to ovarian cells. *Biol. Reprod.*, 10:179-198.

Channing, C. P., and Ledwitz-Rigby, F. (1975): Methods for assessing hormone-mediated differentiation of ovarian cells in culture and in short-term incubations. In: *Methods of Enzymology* (J. G. Hardman and B. W. O'Malley, eds.), Part D, pp. 183-230. Academic Press, New York.

Channing, C. P., and Tsafriri, A. (1977): Mechanism of action of luteinizing hormone and follicle stimulating hormone on the ovary *in vitro*. *Metabolism*, 26:413-468.

ElFouley, M., Cook, B., Nekola, M., and Nalbandov, A. V. (1970): Role of the ovum in follicular luteinization. *Endocrinology*, 87:288-293.

Ledwitz-Rigby, F. I. (1974): Possible site of inhibitory action of follicular fluid (FFl) upon porcine granulosa cell (GC) luteinization. Abstract 16 presented at the 7th Annual Meeting of the Society for the Study of Reproduction, held at Ottawa, Canada on August 19-22, 1974.

Ledwitz-Rigby, F. I., Rigby, B. W., Gay, V. L., Young, J., Stetson, M., and Channing, C. P. (1977): Inhibitory action of porcine follicular fluid upon granulosa cells luteinization in vitro: assay and influence of follicular maturation. *J. Endocrinol.* 74:175-184

Ledwitz-Rigby, F., Stetson, M., and Channing, C. P. (1973): Follicular fluid (FFl) inhibition of LH stimulation of cyclic-AMP levels in porcine granulosa cells (GC). *Biology of Reproduction,* 9:94-95, Abstract 85.

Louvet, J., Harmon, S. M., Schreiber, J. R., and Ross, G. T. (1975): Evidence for a role of androgens in follicular maturation. *Endocrinology,* 97:366-376.

Nekola, M., and Nalbandov, A. V. (1971): Morphological changes of rat follicular cells as influenced by oocytes. *Biol. Reprod.,* 4:154-160.

Nicosia, S. V. (1972): Luteinization of rabbit preovulatory granulosa cells cultured in vitro in presence of follicular oocytes. l. Growth characteristics and progestin biosynthesis. *Fert. Steril.,* 23:791-801.

Reichert, L. (1976): Characteristics of gonadotropin receptors in seminiferous tubules. In: *Receptors for Reproductive Hormones* (L. Birnbaumer, ed.). Volume II: Peptide Hormones. Academic Press, New York *(in press).*

Tsafriri, A., and Channing, C. P. (1975): An inhibitory influence of granulosa cells and follicular fluid upon porcine oocyte meiosis *in vitro.* *Endocrinology,* 96:922-927.

Tsafriri, A., and Channing, C. P. (1977*a*): Lack of an inhibitory influence of oocytes upon luteinization of porcine granulosa cells in culture. *J. Reprod. Fert.,*50:103-105.

Tsafriri, A., and Channing, C. P. (1977*b*): Regulation of ovulatory processes: ovum maturation, follicular rupture and luteinization. In: *Advances in Fertility Regulation through Basic Research* (W. A. Sadler and S. Segal, eds.). Raven Press, New York *(in press).*

Tsafriri, A., Pomerantz, S. H., and Channing, C. P. (1976): Inhibition of oocyte maturation by porcine follicular fluid: partial characterization of the inhibitor. *Biology of Reproduction,* 14:511-516.

Yang, K., Samaam, N., and Ward, D. N. (1976): Characterization of an inhibitor for luteinizing hormone receptor site binding. *Endocrinology,* 98:233-241.

Ovarian Follicular Development and Function,
edited by A. R. Midgley and W. A. Sadler.
Raven Press, New York © 1979.

In Vivo Identification and Structure of Atretic Murine Ovarian Follicles

Richard F. Wilkinson, Jr., Anne Grete Byskov, and Everett Anderson

Laboratory of Human Reproduction and Reproductive Biology and the Department of Anatomy, Harvard Medical School, Boston, Massachusetts 02115

All discussions on the morphological characteristics of atretic follicles have relied on the interpretation of sectioned ovaries (e.g., see Zuckerman et al., 1962). If, on the other hand, one were able to identify atretic follicles prior to the drastic measures required to observe sectioned ovaries (i.e., fixation, dehydration, embedding, etc.), experiments could be designed with objectives in mind that would distinguish and compare particular features of healthy and atretic follicles. For example, the hypothesis recently advanced by Richards and Midgley (1976) that suggests differential quantities of luteinizing hormone receptors between cells of atretic and healthy follicles could be more confidently tested. Other similarly designed experiments could also be planned in the same comparative context. In this chapter, which precedes a more detailed description (Wilkinson et al., *in preparation*), we describe some morphological criteria that will be of value in distinguishing healthy from atretic murine ovarian follicles *in vivo*. Additionally, some descriptions of PMSG-treated follicles are presented.

MATERIALS AND METHODS

While dissecting ovaries from euthanized 18- to 28-day postpartum mice, we noticed that the medium-sized pre- and early antral follicles (about 400 μm in diameter) could be placed in one of two classes based on gross morphology alone. Follicles of homogeneous appearance were pooled and processed for light and electron microscopy. In order to study the effect of gonadotropins on follicular morphology, 5 IU of pregnant mare serum gonadotropins (PMSG) were injected

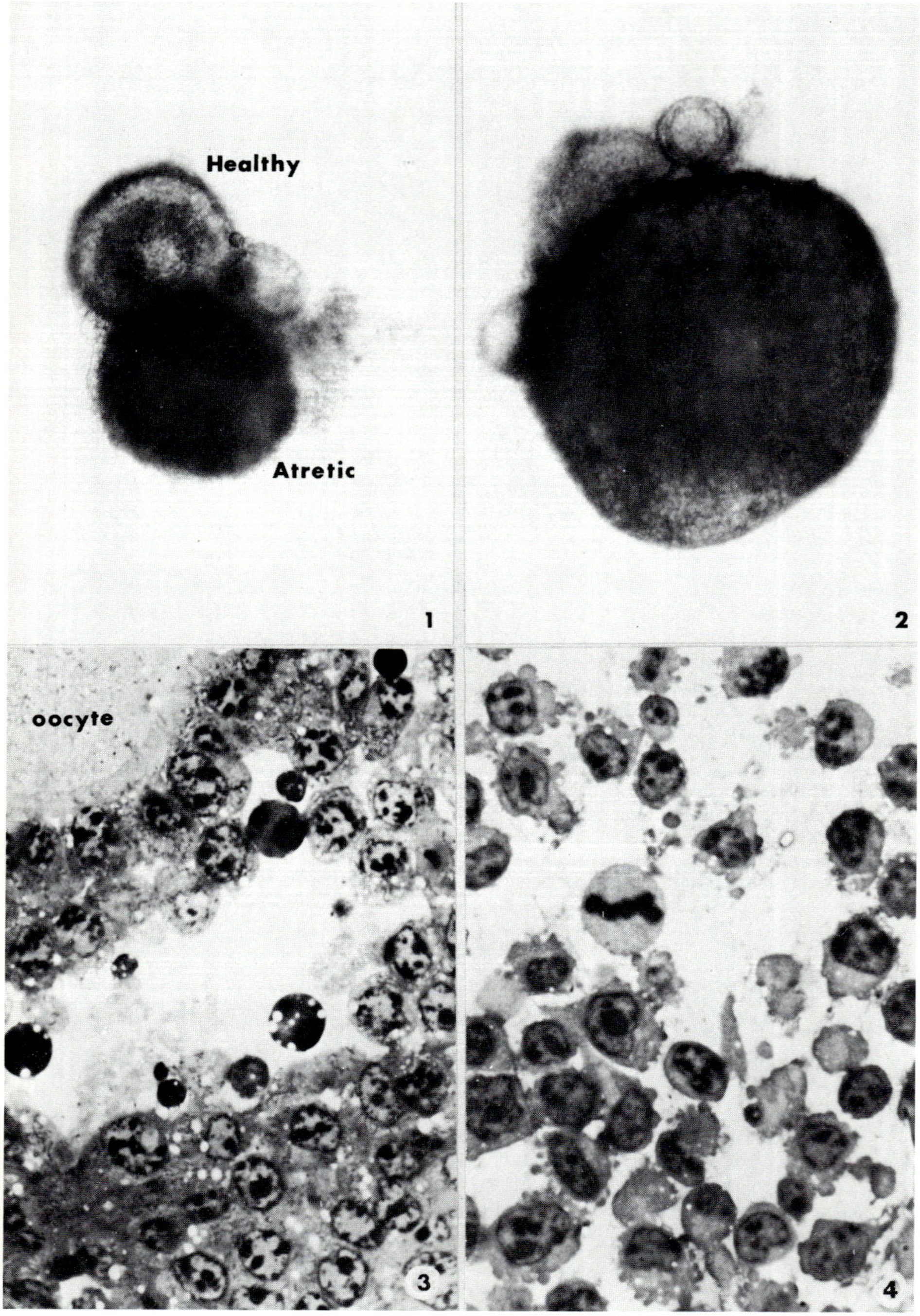
Healthy
Atretic
1
2
oocyte
3
4

intraperitoneally 24 hr prior to dissection. The follicles from these PMSG-treated animals were also processed for microscopy.

RESULTS AND DISCUSSION

After microscopical examination, it was found that one of the two follicular "classes" was composed of healthy follicles and the other of follicles in early stages of atresia (as described by Byskov, 1974). Follicles in the healthy "class" appeared lucent and spherical in outline, which was in contrast to follicles in the atretic "class," which were more opaque, pink in color, and often more oval (Fig. 1). Microscopically, the atretic follicles contained granulosa cells that were degenerating (Fig. 3) and characteristically had pyknotic nuclei and dense-staining ribosome-like particles filling the cytoplasm (Figs. 5 and 6). Additionally, these degenerating cells had spherical intracellular structures that seemed to be remnants of or were derived from rough endoplasmic reticulum (Fig. 6). These spherical bodies were frequently located close to the plasma membrane, often appearing to have fused with the plasmalemma and to have released their contents into the intracellular space (Fig. 6). Granulosa cells in healthy follicles were isomorphous and showed none of the changes associated with degenerating granulosa cells.

We also studied the effects of PMSG treatment and follicular "rescue" as described by Peters et al. (1975). As shown in Fig. 2, which is at the same magnification as Fig. 1, the volume of PMSG-treated follicles increased about seven times. Many granulosa cells were in mitotic division, cells were widely spaced, and the interphase cells were highly pleomorphic (Fig. 4). In addition to the pleomorphic shape, manifested by numerous microvillar and pseudopodial processes (Fig. 7), the endoplasmic reticulum was

←

FIG. 1. Photomicrograph of an atretic and healthy ovarian follicle. ×126

FIG. 2. Photomicrograph of a follicle from an animal that had been given 5 IU PMSG 24 hr prior to dissection. ×126

FIG. 3. One-micrometer plastic section through a 21-day atretic follicle. Toluidine blue stain. ×720

FIG. 4. One-micrometer plastic section through a follicle from a PMSG-treated animal. Toluidine blue stain. ×720

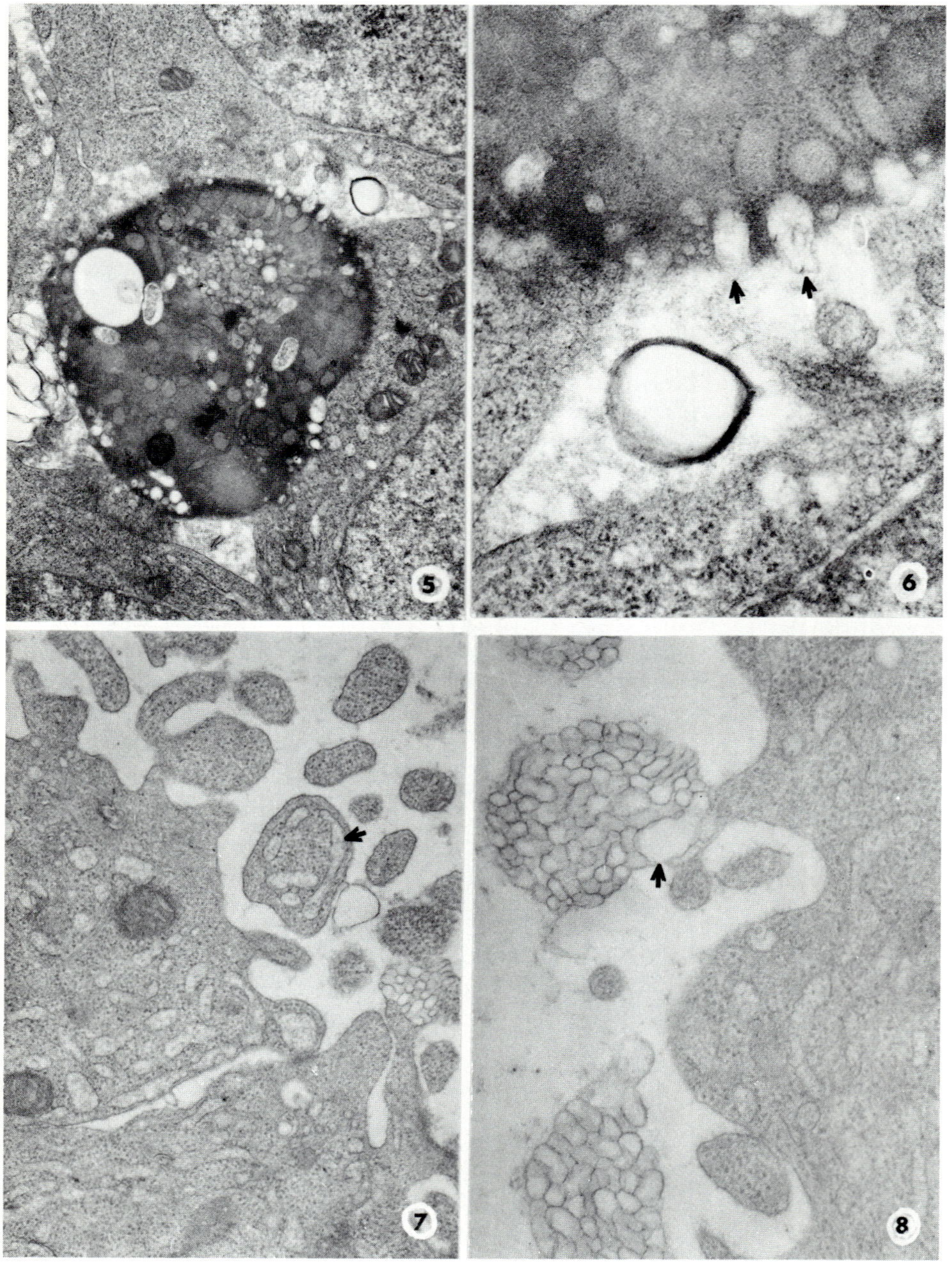
5
6
7
8

greatly distended and often was located in attenuated surface projections (Fig. 7). Often single or small groups of mitochondria were found in the interstitial spaces, suggesting that some granulosa cells lysed as a result of exposure to PMSG. A curious feature of cells in the treated follicles was the consistent presence of membranous structures at the surface of granulosa cells in treated animals (Figs. 7 and 8). These "foamy-looking" membranous structures appear to be derived from some pseudopodial processes. Whether or not this phenomenon is related to the processes noted in degenerating granulosa cells (the spherical bodies noted in Fig. 6) remains to be investigated further.

To the authors' knowledge, this is the first description of a reliable *in vivo* method that will aid in distinguishing healthy from atretic murine follicles. The use of this method should aid in the clarification of differences and similarities that exist among the various cellular compartments of ovulatory and atretic follicles. Atresia presents for the anatomist investigations on cellular differentiation or dedifferentiation and cell death. In addition, this system suggests potential in correlating exogenous hormone treatment (such as the PMSG described herein) and morphological change in granulosa cells. Since atretic follicles can be "rescued" by PMSG treatment (Peters et al., 1975), the change in cellular morphology probably reflects a change in function and therefore is a system worthy of continued attention.

←

FIG. 5. Electron micrograph of a degenerating granulosa cell in an atretic follicle. ×2700

FIG. 6. Higher power electron micrograph of the degenerating cell shown in Fig. 5. ×8100

FIG. 7. Electron micrograph of a granulosa cell from a follicle that was removed from a PMSG-treated animal. ×4500

FIG. 8. Higher magnification of the membranous structures found associated with the surface of granulosa cells from PMSG-treated animals. ×8100

ACKNOWLEDGMENT

This work was supported by NIH grant HD06822.

REFERENCES

Byskov, A. G. S. (1974): Cell kinetic studies of follicular atresia in the mouse ovary. *J. Reprod. Fertil.*, 37:277-285.

Peters, H., Byskov, A. G., Himelstein-Braw, R., and Faber, M. (1975): Follicular growth: The basic event in the mouse and human ovary. *J. Reprod. Fertil.*, 45:559-566.

Richards, J. S., and Midgley, A. R., Jr. (1976): Protein hormone action: A key to understanding ovarian follicular and luteal cell development. *Biol. Reprod.*, 14:82-94.

Zuckerman, S., Mandl, A. N., and Eckstein, P. (eds.) (1962): *The Ovary, Vols. I and II*. Academic Press, New York.

Ovarian Follicular Development and Function,
edited by A. R. Midgley and W. A. Sadler.
Raven Press, New York © 1979.

Estrogen-Induced Follicular Atresia in Rhesus Monkeys

James R. Clark*, Donald J. Dierschke, and Richard C. Wolf

Wisconsin Regional Primate Research Center, University of Wisconsin, Madison, Wisconsin 53706

The follicle destined to ovulate can be identified, using laparoscopy, by day 6 of the menstrual cycle in rhesus monkeys (Clark and Dierschke, 1975). Yamaji et al. (1971) observed that injections of estradiol benzoate administered at about this stage of the menstrual cycle in monkeys induced surges of gonadotropic hormones, but ovulation did not ensue. Rather, peripheral serum concentrations of estrogen were low for several days after the induced surges, suggesting a disruption of the normal sequence of follicular maturation. Therefore it would appear that the exogenous estrogen interfered with the continuing development of the follicle, although it was not clear in their study (Yamaji et al., 1971) if the elevated progesterone secretion, which was eventually noted in the same cycle, resulted from ovulation of the same or a newly developed follicle. We have used this experimental situation as a possible model system in which to investigate the howmonal requirements for the initiation or reinitiation of follicular development in this species. This chapter presents preliminary data regarding the fate of the preovulatory follicle in the presence of elevated concentrations of circulating estrogen.

The preovulatory follicle (3 to 5 mm diameter) was identified by laparoscopy on day 6 of the menstrual cycle and Silastic capsules containing crystalline 17 -estradiol were implanted subcutaneously for 48 hr in 8 monkeys during 12 cycles. Daily blood samples were obtained for analysis of serum concentrations of luteinizing hormone (LH) and estrogen by radioimmunoassay. Laparoscopy was performed at 3 to 5 day intervals to determine ovarian response to the treatment.

---

*Present address: Department of Animal Science, Texas Tech University, Lubbock, Texas 79409.

After implantation of the capsules, the mean serum concentrations were increased by 434 pg/ml above the previous base line (Table 1). Within 24 hr after removal of the exogenous steroid, the mean serum estrogen concentrations

TABLE 1. *Effects of exogenous estrogen*[a] *on serum concentrations of estrogen and follicular atresia in rhesus monkeys*

Animal number	Mean serum estrogen concentrations (pg/ml): Prior to treatment	During treatment	Following treatment	Gross morphology of follicle following treatment[b]
1511	123.4	633.6	74.6	Reduced translucence
1511	180.6	577.7	71.3	Hemorrhagic
1651	129.4	442.0	67.9	Hemorrhagic
2575	230.4	581.5	92.4	Reduced translucence
2575	165.0	601.5	82.4	Hemorrhagic
4862	110.9	491.7	67.6	Hemorrhagic
1287	157.4	677.9	80.5	Follicular cyst
1287	157.8	535.4	122.7	Follicular cyst
F77	176.1	705.1	82.6	Reduced translucence
F77	130.5	561.1	118.2	Hemorrhagic
31Z	136.7	677.1	79.2	Reduced translucence
H47	113.2	539.1	82.7	Reduced translucence
Mean	151.0	585.3	85.2	
SEM	9.9	22.8	5.2	

[a]17β-Estradiol was administered via Silastic capsules placed subcutaneously on day 6 of the menstrual cycle and removed 48 hr later.

[b]Based on observations made at laparoscopy.

were significantly decreased relative to pretreatment concentrations (85 vs. 151 pg/ml, respectively), thus providing endocrinological evidence for the cessation of follicular maturation. In untreated animals, circulating concentrations of estrogen would be near peak concentrations at this stage of the menstrual cycle (Hotchkiss et al., 1971). During laparoscopic examinations after estrogen treatment, the first distinct morphological changes from the pretreatment condition, assumed to be indicative of regression, were noted for each follicle and classified as follows: (1) reduced translucence--cloudy or opaque appearance of transilluminated follicular fluid; (2) hemorrhagic--dark red appearance of follicular contents as viewed through the follicle wall; and (3) cystic degeneration--large fluid-filled structures (>10 mm diameter) with extensive vascularization of the follicle surface (Table 1). All of the above changes were noted within 5 to 9 days from the beginning of estrogen treatment. Subsequently, the diameter of each follicle decreased and the follicle disappeared by twelve to seventeen days after treatment. In two instances, the ovary containing the developing follicle was removed 5 to 6 days after beginning steroid treatment and histological examination revealed the follicle to be in the beginning stages of early atresia according to the criteria of Koering (1969). Associated with the above treatment effects, the exogenous estrogen induced a surge of LH within 48 hr, which was similar to that observed by Yamaji et al. (1971).

Subsequent to regression of the original preovulatory follicle, a different follicle (previously unidentified by laparoscopy) developed in four of the animals. This was followed by a spontaneous LH surge and ovulation. The extended follicular phase averaged 26.2 ± 3.1 days in length and was followed by a luteal phase that varied from 9 to 19 (average, 15.2 ± 2.4) days in length. In these animals, the average cycle length was increased (41.5 ± 4.3 days) when compared to the pretreatment cycle length (29.2 ± 4.0 days). In the eight remaining cycles in which a spontaneous LH surge and ovulation were not observed, the average cycle length was 23.2 2.0 days. Immediately after the cycle of steroid treatment, normal ovulatory cycles were observed in 11 of 12 cases (average, 29.9 1.5 days).

These results indicate that high concentrations of estrogen induced rapid regression of the follicle destined to ovulate. The earliest indicator of regression was the low circulating concentrations of estrogen immediately after treatment. Later, histological evidence of follicles in early atresia was observed, and this was followed by gross morphological changes noted by laparoscopy. These results, however, do not allow us to conclude whether the exogenous estrogen acts directly on the ovary or at the hypothalamo-hypophyseal level.

ACKNOWLEDGMENT

This is Publication No. 17-008 of the Wisconsin Regional Primate Research Center. This research was supported by the Wisconsin Alumni Research Foundation, National Institutes of Health (Grant RR-00167), and National Institute of Child Health and Human Development (Grant HD-01816).

REFERENCES

Clark, J. R., and Dierschke, D. J. (1975): Follicle/corpus luteum relationships in rhesus monkeys. In: *Proceedings of 8th Annual Meeting of the Society for the Study of Reproduction,* Abstract 45, pp. 48-49.

Hotchkiss, J., Atkinson, L. E., and Knobil, E. (1971): Time course of serum estrogen and luteinizing hormone concentrations during the menstrual cycle of the rhesus monkey. *Endocrinology*, 89:177-183.

Koering, M. J. (1969): Cyclic changes in ovarian morphology during the menstrual cycle in *Macaca mulatta*. *Am. J. Anat.*, 126:73-101.

Yamaji, T., Dierschke, D. J., Hotchkiss, J., Bhattacharya, A. N., Surve, A. H., and Knobil, E. (1971): Estrogen induction of LH release in the rhesus monkey. *Endocrinology*, 89-1034-1041.

Ovarian Follicular Development and Function,
edited by A. R. Midgley and W. A. Sadler.
Raven Press, New York © 1979.

Follicular Atresia in the Mare

W. A. Condon, V. K. Ganjam, R. M. Kenney, and
*C. P. Channing

University of Pennsylvania, New Bolton Center,
Section of Reproductive Studies, Kennett Square,
Pennsylvania 19348 and *Department of Physiology,
University of Maryland, School of Medicine,
Baltimore, Maryland 21201

Any follicle that is destined to undergo ovulation must first escape the normal degenerative process of atresia. The biochemical events involved in determining which follicles will escape atresia and remain viable during their growth from primary to graafian follicles are unknown. The following study was undertaken to determine how a viable follicle differs histologically and biochemically from an atretic follicle.

A total of 41 antral follicles from 12 mares has been examined so far in this study. Follicular fluid was aspirated from all follicles immediately on removal of the ovary for quantification of total estrogens and progesterone. An aliquot of the fluid was also added to indomethacin for prostaglandin F analysis. All fluids were stored at -20°C. Both steroid hormones and prostaglandin F were quantitated by radioimmunoassays utilizing highly specific antibodies. Half the follicle wall was then processed for histological examination and half was used as a source of granulosa cells; attempts were made to determine the ability of these granulosa cells to bind ^{125}I-HCG.

The follicles were classified as viable or atretic based on histological evaluation. Pyknosis in the mural granulosa was the earliest evidence used to classify a follicle as atretic.

A summary of the biochemical determinations made on the follicular fluid from each follicle is shown in Fig. 1. Viable follicles contained significantly higher estrogen ($p < 0.01$) and progesterone ($p < 0.025$) levels than did atretic follicles, although prostaglandin F values were significantly higher ($p < 0.05$) in the latter.

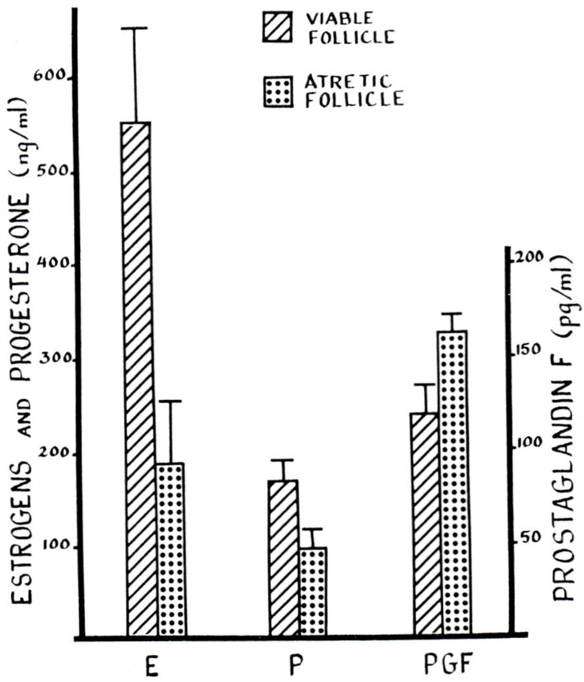

FIG. 1. Concentrations of estrogens (E), progesterone (P), and prostaglandin F (PGF) in follicular fluid of viable and atretic mare follicles.

Of these three parameters, estrogen concentration was the best indicator of viability, and it correlated well with other parameters such as gross vascularity, follicle diameter, and volume of follicular fluid (r = 0.613) (Fig. 2).

When follicles (n = 14) with an estrogen concentration greater than 200 ng/ml were correlated with the histological observations of their corresponding follicle wall, it was realized that only two of them contained normal theca epithelioid cells. In the remaining follicles the thecal cells had either completely atrophied or had undergone extensive luteinization, as summarized in Table 1. In all cases, however, the granulosa cells were either normal, partially luteinized, or in the earliest stages of atresia (i.e., exhibiting a small number of pyknotic nuclei). The cell type that synthesized the estrogens in these follicles is, therefore, questionable.

In summary, although none of the parameters used in this study proved to be absolute indicators of follicle viability or atresia, atretic follicles tended to have

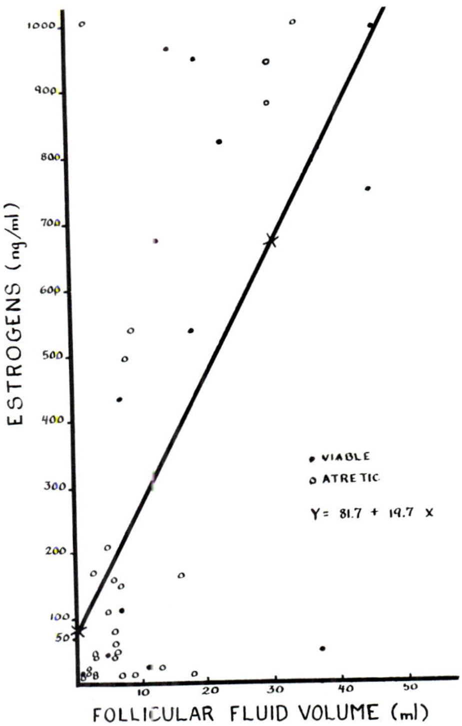

FIG. 2. Correlation of follicular fluid volume with estrogen concentration in viable and atretic mare follicles.

estrogen concentrations ≤ 200 ng/ml, follicular fluid volumes ≤ 10 ml, and very little or no gross vascularity.

TABLE 1. *Histological observations of granulosa and thecal cells in follicles with high levels of estrogen*

Animal number	Follicle number	Estrogens (ng/ml)	Granulosa cells	Thecal cells
1272	L1	495	Incipient atresia	None
1530	L2	1150	Incipient atresia	None
336	L2	540	Incipient atresia	None
336	R1	1150	Incipient atresia	None
1511	L1	882	Incipient atresia	Luteinized
1554	R1	945	Incipient atresia	Luteinized
1272	R1	1000	Partial luteinization	Luteinized
336	L1	750	Partial luteinization	Luteinized
1284	L1	825	Partial luteinization	Luteinized
1511	L2	435	Partial luteinization	Luteinized
1757	R1	950	Active	Luteinized
1757	R2	965	Active	Luteinized

ACKNOWLEDGMENTS

The authors wish to thank Dr. David T. Armstrong for analysis of prostaglandin F, and Drs. Roy Bergman, Wendell Cooper, and Ben Baker for help with the experimental animals.

Ovarian Follicular Development and Function,
edited by A. R. Midgley and W. A. Sadler.
Raven Press, New York © 1979.

A Possible Role for Follicular Fluid in Atresia

Florence Ledwitz-Rigby

Department of Biological Sciences, Northern Illinois University, DeKalb, Illinois 60115

Porcine granulosa cells from preovulatory follicles luteinize *in vitro* in a variety of culture media. However, progesterone secretion (Table 1) and LH-induced cyclic AMP accumulation (Table 2) are suppressed by 48-hr incubations in a medium composed of equal parts of fluid from small (1- to 2-mm) porcine follicles (SFFl) and medium 199. Forty-eight-hour incubations in 50% SFFl had only a slight influence on granulosa cell morphology. The cells appeared healthy but contained fewer lipid droplets and were less epithelioid than cells grown in control media. By 6 days of monolayer culture, however, when control cultures were well luteinized morphologically, SFFl treated cells had failed to luteinize, exhibited membrane blebbing, and detached from the coverslips. Culture media containing fluid from medium-sized (3- to 5-mm) follicles (MFFl) was only partially inhibitory of luteinization, while fluid from large (6- to 12-mm) follicles was as permissive as other control culture media containing either porcine serum or serumless medium 199D by all of the criteria mentioned above. Three millimolar theophylline or 0.2 mM methylisobutyl xanthine enhanced LH-induced cyclic AMP accumulation in SFFl-treated cells but had no significant effect on cells in the other treatment groups (Table 3). There was no difference in the ability of granulosa cells to specifically bind ^{125}I-HCG after 2 days of incubation in 50% SFFl, MFFl, LFFl, serum, or the serumless medium 199D (Table 4).

These observations may indicate the presence of a luteinization inhibitor in fluid from small follicles, which is gradually lost as the follicle matures. Follicles that do not experience such a change in fluid composition may be unable to respond to LH stimulation and may be more susceptible to other factors that initiate atresia.

TABLE 1. *Influence of follicular fluid on progesterone secretion during 48-hr incubations*

Experiment No.	Type of culture	Media:	Progesterone secretion (ng/5 × 10^7 cells/day) Mean ± SEM		
			50% SFFl	50% LFFl	Other
1	Monolayer		4,700 ± 400 (3)	>120,000 (3)	--
2	Monolayer		1,100 ± 43 (3)	2,334 ± 90 (3)	2,469 ± 156 (2) (50% serum)
3	Suspension		36 ± 6 (5)	74 ± 8 (4)	--
4	Suspension		81 ± 5 (4)	202 ± 39 (3)	--
5	Suspension		5,656 ± 548 (4)	40,121 ± 2,167 (4)	18,972 ± 4,620 (6) (199D)

Granulosa cells obtained from large (6- to 12-mm) follicles were incubated for 48 hr. The progesterone secreted into the culture media was determined by radioimmunoassay. Follicular fluids and sera were pretreated to remove endogenous progesterone. The number of cultures used to obtain the mean is indicated in parentheses after the mean.

TABLE 2. *Influence of follicular fluid on LH-stimulated cyclic AMP accumulation*

48-Hr preincubation culture medium	Cyclic AMP accumulated (% of control) Mean ± SEM	
	Basal	LH-stimulated
50% SFFl	7.9 ± 0.8 (59)	29.0 ± 2.6 (57)
50% MFFl	23.3 ± 7.3 (9)	75.0 ± 11.3 (14)
50% LFFl	3.4 ± 1.2 (48)	100.0 ± 2.0 (56)
50% Porcine serum	3.3 ± 1.1 (10)	112.4 ± 9.5 (14)
Medium 199D	4.0 ± 1.4 (5)	94.0 ± 8.0 (6)

Granulosa cells harvested from large porcine follicles were preincubated for 48 hr. Subsequently, the cells were incubated for 20 min with or without 1.0 μg/ml ovine LH. Cyclic AMP accumulation was estimated at the end of the 20 min incubation using the Gilman competitive protein binding assay. To compare data obtained in several experiments, the average amount of cyclic AMP accumulated in response to LH by cells preincubated in 50% LFFl has been defined as the control in each experiment, and all other values are expressed as a percent of this value. The number of cultures used to obtain the mean is indicated in parentheses after the mean.

TABLE 3. *Influence of phosphodiesterase inhibitors on LH stimulation of cyclic AMP accumulation*

48-Hr culture medium	Phosphodiesterase inhibitor	No. of experiments	Cyclic AMP accumulation (% of control) Mean ± SEM	
			Basal	LH-stimulated
50% SFF1	None	8	6.6 ± 2.4	26.9 ± 7.4
	3 mM Theophylline	8	6.9 ± 1.6	53.8 ± 13.5
	0.2 mM Methylisobutyl xanthine	1	5.5	46.8
50% LFF1	None	8	7.8 ± 2.6	100.0
	3 mM Theophylline	8	19.0 ± 8.1	106.6 ± 10.6
	0.2 mM Methylisobutyl xanthine	1	3.7	70.6
50% Serum	None	3	7.0 ± 0.5	135.7 ± 26.9
	3 mM Theophylline	3	5.0 ± 2.4	113.9 ± 17.0

Granulosa cells from large follicles were incubated for 48 hr in suspension cultures. Subsequently, they were incubated with or without 1 μg/ml of ovine LH in the presence or absence of a phosphodiesterase inhibitor for 20 min. Cyclic AMP accumulation was estimated at the end of the 20-min incubation. To compare data obtained in several different experiments, the mean cyclic AMP accumulated in response to LH in the absence of an inhibitor by cells preincubated in 50% LFF1 was defined as the control for each experiment. All other data are expressed as a percent of this value.

TABLE 4. *Effects of 48-hr incubations on the binding of ^{125}I-HCG to granulosa cells*

	CPM of ^{125}I-HCG bound per 5×10^6 cells ($\bar{x} \pm$ SEM)		
Treatment	Experiment 1	Experiment 2	Experiment 3
0-Hr control	30,434 ± 3,278	66,553 ± 2,214	34,740 ± 1,462
48-Hr incubation			
50% Small Follicular fluid	9,800 ± 271	11,894 ± 342	9,060 ± 922
50% Medium Follicular fluid	8,423 ± 654	12,916 ± 724	7,942 ± 592
50% Large Follicular fluid	7,430 ± 495	6,991 ± 962	9,952 ± 1,255
50% Pig serum	--	12,858 ± 4,946	8,033 ± 589
50% Pig serum	--	--	9,055 ± 214
Medium 199D	8,447 ± 1,453	8,687 ± 1,349	14,534 ± 1,173
Nonspecific binding	--	10.3 ± 4.9	16.2 ± 2.0%

Binding of ^{125}I-HCG to granulosa cells was estimated after no incubation (0-hr control) or after a 48-hr incubation in either 50% follicular fluid from small (1- to 2-mm), medium (3- to 5-mm), or large (6- to 12-mm) porcine follicles, 50% pig serum in a balance of medium 199, or chemically defined medium 199D. Nonspecific binding was estimated in the presence of 1.0 μg/ml unlabeled HCG.

Ovarian Follicular Development and Function,
edited by A. R. Midgley and W. A. Sadler.
Raven Press, New York © 1979.

In Vitro Maturation of Oocytes Collected from Atretic Follicles

John J. Peluso

Reproductive Physiology Laboratories and C. S. Mott Center for Human Growth and Development, Wayne State University School of Medicine, Detroit, Michigan 48201

As preovulatory follicles become atretic, acid phosphatase activity appears in the granulosa cell layer, then pyknotic nuclei are observed within the antrum of the follicle (Guraya and Greenwald, 1964; Peluso et al., 1977). Although the follicle undergoes these degenerative changes, the oocyte remains intact and possesses a distinct germinal vesicle and nucleolus. However, the viability of the oocyte may be altered as the follicle degenerates. One indication of the viability of an oocyte is its ability to resume meiosis and undergo germinal vesicle breakdown (GVB) *in vitro*. Thus, an experiment was designed to determine the rate of germinal vesicle breakdown in oocytes collected from follicles in various stages of atresia.

METHODS AND MATERIALS

Thirty-five Wistar rats, 24 days of age, received 5 IU PMSG at 9 A.M. Groups of 10 to 12 rats were killed at 48, 72, and 96 hr after PMSG administration. The ovaries were removed and either fixed in Bouin's solution for histological examination or placed in Brinster's media for ova culture (BMOC-3). The six largest preovulatory follicles from each ovary, which was placed in BMOC-3, were punctured with a needle, releasing the oocyte surrounded by cumulus cells. After the oocytes were washed, they were incubated in 0.1-ml drop of BMOC-3 under paraffin oil at 37°C in a humidified 5% CO_2 and air atmosphere (Biggers et al., 1971). The interval between the time the rats were killed and the time that the oocytes were placed in culture (T = 0) was approximately 45 min. At 0, 1, 2, and 3 hr after culture, oocytes were taken out of culture, incubated with hyaluronidase (350 IU/ml) to remove the cumulus cells, and examined under Nomarski interference optics. The percentage of oocytes undergoing GVB was analyzed by Chi square with p values less than 0.05 considered to be significant.

RESULTS

Histology

Forty-eight hours after PMSG treatment, the ovaries of 26-day-old rats had many large preovulatory follicles. The oocytes in these follicles were intact and possessed distinct germinal vesicles and nucleoli. Seventy-two hours after PMSG injection, the granulosa cell layer of the largest follicles was noticeably thinned and a few pyknotic nuclei were present. By 96 hr after PMSG injection, granulosa cells ceased dividing, and many pyknotic nuclei were observed. Although the granulosa cell layer had deteriorated, the oocyte still possessed an intact germinal vesicle.

Germinal Vesicle Breakdown

Most of the oocytes collected at 48 hr after PMSG injection possessed intact germinal vesicles through the first hour of culture. The percentage of oocytes undergoing GVB increased to 67% after 2 hr in culture with a further increase to 92% after 3 hr of culture. Within 45 min after oocytes were removed from follicles, 72 hr after PMSG administration, 36% had undergone GVB. This percentage of GVB was significantly greater than the preculture percentage of GVB of oocytes collected 48 hr after PMSG treatment. Although the percentage of GVB did not increase after 1 hr in culture, an increase in GVB occurred at 2 hr but then remained constant. In 55% of the oocytes 96 hr after PMSG treatment, the germinal vesicle was broken down. This percentage of GVB was significantly higher than the preculture percentage for oocytes collected from follicles 48 and 72 hr after PMSG injection. In addition, the cytoplasm of these oocytes was very granular. When these oocytes were cultured, the percentage of GVB gradually increased, reaching 88% after 3 hr of culture. The percentage of oocytes to undergo GVB was less for oocytes collected 96 hr than for those collected 48 hr after PMSG treatment (see Fig. 1). In addition, 15 of 26 oocytes were either fragmented or had a very granular cytoplasm 3 hr after culture.

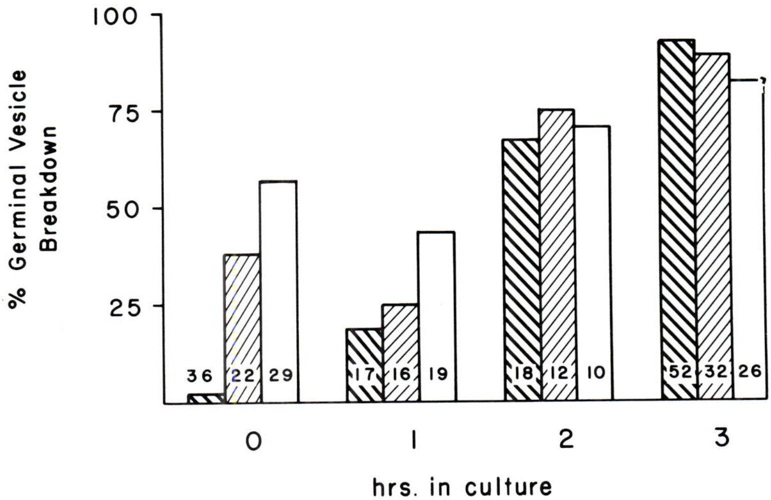

FIG. 1. The rate of germinal vesicle breakdown in oocytes collected from follicles at 48 hr (▧), 72 hr (▨), and 96 hr (□) after PMSG treatment.

DISCUSSION

In atretic oocytes, the mitochondria and other organelles cluster together leaving areas of cytoplasm free of organelles (Vazquez-Nin and Sotelo, 1967). The redistribution of organelles could account for the granular nature of oocytes collected at 96 hr after PMSG treatment and could indicate that these oocytes are beginning to degenerate. In addition, these oocytes undergo GVB *in vitro* more rapidly than those of healthy follicles and fragment when cultured for 3 hr. Similarly, fragmented oocytes are often observed within follicles that are in the final stages of degeneration (Ingram, 1962). These observations suggest that once an atretic oocyte is removed from the influence of the follicle, whether by being placed in culture or by the degeneration of its follicles, it spontaneously fragments.

By 72 hr after PMSG treatment, the follicle begins to degenerate as indicated by the appearance of acid phosphatase activity within the granulosa cell layer (Peluso et al., 1977). However, degenerative changes within the oocyte occur 96 hours after PMSG injection. Since the granulosa cells are thought to be a nutrient source for the oocyte (Zamboni, 1974), any degenerative changes within the granulosa cells could be detrimental to the oocyte. Thus, the temporal relationship between degenerative changes within the follicle and the oocyte suggest that the follicle deteriorates prior to the oocyte.

ACKNOWLEDGMENT

This work was supported by Ford Foundation grant 710-0287A.

REFERENCES

Biggers, J. D., Whitten, W. K., and Whittingham, D. G. (1971): The culture of mouse embryos *in vitro*. In: *Methods in Mammalian Embryology*, edited by J. C. Daniel, Jr., pp. 86-116. Freeman, San Francisco.

Guraya, S. S., and Greenwald, G. S. (1964): A comparative histochemical study of interstitial tissue and follicular atresia in the mammalian ovary. *Anat. Rec.*, 149:411-434.

Ingram, D. L. (1962): Atresia. In: *The Ovary, Vol. 1,* edited by S. Zuckerman, pp. 247-273. Academic Press, New York.

Peluso, J.J., Steger, R.W., and Halez, E. (1977): Sequential changes associated with the degeneration of preovulatory rat follicles. J. Reprod. Fert. 49:215-218.

Vazquez-Nin, G.H., and Sotelo, J.R. (1967): Electron microscope study of the atretic oocytes of the rat. Zellforsch., 80:518-533.

Zamboni, L. (1974): Fine morphology of the follicle wall and follicle cell-oocyte association. Biol. Reprod. 10:125-149.

Ovarian Follicular Development and Function,
edited by A. R. Midgley and W. A. Sadler.
Raven Press, New York © 1979.

Discussion Summary: Session on Atresia

A. G. Byskov

The introductory remarks emphasized the complexity of follicular growth as well as atresia: that many components not only in the follicle itself but also in its proximate surroundings may be involved in regulating the fate of the follicles. The chairman expressed the need for further knowledge of these events, particularly in the primate ovary.

Morphological descriptions to distinguish large atretic follicles from healthy ones *in vivo* were presented for both the mouse and the rhesus monkey. The atretic follicles were less translucent than the healthy follicles, which also had a richer blood supply. In the mouse, the validity of these criteria was supported by ultrastructural studies.

The importance of the follicular integrity was mentioned in relation to onset of atresia. It was suggested that atresia might be associated, for example, with the permeability of the basement membrane or with the dissociation of the oocyte and cumulus cells. The importance of receptors for FSH and LH in developing and maturing follicles was discussed. It was proposed that the first steps leading to atresia could be a loss of receptors for gonadotropins.

In some studies, the intact PMSG-injected mouse was used as a model to study follicular atresia. The convenience of using PMSG-treated animals for studying follicular atresia was questioned, and it was suggested to use hypophysectomized animals given PMSG followed by anti-PMSG.

In light of studies on hypophysectomized immature rats the function of gonadotropins was evaluated. Treatment of these animals with gonadotropins as well as with estrogens decreased the number of atretic follicles. However, results of combined treatment with antiestrogens indicated that the antiatretic effect of gonadotropins most likely is mediated by a local stimulation of estrogen synthesis. Finally, the atretogenic effect of androgens was tested in a system in which estrogens and HCG with and without antiandrogens were given. The results showed that the additional treatment with antiandrogens inhibited the onset of atresia. Thus, it appears that the balance between estrogen and androgen plays an important role in determining whether a follicle becomes atretic or not.

That exogenous estrogens interfered with the follicular development was shown in a study of the rhesus monkey. Silastic capsules with 17-β-estradiol left subcutaneously for 2 days resulted in regression of the preovulatory follicle, development of a new ovulatory follicle, and a prolongation of the following cycle.

In a study of large atretic follicles in the mare it was shown that in the early stages of atresia the follicle fluid contained estrogens, progesterone, and prostaglandin F within the levels of healthy follicles, although the atretic ones tended to have lower concentrations of estrogens and larger amounts of prostaglandin F than the healthy follicles.

Much interest was shown in the possible regulatory mechanisms of follicular growth and atresia by inhibitors. Such intrafollicular inhibitors might regulate the cellular response to LH and FSH. Studies of cultures of granulosa cells from preovulatory porcine follicles showed that their luteinization could be inhibited in the presence of fluid from small and medium follicles. This indicated that the follicle fluid from these follicles contained a luteinization inhibitor. It was pointed out that LH inhibitors could prevent *in vitro* luteinization of *in vivo* LH-primed mature granulosa cells. Furthermore, it was reported that estrogens had no influence on the activity of the inhibitor, but that steroids may act on the secretion of the inhibitor.

The discovery of polyploidy in the granulosa cells can represent a new parameter that may be involved in follicular development and atresia. Granulosa cells from small follicles showed the highest values of ploidy, whereas the values were very near to 2 N in large follicles.

In closing, it was pointed out that the atretic follicles are not only "used follicles," but that they produce a significant amount of sex steroid hormones.

FOLLICULAR MORPHOLOGY

Ovarian Follicular Development and Function,
edited by A. R. Midgley and W. A. Sadler.
Raven Press, New York © 1979.

Follicular Morphology

Everett Anderson

Department of Anatomy and Laboratory of Human
Reproduction and Reproductive Biology,
Harvard Medical School, Boston, Massachusetts 02115

In the study of the ultrastructure of cellular components of the ovarian follicle, it has become clear, as it has in other cell systems, that the plasma membrane, which may be specialized for many biochemical events, delineates a cytoplasmic matrix richly compartmentalized by a system of membranes. Most of the intracellular membranes are either morphologically highly organized with many complex functions, such as mitochondria, or are not so highly structurally organized but with nevertheless varied biochemical functions, such as the endoplasmic reticulum. This chapter will not direct attention to the ultrastructural organization of the cellular constituents of the differentiating follicle. Since the plasma membrane is a substrate for a multiplicity of symphonic cellular episodes, such as antigen-antibody reaction and hormone stimulation, to name two, this chapter will focus attention on certain specialized regions of the plasma membrane of granulosa cells whose organization has been elucidated by the explication of new methodology.

It has become abundantly clear in recent years that the activation of target cells by protein hormones is mediated by events occurring at the level of the plasma membrane (see Catt and Dufau, 1976). The granulosa cells of the follicular compartment of the mammalian ovary respond to tropic stimuli from the pituitary, thereby making their plasma membranes and their specializations an appealing model. By use of colloidal lanthanum as a tracer, investigators have studied the gap junction, one specialized region of the plasma membrane of granulosa cells (Anderson, 1971; Fletcher and Everett, 1973; Merk et al., 1973; Albertini and Anderson, 1974). Gap junctions have been implicated as sites of molecular and ionic exchange and electrical and metabolic coupling between cells which serve

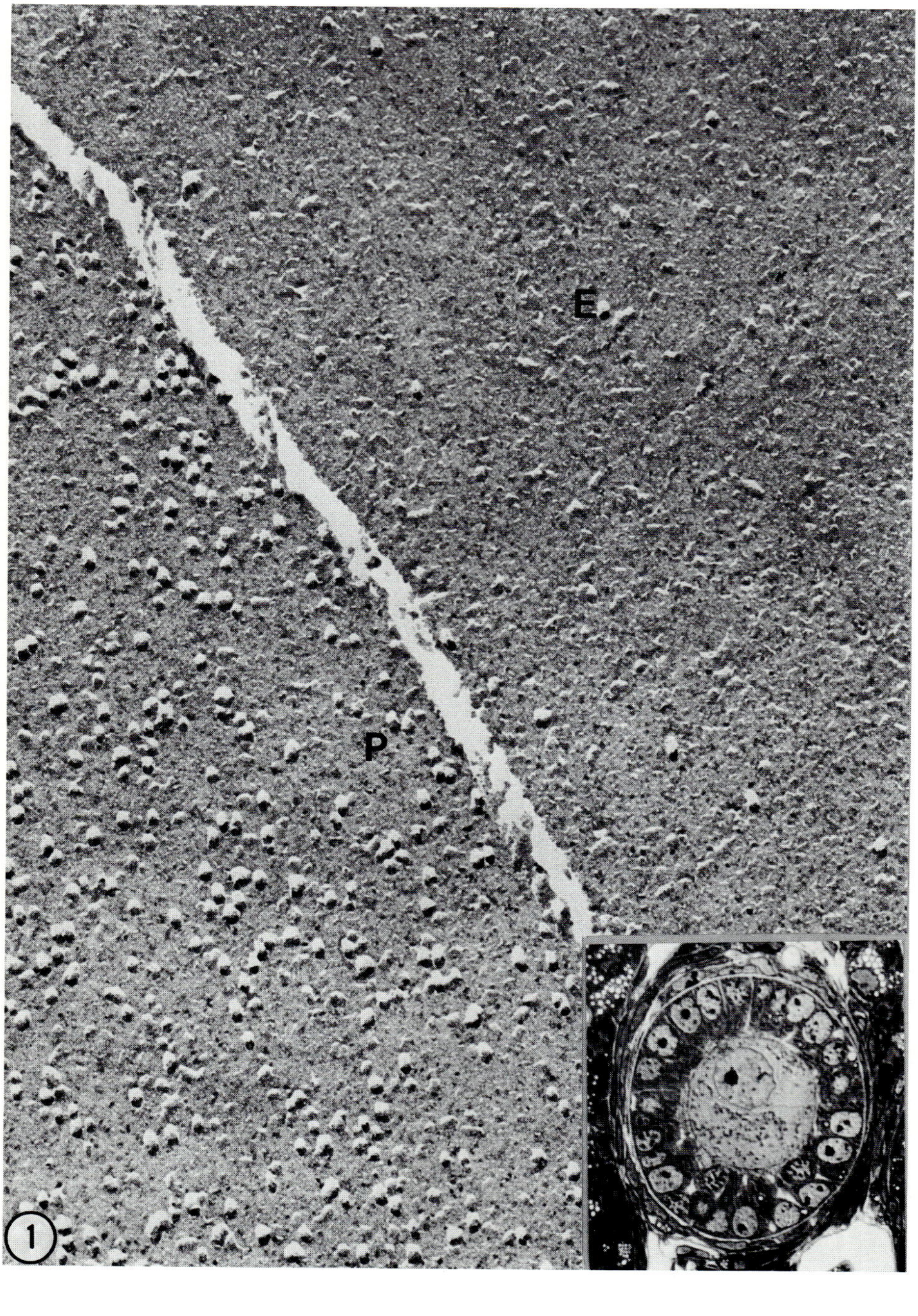
E
P
1

to coordinate cell function in many tissues (Furshpan and Potter, 1968; McNutt and Weinstein, 1970; Friend and Gilula, 1972; Bennett, 1973).

Since follicular development can be hormonally manipulated and follicle size related to the stage of development, the granulosa cellular layer can be readily analyzed with respect to the assembly of gap junctions. The morphological analysis of gap junction evolvement could not have been possible were it not for the development of the freeze-fracture technique (Branton, 1966). This technique has become very useful in analyzing the internal construction of cellular membranes because it allows for direct visualization of intramembranous differentiations on platinum-carbon replicas of frozen and cleaved biological specimens.

In collaboration with Dr. David Albertini we have studied the development of ovarian gap junctions in a number of mammalian species (mice, rat, rabbit) primarily by the freeze-fracture technique. The nomenclature as suggested by Branton et al. (1975) will be used in this chapter. For this investigation we divided the follicles as follows: preantral, early antral, and graafian or preovulatory follicle.

PREANTRAL FOLLICLES

An early preantral follicle consists of an oocyte surrounded by a layer of squamous-shaped granulosa cells. As the follicle grows, the granulosa cells change shape from a simple squamous to one of either cuboidal or short columnar cells (Fig. 1, inset). Our observations of rabbit preantral follicles reveal no gap junctions between cells; however, the protoplasmic face (P-face) is distinguished by many randomly scattered 60 to 100 Å particles (Fig. 1). On both the P-face and E-face (E-face is adjacent to the extracellular space) are particles and pits. There are no highly ordered regions of particles on either face. The granulosa cells of early preantral follicles of rabbits are associated with each other by *maculae adherentes*. In the rat they are associated with each other by both gap junctions and maculae adherentes.

←

FIG. 1. The inset is an early preantral follicle containing an oocyte encompassed by a layer of cuboidal granulosa cells. The freeze-fracture replica (main portion of figure) shows the particle-rich P-face and particle-poor E-face of granulosa cells from a preantral follicle. ×193,600 Rabbit. Inset, toluidine blue stained, ×352 Mouse.

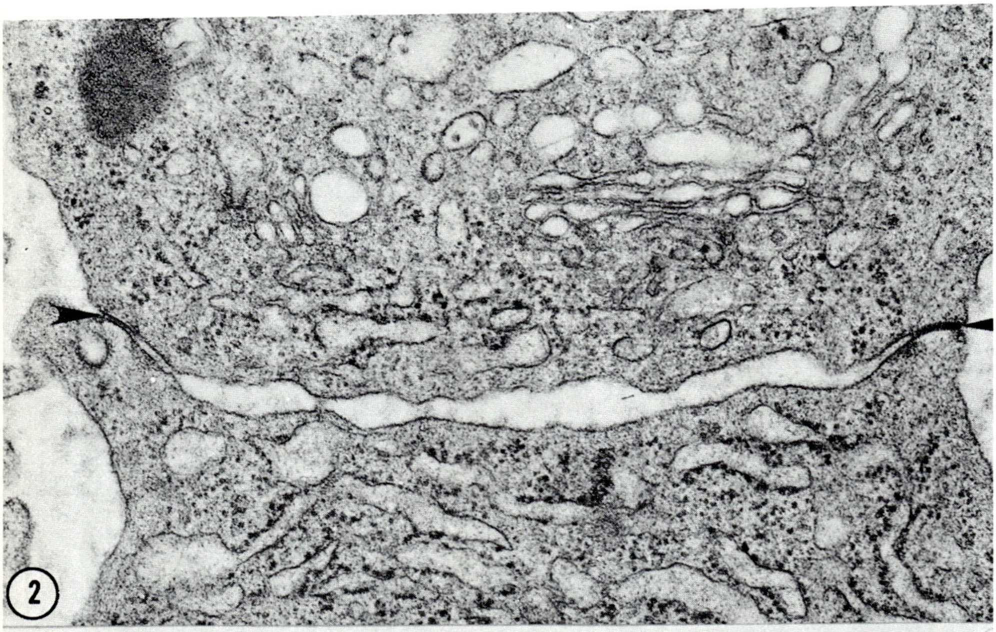

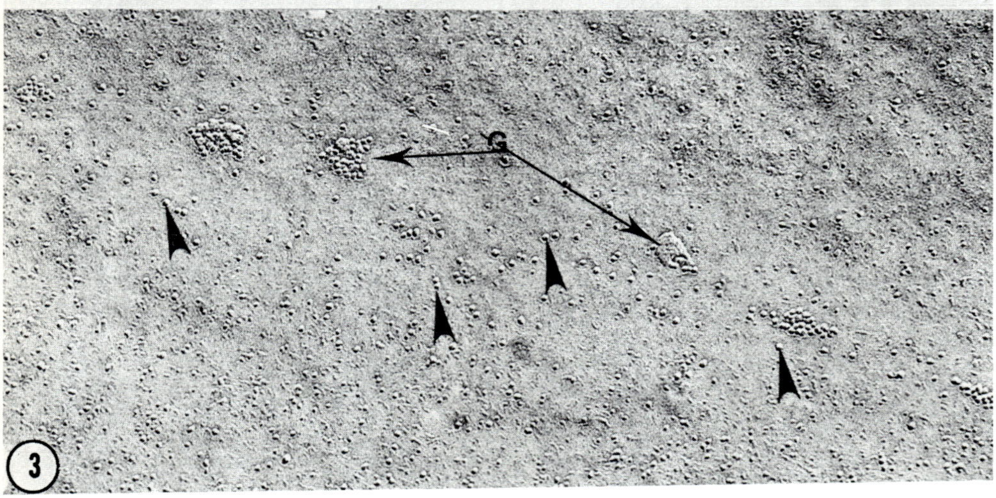

FIG. 2. Early antral follicle depicting (*arrows*) small gap junctions between two granulosa cells. × 52,800 Rabbit.

FIG. 3. Freeze-fracture replica between granulosa cells of an early antral follicle illustrating small gap junctions (G) and large intramembrane particles (*arrows*). × 61,600 Rabbit.

EARLY ANTRAL FOLLICLES

Before an antrum is formed there appears at the limits of the *membrana limitans* the theca interna (see below). From the formation of the antrum onward (Fig. 4, inset) the growth and function of the follicle is under the control of pituitary gonadotropic hormones (Hisaw, 1947; Greenwald, 1974; Richards and Midgley, 1976).

Gap junctions in all species examined appear as minute attachment sites that are frequently found at the edges of granulosa cells (Fig. 2). Freeze-fracture replicas of early antral follicles exhibit a number of clustered aggregates or plaques of 90 Å particles on the P-face and corresponding pits on the E-face (Fig. 3). These aggregates are separated from each other by a zone of particle depletion (Albertini et al., 1975). We view these small plaques as early formative stages of the gap junction of granulosa cells. As differentiation proceeds, large gap junctions may be found occupying rather large areas of the membranes (Figs. 4 to 6).

By using lanthanum as a tracer, it can be shown that the substance percolates among the intercellular spaces. In some regions, the lanthanum accentuates the 30 to 40 Å gap (Fig. 7). En face view of the junction reveals hexagonal units (Fig. 8); in the center of each is a dot. The hexagonal pattern results from the organization of the subunits comprising each particle (Paracchia, 1973). Goodenough (1974) has biochemically characterized the principal protein of gap junctions and has named it *connexin.*

GRAAFIAN OR PREOVULATORY FOLLICLES

Freeze-fracture replicas of large graafian or preovulatory follicles (mice, rats, rabbits) reveal the equivalent particle composition as the mid-to-late antral follicles. The junctions, however, appear to be larger than the mid-to-late antral ones and the particles show a preferred ordering, i.e., most are aggregated in either rectangles or hexagons.

Between granulosa cells, especially those of the graafian or preovulatory follicles, are some annular gap junctions (Fig. 9) (Espey and Stutts, 1972; Merk et al., 1973; Albertini and Anderson, 1974). These annular structures may contain any constituent(s) of the cytoplasm. In freeze-fracture replicas there are both 90 Å particles and pits (Fig. 10).

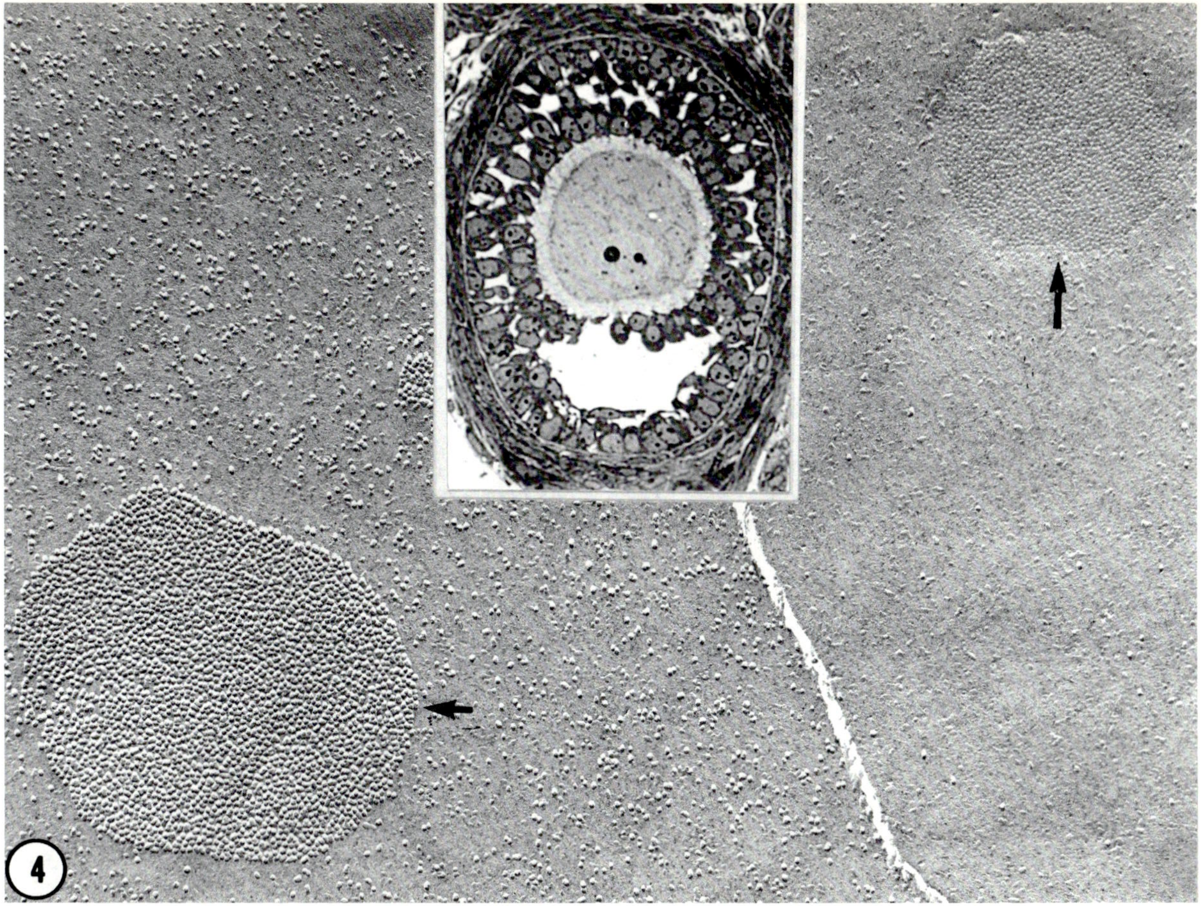

FIG. 4. Inset is an antral follicle of the mouse. The rest shows the organization of particles within a gap junction on the P-face (*short arrow*) and pits of another junction on the complementary E-face (*long arrow*). ×80,000; inset ×500, toluidine blue stained. Rabbit.

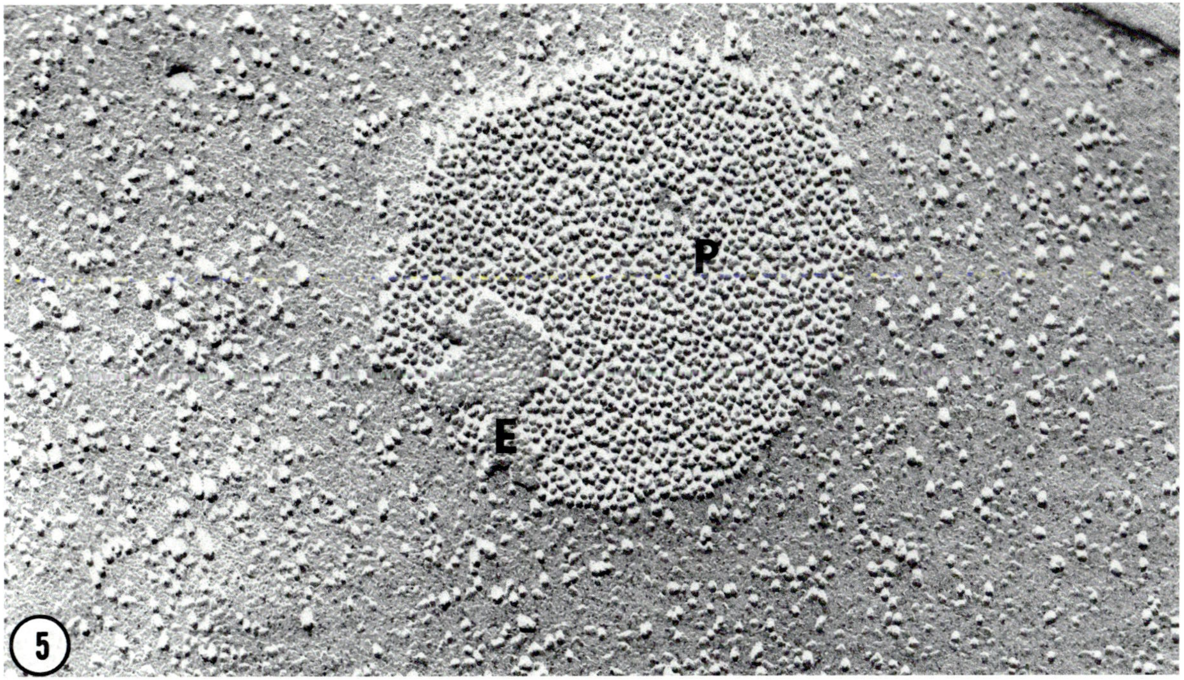

FIG. 5. A small region of an antral follicle illustrating particles comprising a portion of particles of a gap junction on the P-face and pits on the E-face. ×90,000. Rabbit.

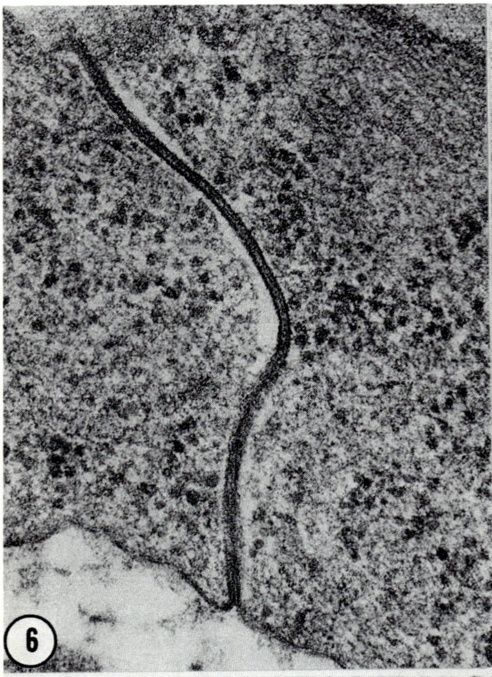
6

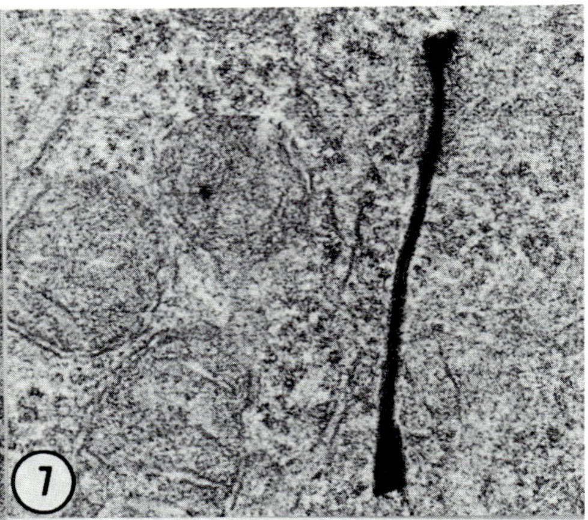
7

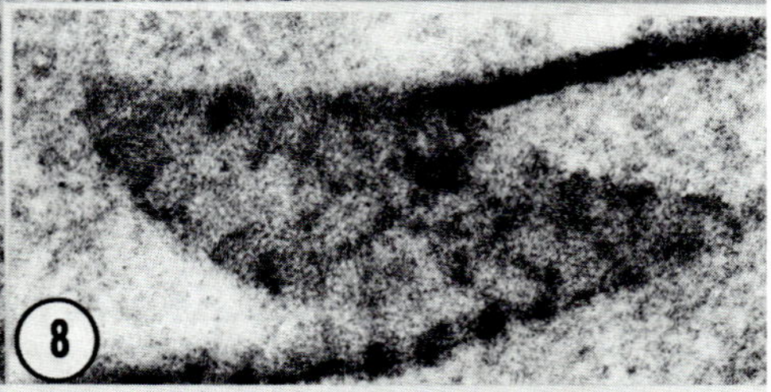
8

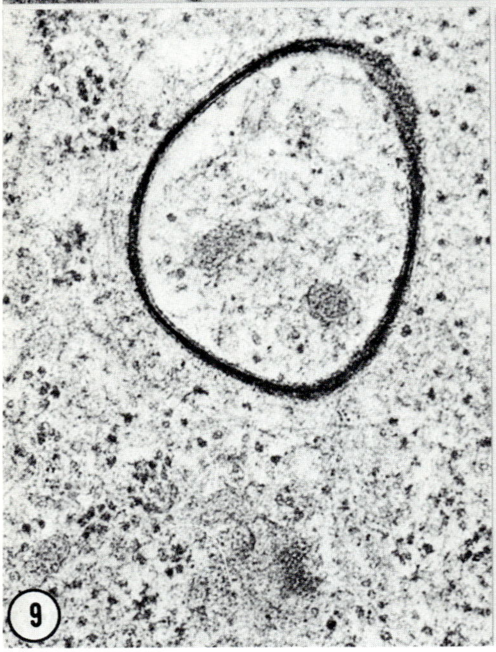
9

10

We have found that annular gap junctions are internalized and their fate is unknown (Albertini and Anderson, 1974). It is possible that they are degraded by lysosomes or some hydrolytic enzyme(s) of the cytoplasm, since we have found similar structures to be acid-phosphatase positive in lutein cells of pregnant rats (Anderson and Albertini, *unpublished observations*).

THECA

As indicated previously, during the development of the associated compartments of the follicle there appears at the limits of the *membrana limitans* (the basal lamina of the outer layer of granulosa cells) a sheath or theca. This layer ultimately differentiates into an inner vascular glandular layer and an outer nonvascular connective tissue-like layer.

We have found that the cells comprising the theca interna are also associated with each other via gap junctions (Fig. 11). Elements of endoplasmic reticulum (ER) of the smooth variety are frequently observed associated with gap junctions between thecal cells of large preovulatory follicles (Fig. 12, 5 to 7 hr postcoitus). In Fig. 12 the cleavage plane leaves the junctional plasma membrane interior to reveal an underlying fenestrated cistern of smooth ER.

We have found, as have others, that prior to ovulation the capillaries of the theca interna become fenestrated (see Byskov, 1969). We (Anderson, et al., 1974) found changes in the tight junctions between endothelial cells following sexual stimulation in rabbits, i.e., 3, 5, 7, 9, and 11 hr postcoitus. When

←

FIG. 6. A gap junction between granulosa cells after staining with uranyl acetate and lead. Note its five-layered structure. ×70,400 Rabbit.

FIG. 7. The 30 to 40 gap of the gap junction is accentuated by the deposition of lanthanum. ×79,200 Rabbit.

FIG. 8. En face view of a gap junction from lanthanum impregnated material. Note the hexagonal lattice. ×110,160 Rabbit.

FIG. 9. An annular gap junction from lanthanum infiltrated material. ×352,000 Rabbit.

FIG. 10. Freeze-fracture replica of an annular gap junction with many clusters of particles. ×44,000 Rabbit.

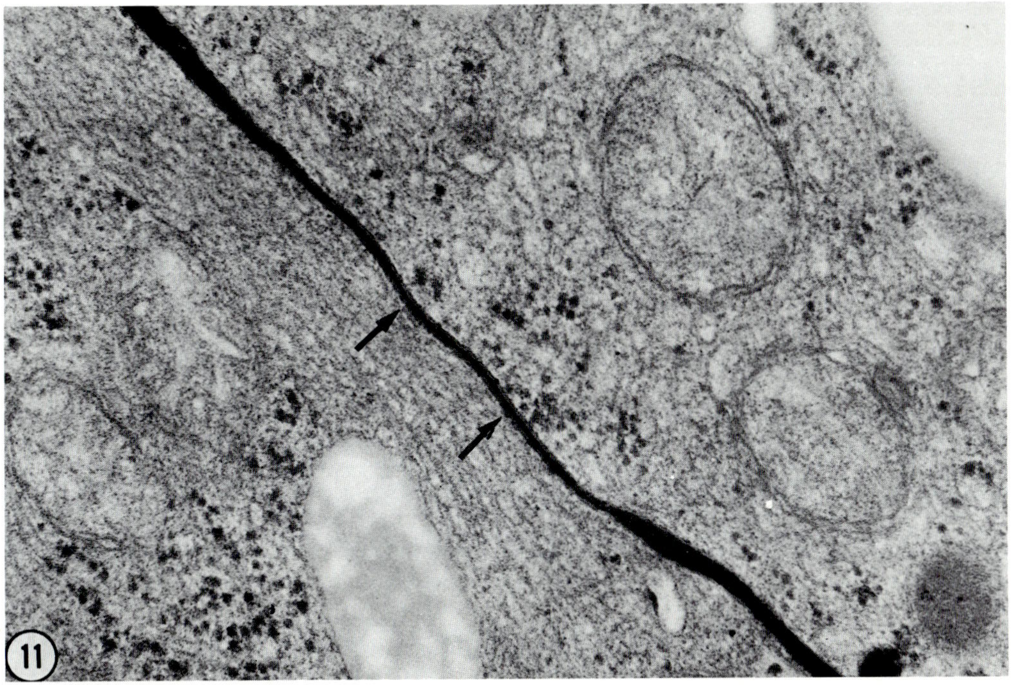

FIG. 11. A gap junction (*arrows*) between two thecal cells of material treated with lanthanum. × 52,800 Rabbit.

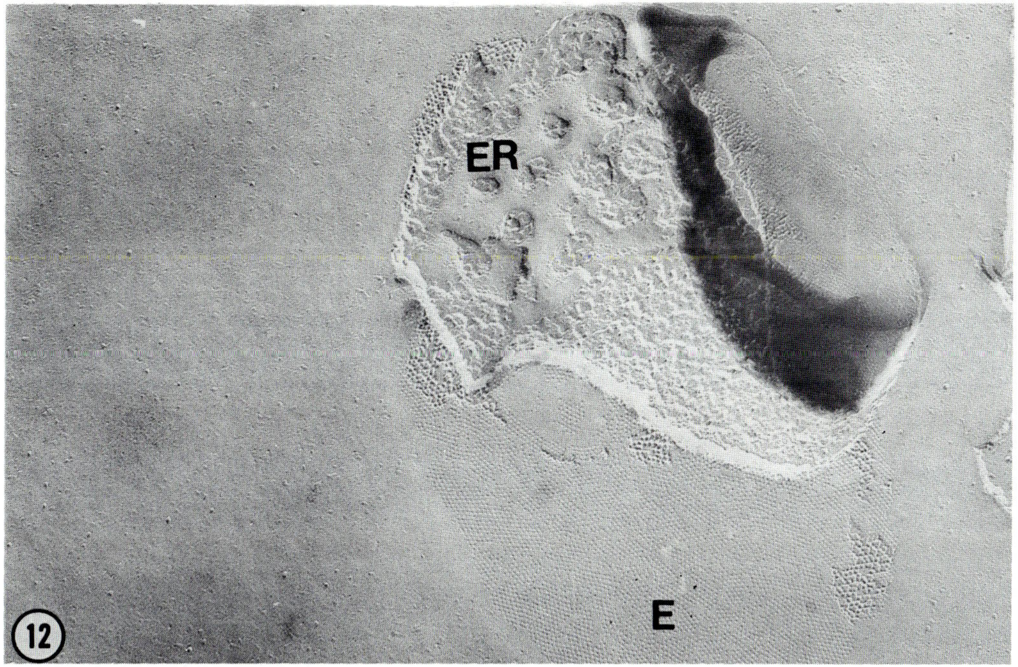

FIG. 12. A freeze-fracture preparation in which the cleavage plane leaves the junctional plasma membrane interior to reveal an underlying fenestrated cistern of smooth ER. Note a position of the E-face. ×616,000 Rabbit.

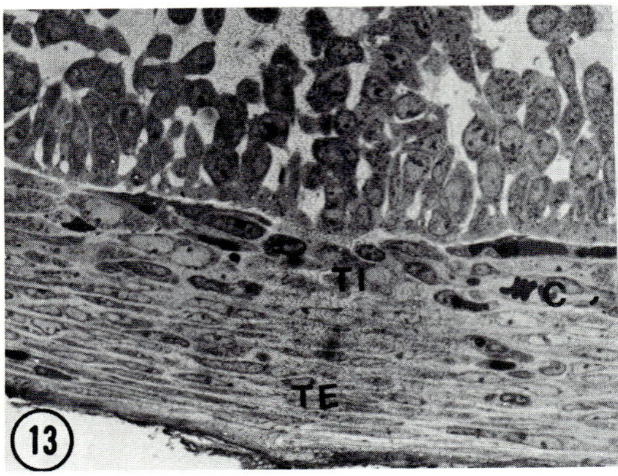

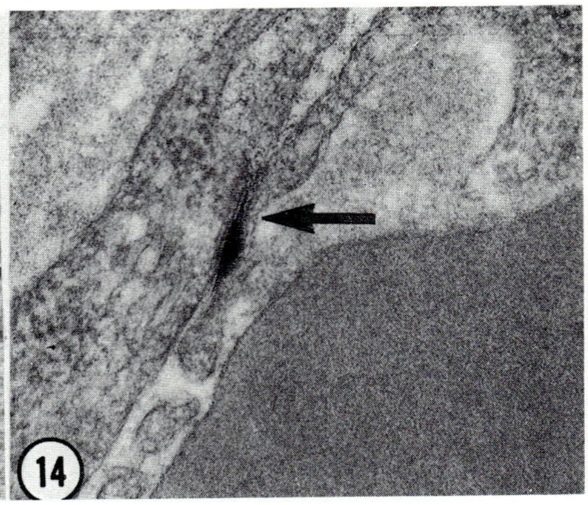

FIG. 13. A small portion of the wall of a graafian or preovulatory follicle 7 hr postcoitus. TI, theca interna; C, capillary; TE, theca externa. ×484 Rabbit.

FIG. 14. A tight junction between capillary endothelial cells from the theca interna. Lanthanum treated. ×44,000 Rabbit.

replicas were examined of material of animals 3 to 5 hr after mating, the tight junctions were composed of a few interconnected linear elements in association with gap junctions (Fig. 16). From 7 to 11 hr postcoitus the interconnected linear elements of the gap junction increased in number and anastomosed (Fig. 15). The function of the theca is not well understood. Hisaw (1947) suggested that the "...theca interna provides the follicle with a blood supply and either this or a stimulus derived from the theca itself may account for the increase in mitotic activity of the granulosa in a growing follicle." The experiments of others indicate that the theca and granulosa cells act synergistically in the production of estrogen (Falck, 1959; Ryan et al., 1968; Makris and Ryan, 1975).

CONCLUSIONS

The populations of granulosa cells in follicles of rabbits are associated with each other by specialized sites of attachment known as *maculae adherentes*. They are also coupled to each other by gap junctions. As indicated before, gap junctions function to electrically and metabolically couple cells in many different tissue types. We feel that gap junctions during folliculogenesis serve a simi-

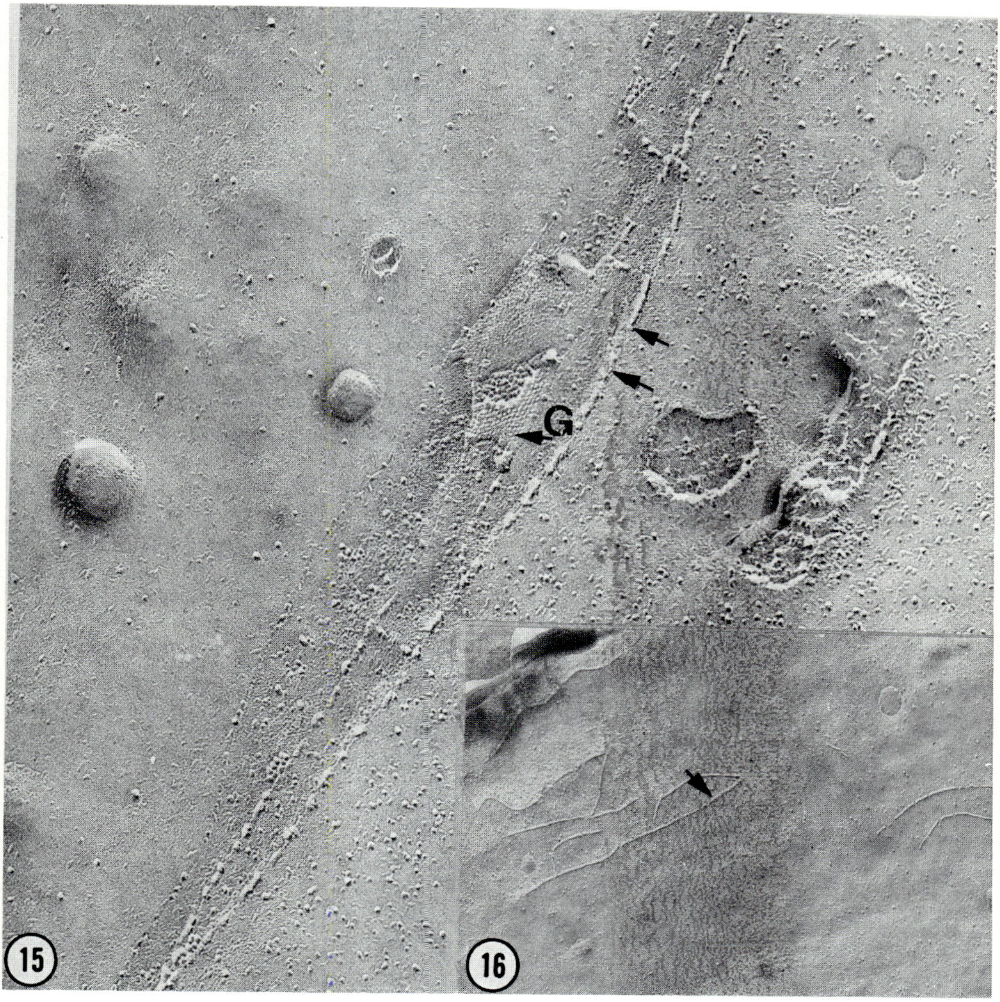

FIG. 15. A tight junction between two endothelial cells from a 9-hr postcoitus thecal capillary. The P-face ridge is frequently particulate (*arrows*), whereas the nearly complete E-face furrow is randomly occupied by particles. Gap junctions (G) are often observed in the interstices of the tight junctional strands. ×45,500. Rabbit.

FIG. 16. A portion of an endothelial cell from a follicle of an unstimulated rabbit. ×30,000.

lar function of coordination and also accommodate many of the biochemical events associated with growth, maturation (meiosis), and ovulation.

ACKNOWLEDGMENT

This study was aided by grant HD06822 and Center grant HD 06645(NICHD).

REFERENCES

Albertini, D. F., and Anderson, E. (1974): The appearance and structure of the intercellular connections during the ontogeny of the rabbit ovarian follicle with particular reference to gap junctions. *J. Cell Biol.*, 63:234-250.

Albertini, D. F., Anderson, E., and Letourneau, R. (1974): The ultrastructural design of the theca interna during the ovulatory process in the rabbit. *J. Cell Biol.*, 63:234-250.

Albertini, D. F., Fawcett, D. W., and Olds, P. J. (1975): Morphological variations in gap junctions of ovarian granulosa cells. *Tissue Cell*, 7:389-405.

Anderson, E. (1971): Intercellular junctions in the differentiating graafian follicle of the mouse. *Anat. Rec.*, 168:473.

Bennett, M. V. L. (1973): Function of electronic junctions in embryonic and adult tissues. *Fed. Proc.*, 32:65-75.

Branton, D. (1966): Fracture faces of frozen membranes. *Proc. Natl. Acad. Sci. U.S.A.*, 55:1048-1056.

Branton, S., Bullivant, S., Gilula, N. B., Moor, H., Muhlethaler, K., Northcote, D. H., Packer, L., Satir, B., Satir, P., Speth, V., Staehlin, L. A., Steere, R. L., and Weinstein, R. S. (1975): Freeze-etching nomenclature. *Science*, 190:54-56.

Byskov, A. G. (1969): Ultrastructural studies on the preovulatory follicle in the mouse ovary. *Z. Zellforsch.*, 100:285-299.

Catt, K. J., and Dufau, M. L. (1976): Basic conceptions of the mechanism of action of peptide hormones. *Biol. Reprod.*, 14:1-15.

Espey, L. L., and Stutts, R. H. (1972): Exchange of cytoplasm between cells of the membrana granulosa in rabbit ovarian follicles. *Biol. Reprod.*, 6:168-175.

Falck, B. (1959): Site of production of oestrogen in rat ovary as studied by microtransplants. *Acta Physiol. Scand.* [Suppl.], 47:163.

Fletcher, W. H., and Everett, J. W. (1973): Ultrastructure reorganization of rat granulosa cells on the day of proestrus. *Anat. Rec.*, 175:320a.

Friend, D. S., and Gilula, N. B. (1972): Variations in tight and gap junctions in mammalian tissues. *J. Cell Biol.*, 53:758-776.

Furshpan, E. J., and Potter, D. D. (1968): Low resistance junctions between cells in embryos and tissue culture. In: *Current Topics in Developmental Biology*, Vol. 3, edited by A. A. Moscona and A. Monroy, pp. 95-127. Academic Press, New York.

Goodenough, D. (1974): Bulk isolation of mouse hepatocyte gap junctions. *J. Cell Biol.*, 61:557-563.

Greenwald, G. S. (1974): Role of follicle-stimulating hormone and luteinizing hormone in follicular development and ovulation. In: *Handbook of Physiology: Section 7: Endocrinology*, edited by S. R. Geiger, pp. 293-323. Waverly Press, Baltimore.

Hisaw, F. L. (1947): Development of the graafian follicle and ovulation. *Physiol. Rev.*, 27:95-119.

Makris, A., and Ryan, K. J. (1975): Progesterone, androstendione, testosterone, estrone and estradiol synthesis in hamster ovarian follicle cells. *Endocrinology*, 96:694-701.

Merk, F. B., Albright, J. T., and Botticelli, C. R. (1973): The fine structure of granulosa cell nexuses in rat ovarian follicles. *Anat. Rec.*, 175:107-125.

McNutt, N. S., and Weinstein, R. S. (1970): The ultrastructure of the nexus. A correlated thin-section and freeze-cleave study. *J. Cell Biol.*, 47:666-688.

Paracchia, C. (1973): Low resistance junctions in crayfish. II. Structural details and further evidence for intercellular channels by freeze-fracture and negative staining. *J. Cell Biol.*, 57:66-76.

Richards, J. S., and Midgley, A. R. (1976): Protein hormone action: A key to understanding ovarian follicular and luteal development. *Biol. Reprod.*, 14:82-94.

Ryan, K. J., Petro, Z., and Kaiser, J. (1968): Steroid formation by isolated and recombined ovarian granulosa and thecal cells. *J. Clin. Endocrinol.*, 28:355-358.

Ovarian Follicular Development and Function,
edited by A. R. Midgley and W. A. Sadler.
Raven Press, New York © 1979.

Preliminary Observations on Time-Dependent Intrafollicular Distribution of ^{125}I-HCG

Seong S. Han

Reproductive Endocrinology Program, Departments of Anatomy and Oral Biology, University of Michigan, Ann Arbor, Michigan 48109

The following is a brief account of recent observations in which the fate of ^{125}I-HCG was followed in follicular cells of the ovary in hypophysectomized rats that were appropriately primed by estrogen and FSH. It will be shown that, during the period or preovulatory LH surge, (1) the peripheral follicle cells bind ^{125}I-HCG initially; (2) there is a very gradual advance of radioactive hormones toward the deeper layers of granulosa cells: (3) the endocytosed ^{125}I-HCG appears to concentrate in lysosomes and multivesicular bodies; and (4) some radioactive grains are localized over nuclear chromatin at later time periods.

Briefly, the experimental approach was as follows: Immature hypophysectomized rats were treated with estradiol (2 mg/day for 4 days), followed by treatment with HFSH (2 μg/day) for 2 days. On day 6 of the experiment, rats were given ^{125}I-HCG (0.3 μCi/g) and killed at 2, 30, 120, and 480 min following the injection of radioactive HCG. Tissues were prepared for light and electron microscopic autoradiography as mentioned elsewhere (*Endocrinology*, 96:589, 1974).

Figures 1 to 3 demonstrate the pattern of distribution of ^{125}I-HCG at the periphery of large follicles as shown at 30 min and 2 and 8 hr after the injection.

Note in Fig. 1 that a small number of thecal and interstitial elements are labeled as well as a moderate labeling of peripheral cells of the follicle. By 2 hr the labeling of three to four cell layers at the periphery of most large follicles was so dense that grains tended to obscure histological details of granulosa cells. It should be noted that the labeling of interstitial elements at this time, however, did not vary significantly from what was observed at 30 min (Fig. 1). The complete masking of granulosa cells by

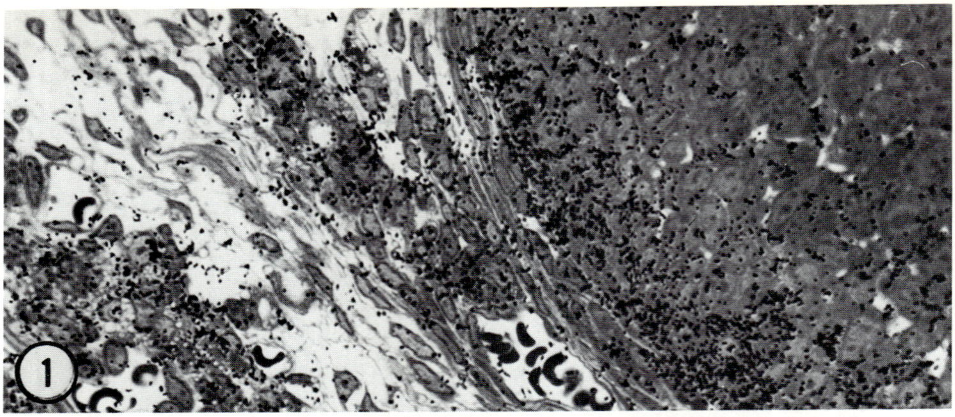

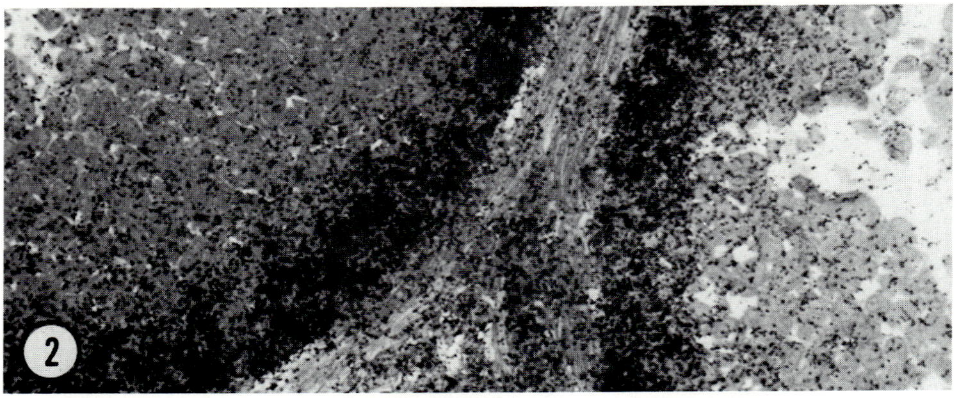

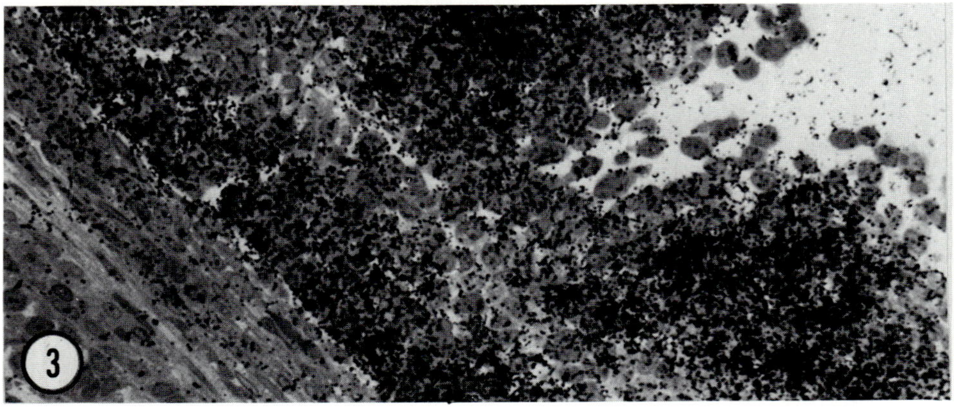

FIG. 1. Radioautographic appearance of silver grains of ^{125}I-hCG.

FIG. 2. Radioautographic appearance of silver grains of ^{125}I-hCG at 2 hours after injection. Note the preferential increase of grains in follicular cells of the interstitial cells. Within the follicles, however, the grains are largely located at the periphery.

FIG. 3. Radioautographic appearance of silver grains of ^{125}I-hCG at 8 hours after injection. Those cells that are close to the follicular lumen appear to have an increased labeling. In contrast the grain density at the follicular periphery is reduced from that seen 2 hours after injection.

silver grains at this time was of interest, since this was in contrast to the somewhat clear delineation of luteal cell periphery when ^{125}I-HCG was localized in the corpus luteum. This might simply be due to the fact that luteal cells were considerably larger than granulosa cells. Or conversely, this might be due to the possibility that indeed radioactive HCG got into the nuclei of granulosa cells.

At 8 hr after the injection of ^{125}I-HCG (Fig. 3), the peripheral labeling of the granulosa cells in large follicles was still heavy but not as heavy as it was at 2 hr after the injection. In contrast, a large number of grains were localized in deeper cells of the follicle. However, only occasional grains were localized over the granulosa cells located at the periantral border. In light of the earlier observations in which small molecular markers such as peroxidase were found to penetrate into the follicular lumen within minutes after intravenous injection, the slow migration of labeled HCG found in our experiments might be taken as indicative of the possibility that a continuous occupancy and saturation of peripheral receptors for the hormone might be necessary before additional free hormone molecules were allowed to migrate further into the deeper granulosa cell layer.

The results from electron microscopic autoradiography show that initially many of the silver grains are localized over the plasma membranes of granulosa cells, as seen in the lower left-hand corner of Fig. 4. They are also present along the surface of some thecal and interstitial elements. While the electron microscopic appearance did not change markedly throughout the period of observation, the appearance of silver grains inside the cells at different time periods was of interest. As early as 30 min after the injection, some of the grains were localized over deep cytoplasm in the vicinity of multivesicular bodies and dense structures that appeared to be primary lysosomes (Figs. 5 and 6). The amount of silver grains concentrated over these structures increased as a function of time.

What was most intriguing was the observation that at 2 hr and onward, a number of granulosa cells contained silver grains that were lying over the nucleus (Fig. 5). The number of nucleus-associated grains was such that it was difficult to regard their localization as accidental or due to background radiation alone. For example, the granulosa cell in Fig. 5 may contain as many as 13 silver grains over or touching the nuclear structure. Furthermore, the location of these silver grains within the nucleoplasm tended to favor the transitional zone between the euchromatin

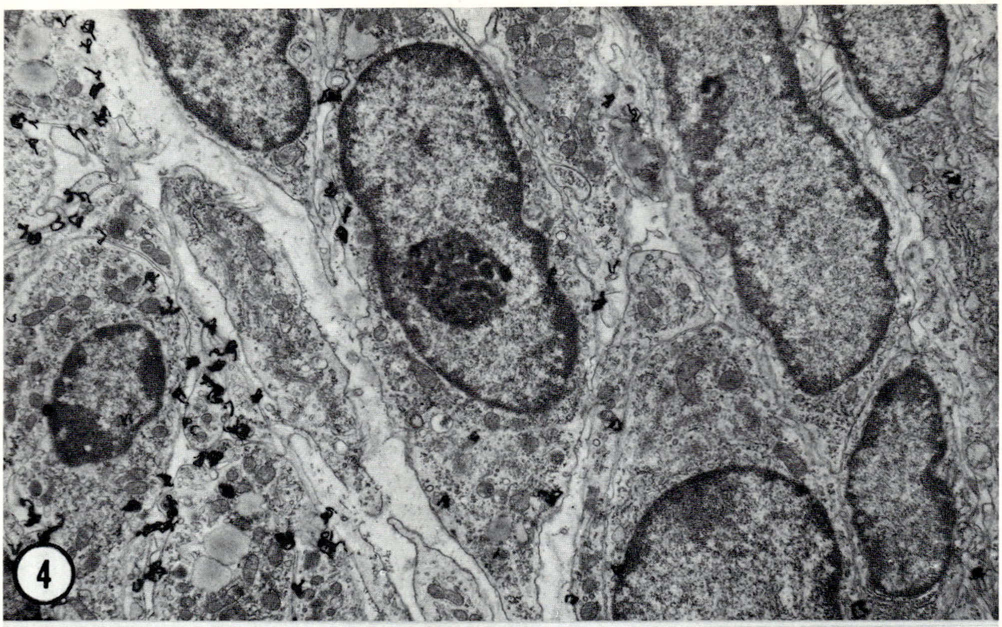

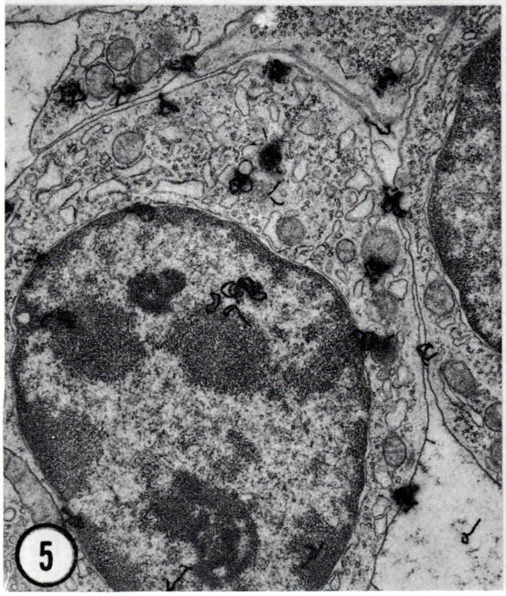

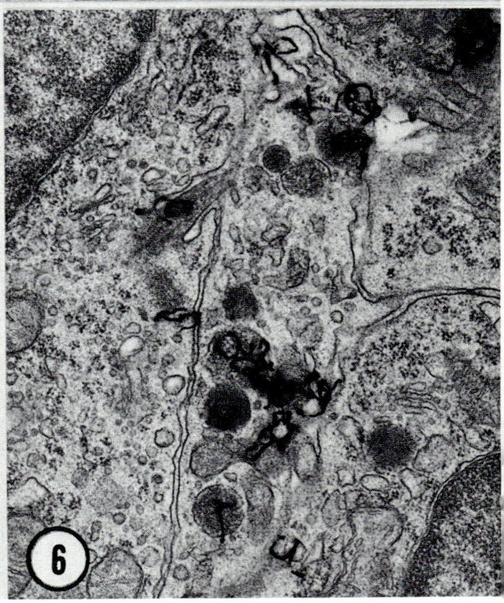

FIG. 4. Low magnification electron microscopic radioautograph at 30 minutes after injection of ^{125}I-hCG. Silver grains are abundant over the granulosa cells at the lower left hand corner. Several grains are also present over thecal cells.

FIG. 5. Electron microscopic radioautograph of a granulosa cell at 2 hours after ^{125}I-injection. A fair number of grains are present over both the cytoplasm and nucleus. This particular nucleus has as many as 13 grains that are within the nucleoplasm or touching nuclear membrane.

FIG. 6. Electron microscopic radioautograph of portion of granulosa cells at 2 hours after ^{125}I-injection. Note that a large number of grains are found over the lysosome.

and heterochromatin. In light of the current understanding of the functional and structural relationship between the two regions of the nucleoplasm, it is of interest that the trace of the hormone be found at the interface where the active and inactive forms of chromatin materials are present side by side.

These preliminary observations allow us to conclude that (1) time-dependent migration of ^{125}I-HCG in large follicles is a slow process that might require gradual saturation of peripheral receptors for the hormone; (2) ^{125}I-HCG is endocytosed by granulosa cells; and (3) that the endocytosed hormone might be sequestered into the lysosomes, multivesicular bodies, and/or transported into nucleus. The significance of the latter observations awaits further investigative efforts.

Ovarian Follicular Development and Function,
edited by A. R. Midgley and W. A. Sadler.
Raven Press, New York © 1979.

Intercellular Junctions in Ovarian Follicles: A Possible Functional Role in Follicle Development*

William H. Fletcher

Department of Anatomy, Duke University Medical Center, Durham, North Carolina 27710

Ovarian follicles are histologically complex but highly ordered tissues whose construction and maturation are paragons of synchronous cellular activity (9). During one such state, luteinization, Bjorkman (3) found rat ovarian granulosa cells were interconnected by "close membrane appositions;" however, the correct identification of these as gap junctions awaited the observations of Anderson (cf. 1). Given the reported function of gap junctions as pathways for intercellular communication (10) it seemed appropriate to test their involvement in coordinating folliculogenesis and luteinization. In collaboration with Drs. John W. Everett, J. David Robertson, and David W. Schomberg, a series of investigations were undertaken to examine the junctions and to assess their possible role in follicle economy.

STRATUM GRANULOSUM

As shown in Figs. 1, 1a, and 2 (10-day-old rat), granulosa cells unite among themselves via gap junctions apparently from the first moment of contact on; similar events occur in ovaries of 3-and 5-day olds. In these early primary follicles, where the granulosa are squamous and directly contact the oocyte without intervention of the zona pellucida, the junctions are usually diminuitive and restricted to the upper or lower margins of apposition. Their morphology is typical whether viewed in thin section (Fig. 1a;

*Dedicated to Dr. John W. Everett on the occasion of his advancement to Emeritus Professor of Anatomy, and in recognition of his numerous contributions to our understanding of ovarian morphology and function.

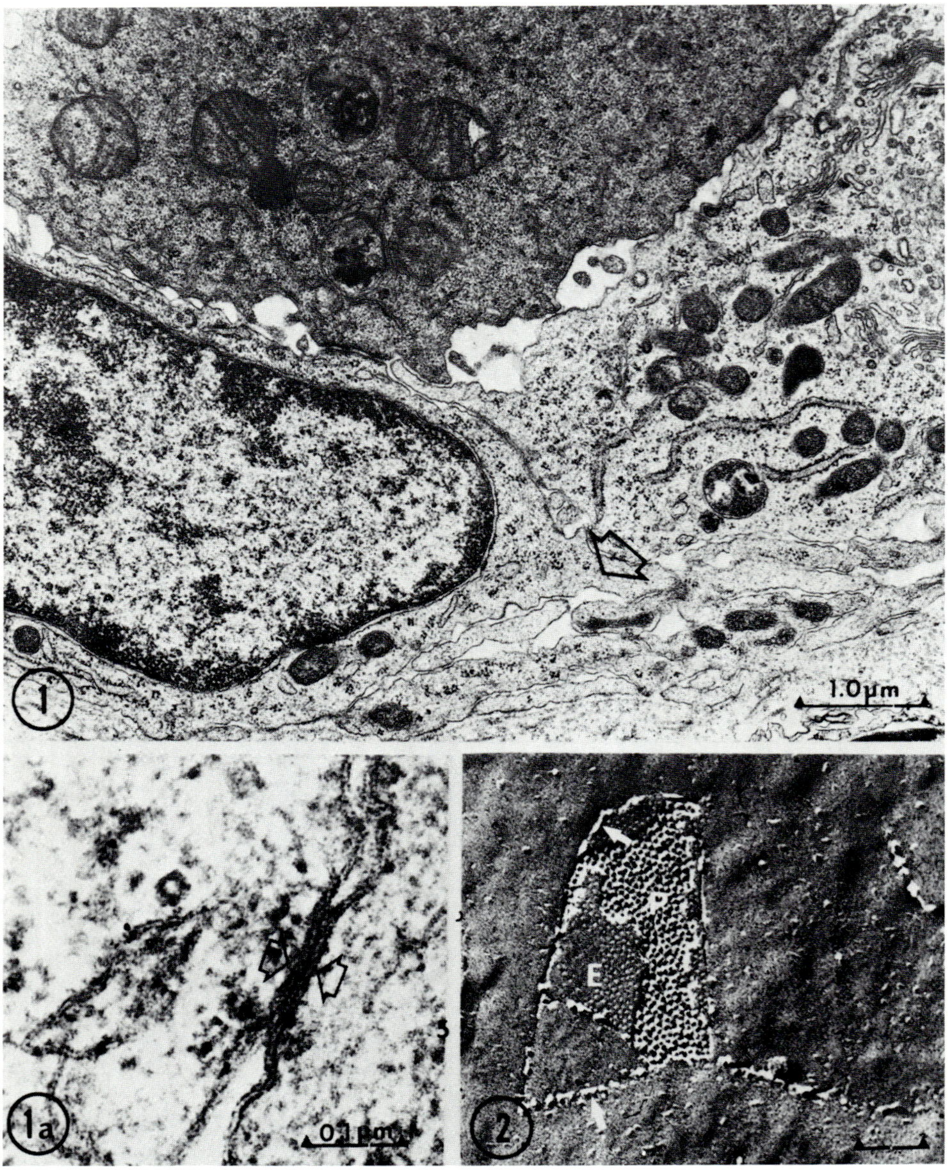
1
1.0 µm
1a
0.1 µm
2
E

open arrows) or freeze-fracture preparations (Fig. 2). The junctional E-face depressions are sometimes circumscribed by a mixed strand of particles and short ridges nestled in a shallow furrow (Fig. 2, *arrows*). This motif is indistinguishable from that of newly assembling gap junctions in small follicles of immature rats treated with estradiol monobenzoate (6), and is consistent with an occurrence between recently apposed granulosa cells. Although these observations contrast markedly with the report that gap junctions do not appear before antrum formation (1), they do support the suggested involvement of the junctions in follicle differentiation (1,7).

Recently Albertini and Anderson (1) speculated that gap junction development may coincide with the appearance of gonadotropin receptors. Our observations, together with the reports (11,12) of gonadotropin binding by the granulosa cells of small follicles, lend credence to their view, but indicate these special properties accrue much earlier than was previously suspected. In rats, the receptors appear in various sized follicles sometime between the 3rd and 5th postnatal days and increase in number thereafter (11). Even in unilaminar follicles of 3-day-old rats, the granulosa cells are joined by small gap junctions. Postreceptor processes also may be active in solid follicles as exogenous gonadotropins elicit specific biochemical responses by immature ovaries containing few, if any, vesicular follicles (4). However, the precise cellular location of this response has not been described.

Given these findings, it is appropriate in this workshop to suggest there may be a correlation between

←

FIG. 1. Primary follicle in the ovary of a 10-day-old rat, with the oocyte lying in the upper left quadrant. Two presumptive granulosa cells nestle close to the oocyte although there is no evidence of zona pellucida formation as yet. Even so, the granulosa cells unite at their bases by a small gap junction (*inside open arrow*).

FIG. 1a. Higher magnification of the junction in Fig. 1. Distance between the arrowheads is 19-20 nm.

FIG. 2. A freeze-fractured primary follicle of a 10-day-old rat. A polygonal aggregate of 7-8 nm P face particles lies near the apical border of the presumptive granulosa cells. At the white arrow is a segment of the E face ridge that circumscribes this definitive junctional domain. Note the tail like continuation of this ridge.

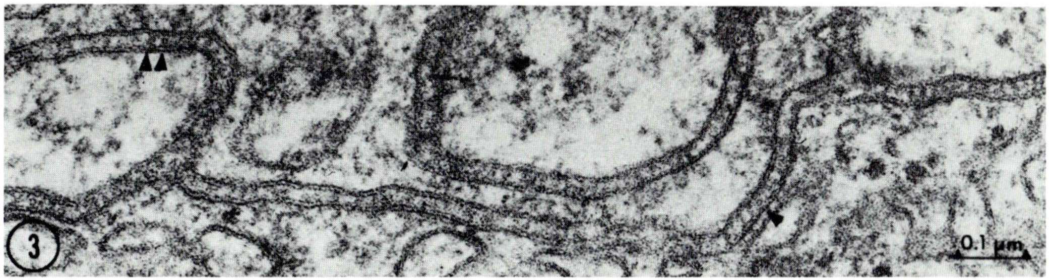

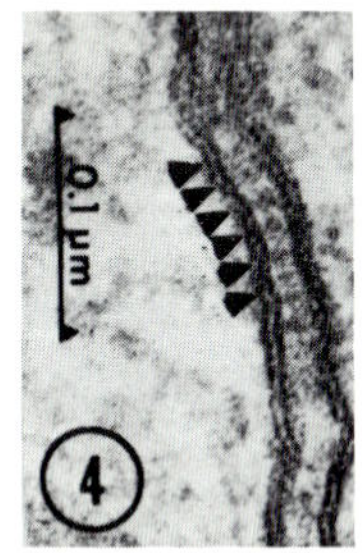

FIG. 3. In mature rats the theca folluculi of large antral follicles often has great spans of the space between neighbor cells differentiated into a septate like contact. Where this occurs the intercellular space is of remarkably constant width, slightly electron opaque, and is occasionally punctuated by 6-7 nm electron dense dots. These features are shown to better advantage in Fig. 4.

FIG. 4. The periodic intercellular densities (black triangles) are sometimes regularly spaced and lie midway between the apposed outer leaflets of the adjacent cell membranes.

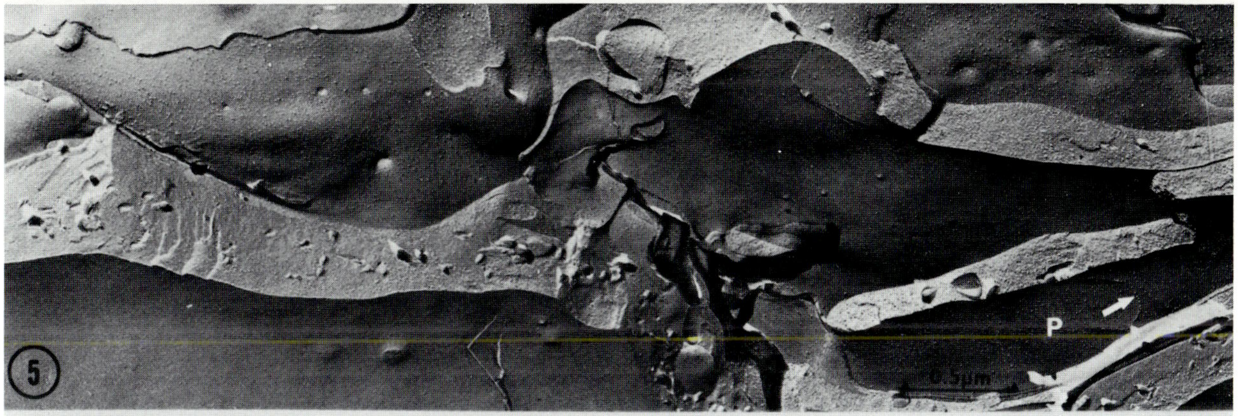

FIGS. 5 and 6. Freeze fracture replicas of the thecal layer of antral follicles in mature rats frequently revealed gap junction specializations. Occasionally these appeared as simple or complex linear strands of 7-8 nm P face particles (white arrow, Fig. 5), although most often the particles clustered polygonally forming typical gap junctions (Fig. 6, gj). In addition, long straight or branched chains of gap junction like particles occurred on the E fracture face (Fig. 6, white arrows). Sometimes, as here, these particle chains were associated with gap junctions on the apposing fracture face. This observation diminishes the possibility that tissue preparation artifacts gave rise to the E face particle rows.

the development of gap junctions and the capacity of hormonal stimuli (LH?; FSH?) to elicit the production or unmasking of "signals" (cyclic AMP? GMP?; ions?) that distribute among contiguous cells, coordinating the construction and maturation of follicles. A precise characterization of the temporal appearance of junctions, receptors, and postreceptor processes would prove helpful in testing this notion.

THECA FOLLICULI

Cells identified as presumptive theca begin to appear in follicles with a minimum of a single layer of "plump" granulosa. The harbinger of this event seems to be basal lamina deposition external to the granulosa. Whether this forms in reaction to the arrival of thecal cells or precedes and possibly initiates their addition is unknown. Thecal cells quickly unite by gap junctions; however, the contacts do not appear to achieve the number or proportions of those joining granulosa cells. In larger follicles, either of mature or immature rats, the dominant contact between thecal cells is like the septate variety described in adrenal cortex (8). This specialization (Fig. 3) consists of adjacent cell membranes separated by a 15- to 18-nm wide cleft (vs. the 2- to 6-nm spacing at gap junctions). The intercellular space is filled with an electron-dense material and is frequently spanned by 5- to 6-nm wide "bars" (*black triangles*) of even more intensely stained substance. The bars may be artifacts of tilt or overlap. When the apposed membranes are crisply defined in very thin sections, the intercellular cleft is punctuated by 6- to 7-nm diameter densities spaced about 15 nm center-to-center (Fig. 4, *triangles*). Both potassium pyroantimonate and colloidal lanthanum avidly bind to these components, indicating they have an affinity for polyvalent cations. In fortuitous fractures through the theca interna, particularly that of large vesicular follicles, small strands (Fig. 5, *white arrow*) or aggregates (Fig. 6, *gj*) of 7- to 8-nm P-face particles are frequent; both forms appear to be gap junctions. In addition, lengthy chains of E-face particles have been seen (Fig. 6, *white arrows*). The deposition of particles on the E-face and absence of ridges both indicate this structure is not a zonula occludens. The occurrence of the E-face chains, most prevalent in preovulatory follicles, together with their morphology suggest they may signify an intramembrane modification of the septate-like contact seen in thin sections. If so, this would be the first freeze-fracture evidence that the phylogenetically ancient septate junctions have been preserved in a modified form in vertebrate tissues. Further comparative studies possibly will resolve this question. The role, if any, of this unusual contact in ovarian physiology should be examined. In adrenal cortex it may influence steroid secretion either by channeling the hormone into narrow canaliculi or by making available stores of ions necessary for the secretory process (8).

OOCYTE

Heterocellular gap junctions have been seen on granulosa processes that pierce the zona pellucida and contact the egg. There is a possibility that these junctions are formed during the initial apposition of granulosa and egg (as in Fig. 1). In thin sections these contacts are small but clearly recognizable as gap junctions. Freeze-fracture reveals that at least two forms occur. One is composed of short straight or curved rows of 7- to 8-nm P-face particles (similar to the junction of Fig. 5). The other is a small macular aggregate of 7- to 8-nm particles like the gap junctions found elsewhere in follicles. If these provide functional coupling between oocyte and granulosa cytoplasms, they may well mediate the passage either of nutrients to the egg (2) or of factors influencing the resumption of meiosis in preparation for ovulation (5).

ACKNOWLEDGMENT

The author was supported by USPHS grant 6 F02 GM 52636; NIH grant 1 P01 10299; and NSF grant GB 21025.

REFERENCES

1. Albertini, D., and Anderson, E. (1974): The appearance and structure of intercellular connections during the ontogeny of the rabbit ovarian follicle with particular reference to gap junctions. *J. Cell Biol.*, 63:234-250 .
2. Biggers, J. D. (1971): New observations on the nutrition of the mammalian oocyte and the preimplantation embryo. In: *The Biology of the Blastocyst*, edited by R. J. Blandau, pp. 319-327. University of Chicago Press, Chicago.

3. Bjorkman, N. (1962): A study of the ultrastructure of the granulosa cells of the rat ovary. *Acta Anat. (Kbh.)*, 51:125-147.
4. DeAngelo, A. B., Schweppe, J. S., Jungmann, R. A., Huber, P., and Eppenberger, U. (1976): Ovarian cyclic adenosine monophosphate-dependent protein kinase activity: Ontogency and effect of gonadotropins. *Endocrinology*, 97:1509-1520.
5. Erickson, G. F., and Ryan, J. K. (1976): Spontaneous maturation of oocytes isolated from ovaries of immature hypophysectomized rats. *J. Exp. Zool.*, 195:153-158.
6. Fletcher, W. H. (1975): Assembly of an "enclosed gap junction" by the granulosa cells of developing ovarian follicles in sexually immature rats. *J. Cell Biol.*, 67-116a. 1975 (abs.).
7. Fletcher, W. H., and Everett, J. W. (1973): Ultrastructural reorganization of rat granulosa cells on the day of proestrus. *Anat. Rec.*, 175:320.
8. Friend, D. S., and Gilula, N. B. (1972): A distinctive cell contact in the rat adrenal cortex. *J. Cell Biol.*, 53:148-163.
9. Greenwald, G. S. (1974): Role of follicle-stimulating hormone and luteinizing hormone in follicular development and ovulation. In: *Handbook of Physiology, Vol. IV, Sect. 7, Part II*, edited by R. O. Greep and E. B. Astwood, pp. 293-323. American Physiological Society, Washington, D. C.
10. Lowenstein, W. (1973): Membrane junctions in growth and differentiation. *Fed. Proc.*, 30:60-64.
11. Petrusz, P. (1974): Demonstration of gonadotropin binding sites in the rat ovary by an immunoglobulin-enzyme bridge method. *Eur. J. Obstet. Gynecol. Reprod. Biol.*, 4:3-9.
12. Richards, J. S., and Midgley, A. R., Jr. (1976): Protein hormone action: A key to understanding ovarian follicular and luteal cell development. *Biol. Reprod.*, 14:82-94.

Ovarian Follicular Development and Function,
edited by A. R. Midgley and W. A. Sadler.
Raven Press, New York © 1979.

Studies on the Structure of the Extracellular Matrices of the Mammalian Follicles as Revealed by High-Voltage Electron Microscopy and Cytochemistry

Yuan-Hsu Kang, Winston A. Anderson, *S. C. Chang, and *R. J. Ryan

Department of Zoology and Howard University Cancer Research Center, Howard University, Washington, D. C. and *Department of Molecular Medicine, Mayo Medical School, Rochester, Minnesota 55901

The fine structural specialization of the ovarian follicle is correlatable with gonadotropin regulation of follicular development. Increased granulosa cell mitosis and folding of cell surfaces correlate with increased numbers of cell surface receptors for gonadotropins and stimulatable cyclase in growing follicles (Lee, 1975, 1976; Chang et al., 1976, 1977). Increase in follicle size is further marked by hypertrophy of the oocyte and the elaboration of extracellular components including the zona, liquor, and follicular basement lamina. The zona is considered a complex filamentous sheath composed of mucopolysaccharides elaborated by the oocyte and the differentiating granulosa cells. However, fine structural studies consistently reveal that antral spaces contain amorphous material rich in glycoproteins and substances derived as an exudate from the blood (Anderson, 1967, 1972; Gwatkin, 1967; W. Anderson, 1972; Hastings II et al., 1972).

This report, based on observations from transmission and high-voltage transmission electron microscopy and cytochemistry, demonstrates the complex ultrastructure and origin of the extracellular matrix of the rat ovarian follicles.

MATERIALS AND METHODS

Sprague-Dawley rat ovaries were fixed by left ventricular perfusion with glutaraldehyde-paraformaldehyde mixtures in 0.1 M cacodylate buffer at pH 7.2 (Karnovsky, 1965). Aldehyde-fixed tissues were embedded in glycol methacrylate according to Leduc and Bernhard (1967) and in Epon (Luft, 1961). Semithick sections (1 μm thick) and thin sections (900 Å) of glycol-methacrylate-embedded tissue, cut with glass and diamond knives, were stained for mucopolysaccharides by floatation using the chromic acid-phosphotungstic acid (CA-PTA) technique according to Rambourg (1967), Rambourg et al. (1969), and Fléchon (1970). One-micron-thick and thin sections were then mounted on formvar-coated grids to be examined by conventional and high-voltage transmission electron microscopy. Thick sections of Epon-embedded tissue were stained with uranyl acetate and lead citrate (Venable and Coggeshall, 1965).

Horseradish peroxidase (0.1 to 0.5 ml of 10 mg/ml Sigma Type VI horseradish peroxidase) was administered to young female Sprague-Dawley rats by injection into a saphenous vein. Animals were sacrificed by cervical dislocation at 3 to 10 min after injection. The excised ovaries were fixed for 30 to 50 min by immersion in glutaraldehyde-paraformaldehyde fixative (Karnovsky, 1965), rinsed for 12 hr in 0.1 M cacodylate buffer (pH 7.2), and cut into 100- to 200-μm-thick sections using an Oxford vibratome. These sections were incubated in diaminobenzidine (DAB)-H_2O_2 medium (Graham and Karnovsky, 1966) for the cytochemical demonstration of peroxidase activity. The incubated sections were postfixed for 60 min in 2% osmium tetroxide, dehydrated in alcohol-water solutions, and embedded in Epon. One-micron-thick sections, without counterstaining, were examined by high-voltage electron microscopy.

All thick sections were examined with an AEI-EM7 high-voltage electron microscope (University of Wisconsin, Madison, Wisconsin) at 1,000 kV and at ± 15° tilting angles. Thin sections were studied with a Siemens 101 electron microscope.

OBSERVATIONS

Extracellular mucopolysaccharide material including the zona, granulosa cell surface coat, and the follicle basement lamina is intensely stained with the chromic acid-phosphotungstic acid technique (Figs. 1 to 9). Intracellular sites that were also intensely stained include Golgi saccules and vesicles, multivesicular bodies in oocytes and granulosa cells (Figs. 1 and 2). Cortical granules distributed singly, as aggregates and/or in close association to the Golgi complex (Figs. 3 and 4) are also intensely stained.

When thin sections of glycol-methacrylate-embedded ovarian tissues are examined, the zona, cortical granules and Golgi complex of oocytes, and the cell surface coat of granulosa cells are intensely stained by the CA-PTA technique (Fig. 5). The zona consistently appears as a somewhat granular or homogeneously dense sheath. In contrast, in thick sections of similarly processed tissues examined by high-voltage electron microscopy, the zona appears as a coarse, fibrous, and intensely electron-opaque sheath (Fig. 6). Mature follicles possess zona composed of coarse filaments or fibers that obscure the microvilli and cytoplasmic processes of granulosa cells that permeate the zona. In immature follicles (preantral), microvilli and cell processes appear as a complex network in negative contrast against the intensely stained zona material (Figs. 1 to 3).

The antral spaces of mature follicles in tissues fixed in aldehyde and embedded in Epon appear empty containing flocculent electron dense material. Aldehyde-fixed tissues embedded in glycol methacrylate and stained by the CA-PTA technique show antra containing a complex network of coarse fibers (Fig. 7). The coarse fibers are closely apposed to the granulosa cells and are frequently continuous with the zona material that is similarly stained. Corresponding thin sections illustrate that the cell surface coat of granulosa cells of the mature follicle is continuous with filamentous material filling the antral space (Fig. 8). The mucopolysaccharide material in the antrum is highly branched and reticulated and completely fills the antrum of the mature follicle (Fig. 9).

At 3 min following injection, horseradish peroxidase fills the intercellular spaces of the theca externa, theca interna, and the follicle basement lamina (Fig. 10). Exogenous peroxidase permeated the spaces between the follicle granulosa cells, and fills the antral spaces and zona of mature follicles (Figs. 11 to 13). Complex junctions (presumably gap junctions) at the base of the granulosa cells (Fig. 11) and the macula adherens between granulosa cell processes and the oolemma are permeated by exogenous peroxidase. The complex network of microvilli and oolemma are easily discerned within the zona that is saturated with peroxidase (Figs. 12 and 13). Neither the oocyte nor granulosa cells incorporate significant amounts of exogenous peroxidase through pinocytosis during the time of exposure to the tracer (3 to 10 min).

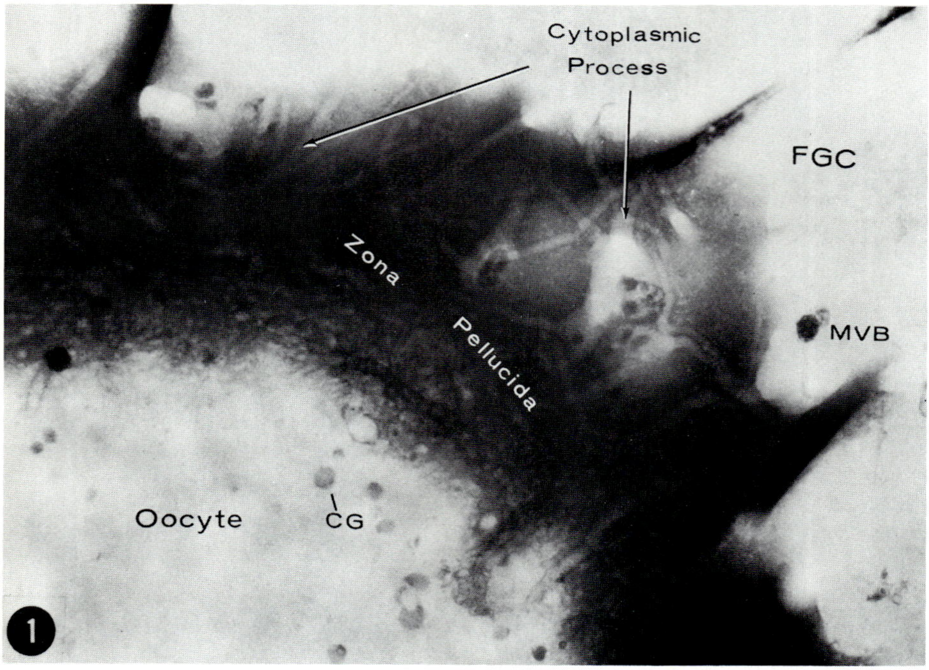
Cytoplasmic
Process
FGC
Zona
Pellucida
MVB
Oocyte
CG
1

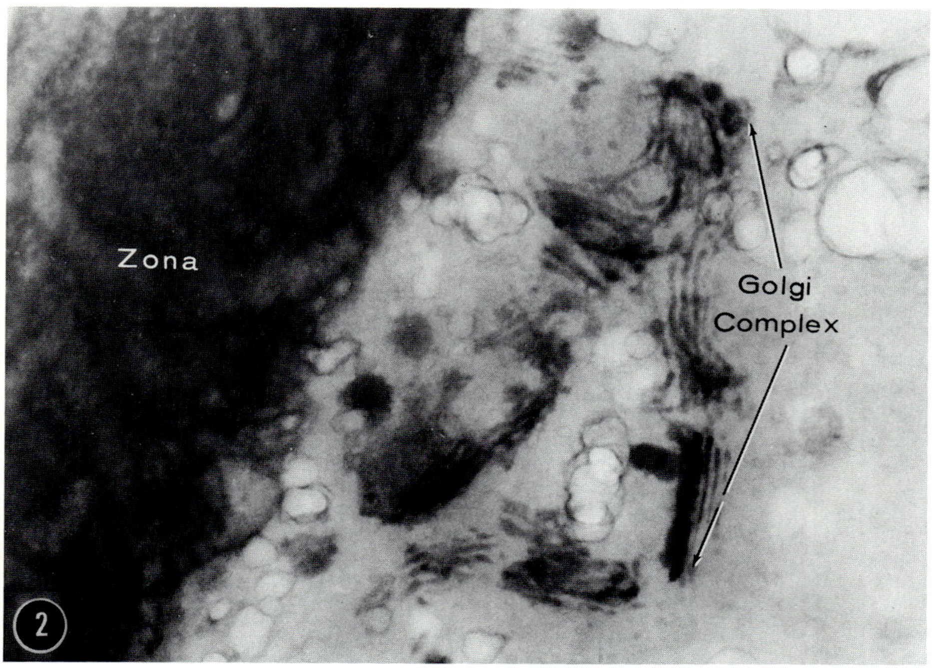
Zona
Golgi
Complex
2

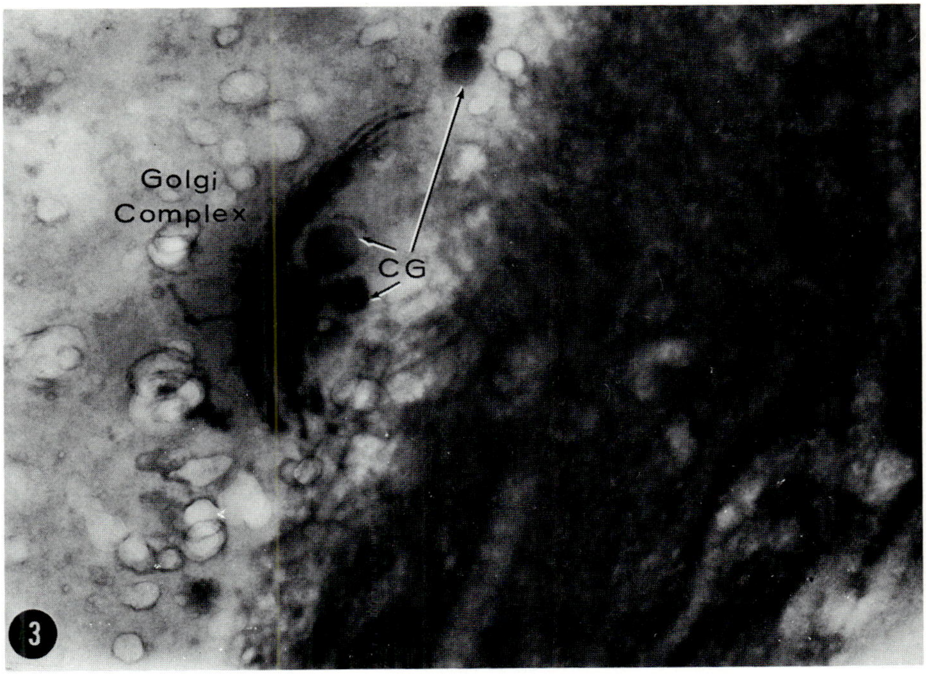

FIGS 1-3. One-micron-thick sections of preantral follicles stained by chromic acid-phosphotungstic acid show that nonfibrous mucopolysaccharide material is present in the intercellular spaces between granulosa cells (FGC) (Fig. 1). The zona, cortical granules (CG), multivesicular bodies (MVB), and Golgi complex are also stained (Figs. 2 and 3). Fig. 1, × 11,700 ; Fig. 2, × 36,000 and Fig. 3, × 36,000

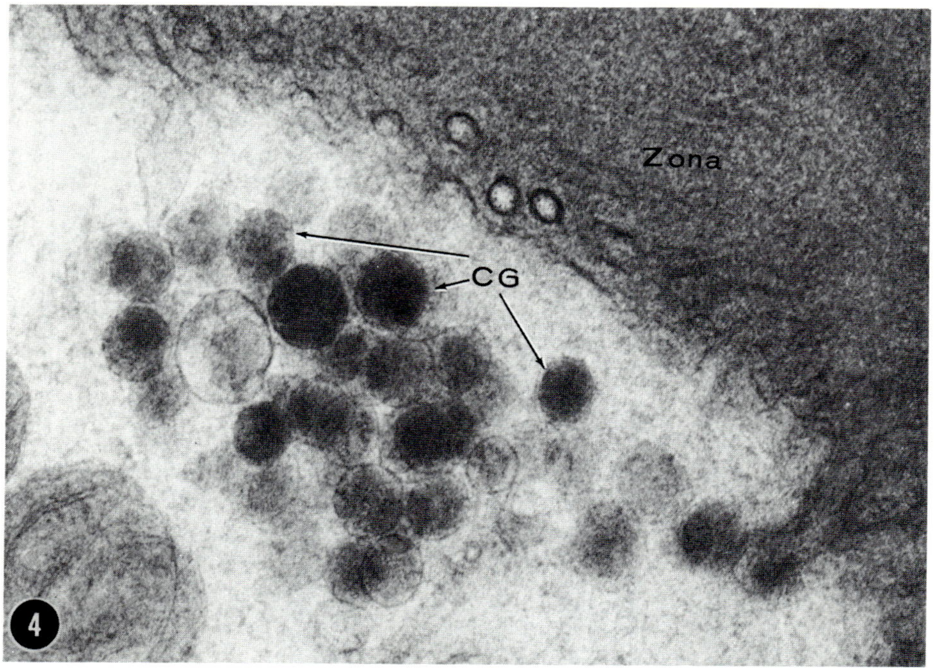

FIG. 4. This high-voltage electron micrograph depicts a three-dimensional view of a cortical granule aggregate in a preantral follicle embedded in Epon and stained with uranyl acetate and lead citrate. ×44,000

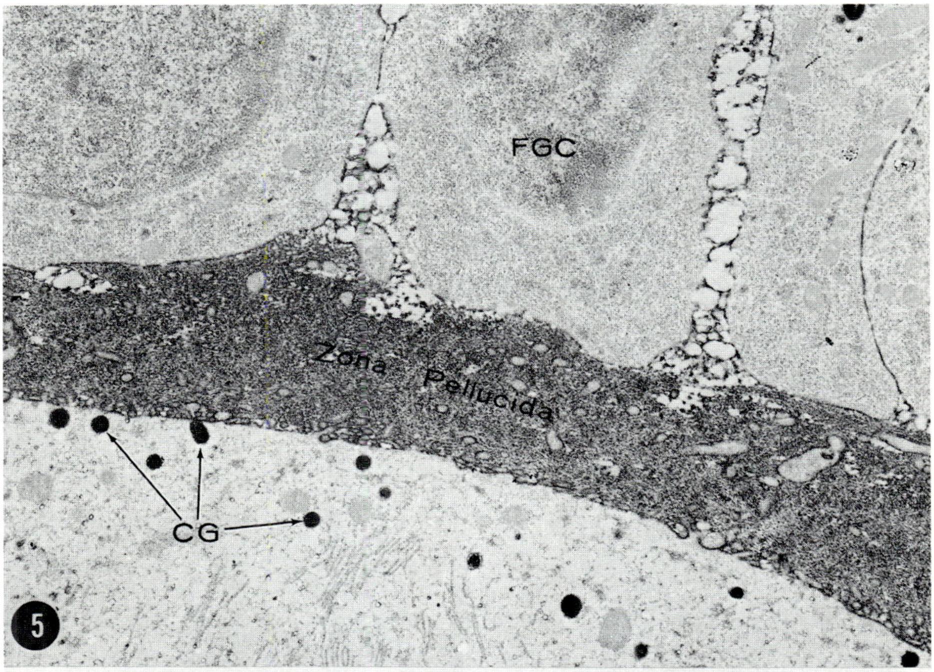
FGC
Zona Pellucida
CG
5

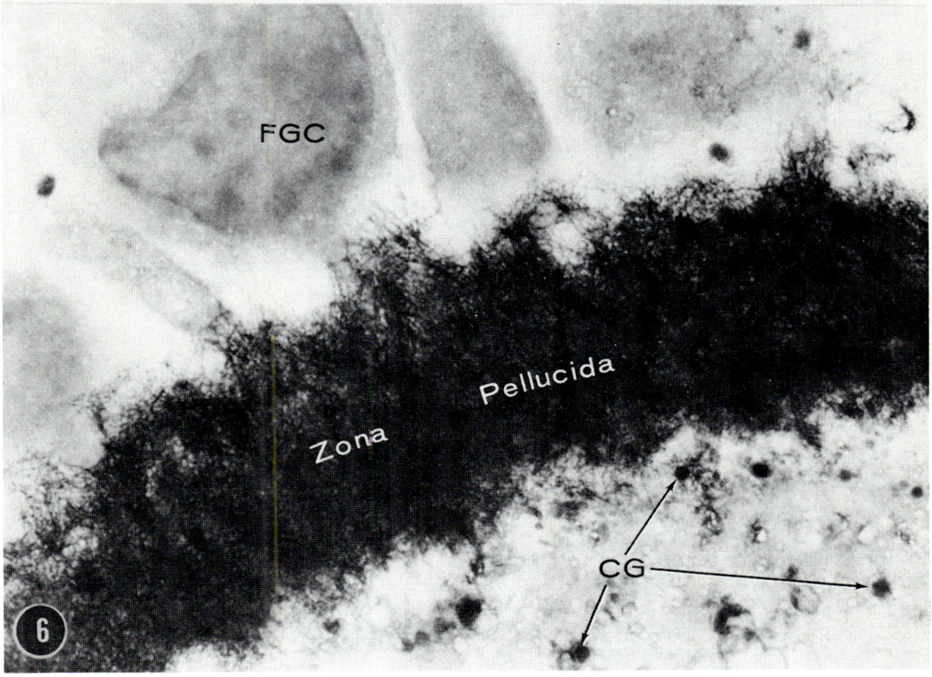
FGC
Zona Pellucida
CG
6

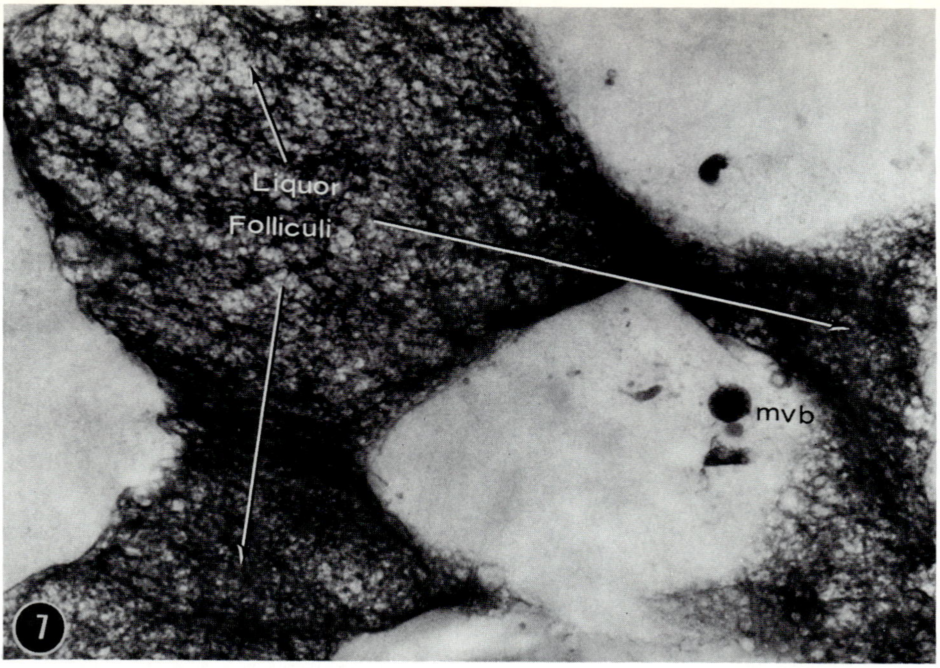

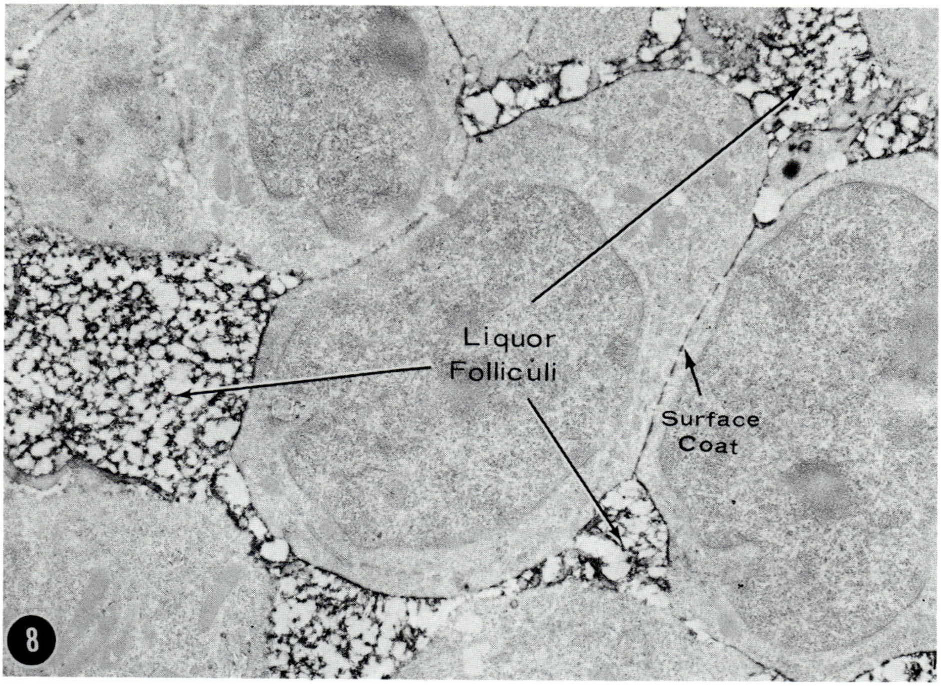

FIGS. 5-9. By using the chromic acid-phosphotungstic acid technique the structure of the zona pellucida and liquor folliculi in the antral follicles is shown respectively in thin sections (Figs. 5, 8, and 9) and in thick sections (Figs. 6 and 7). Zona appears as dense granular mucopolysaccharide material in thin section (Fig. 5), and as coarse fibrous material in the corresponding thick section (Fig. 6). The liquor folliculi is revealed as a complex network structure in thick section (Fig. 7), whereas it shows a fine reticular configuration in the thin sections of antral (Fig. 8) and preovulatory (Fig. 9) follicles. Fig. 5, × 11,700 Fig. 6, × 9000 Fig. 7, × 11,440 and Fig. 9, × 11,700

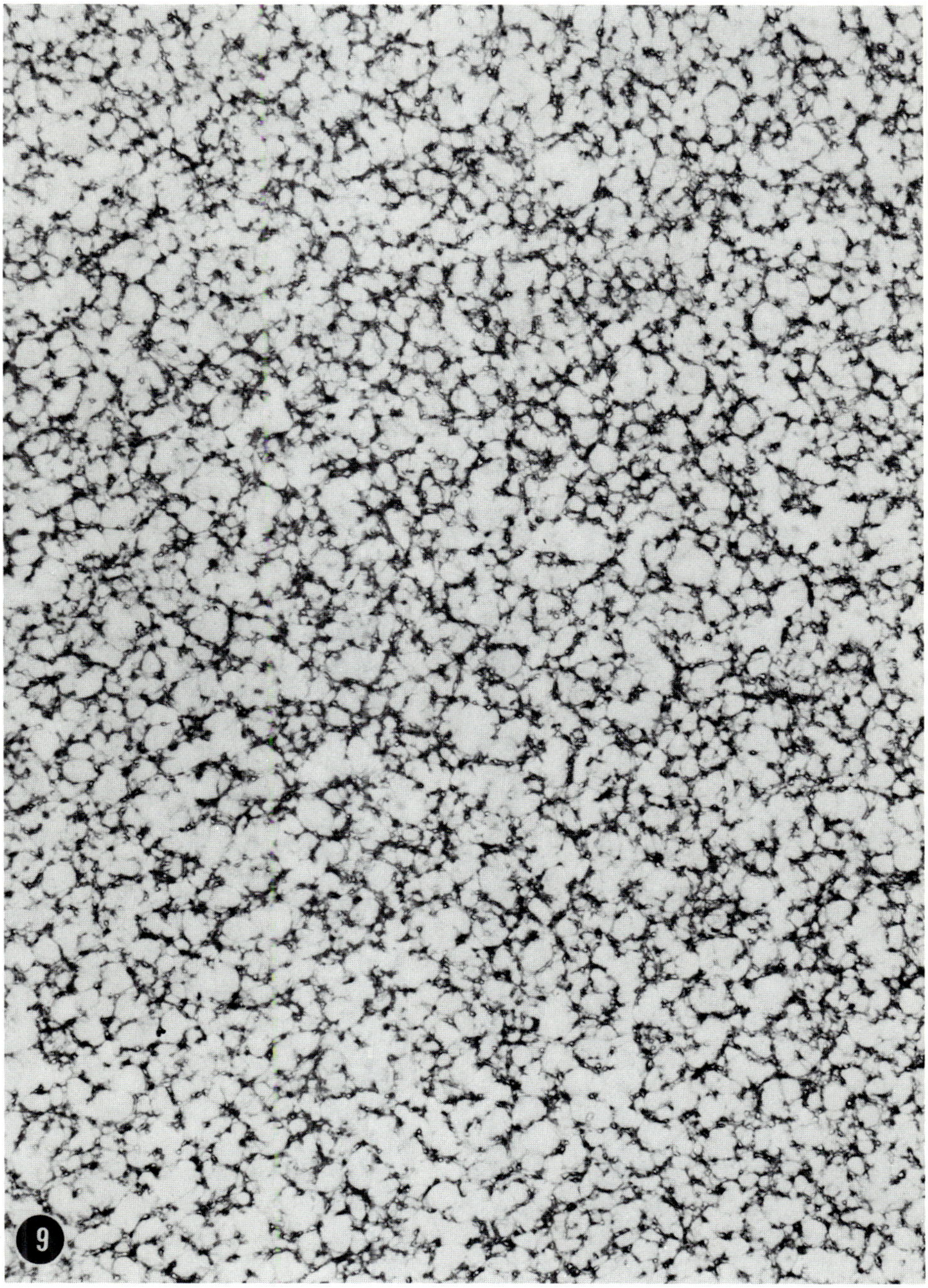
9

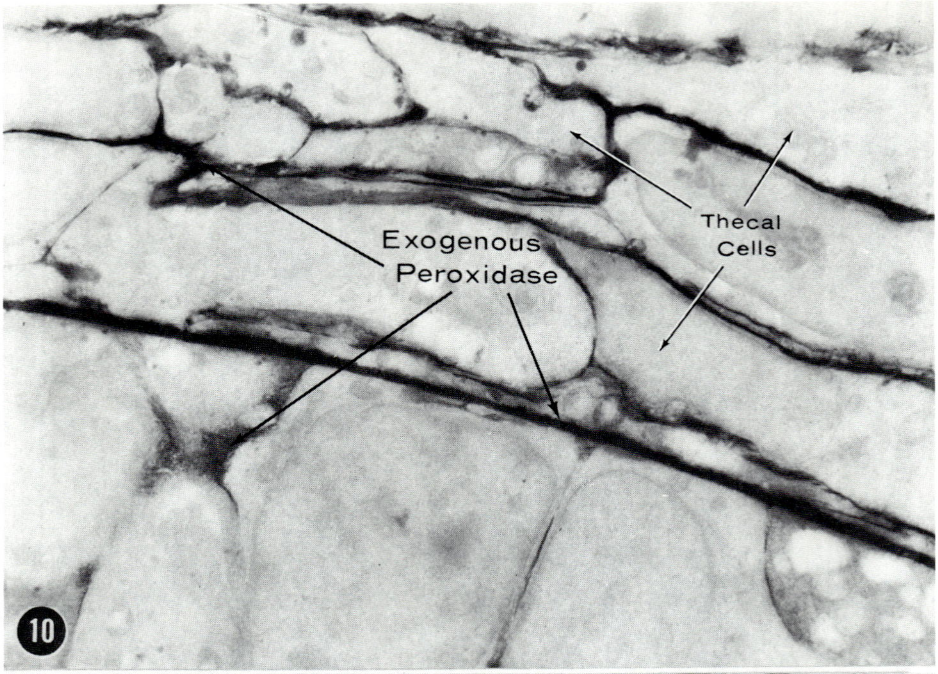

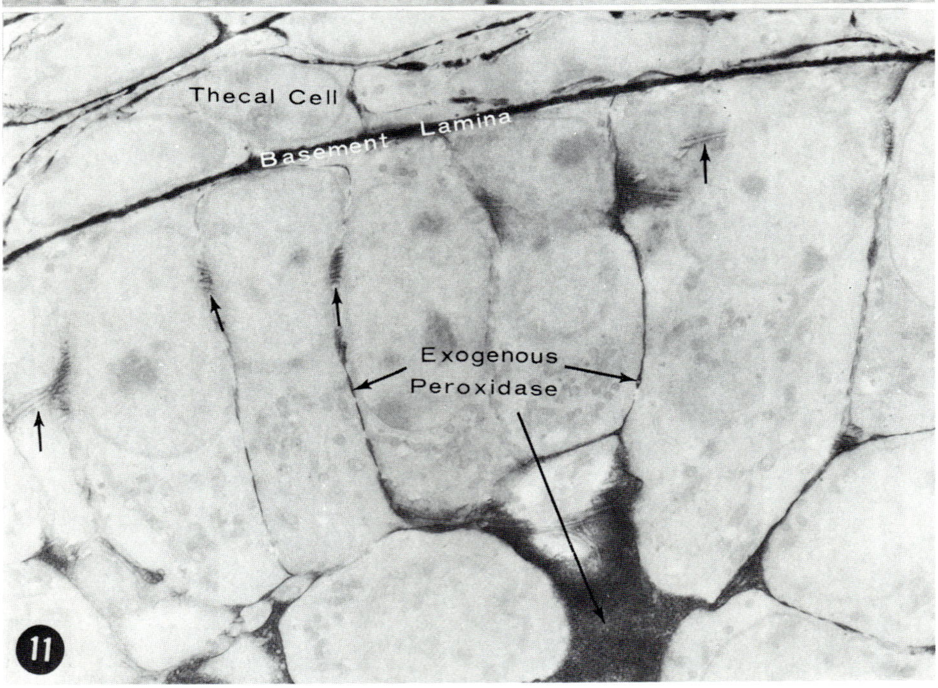

FIGS. 10-13. Three to ten minutes after injection, horseradish peroxidase appears in the intercellular spaces of the theca cell and granulosa cell layers, basement lamina (Figs. 10 and 11) and in the zona pellucida (Figs. 12 and 13). Some complex junctions

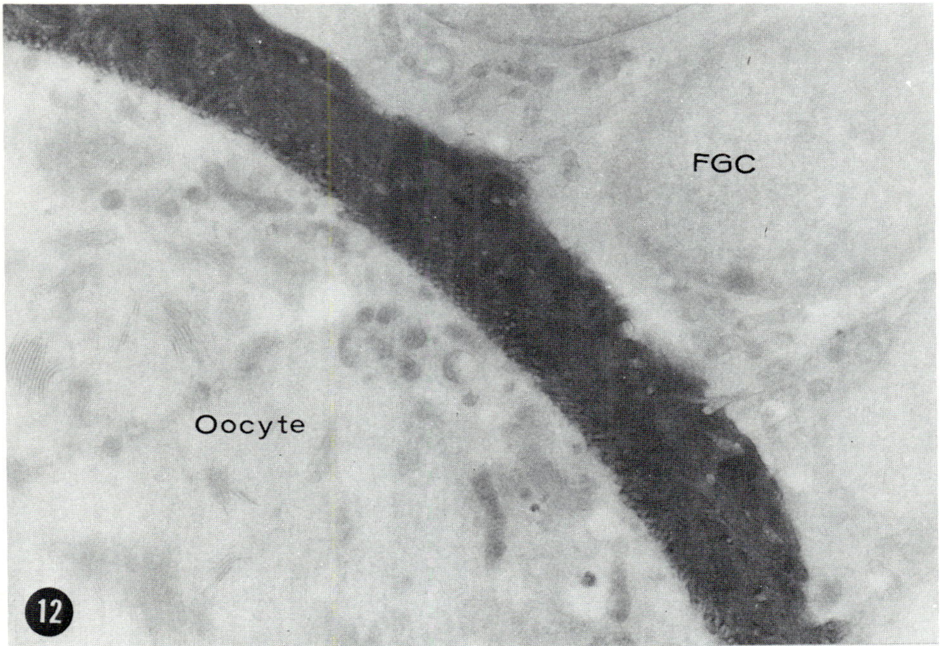

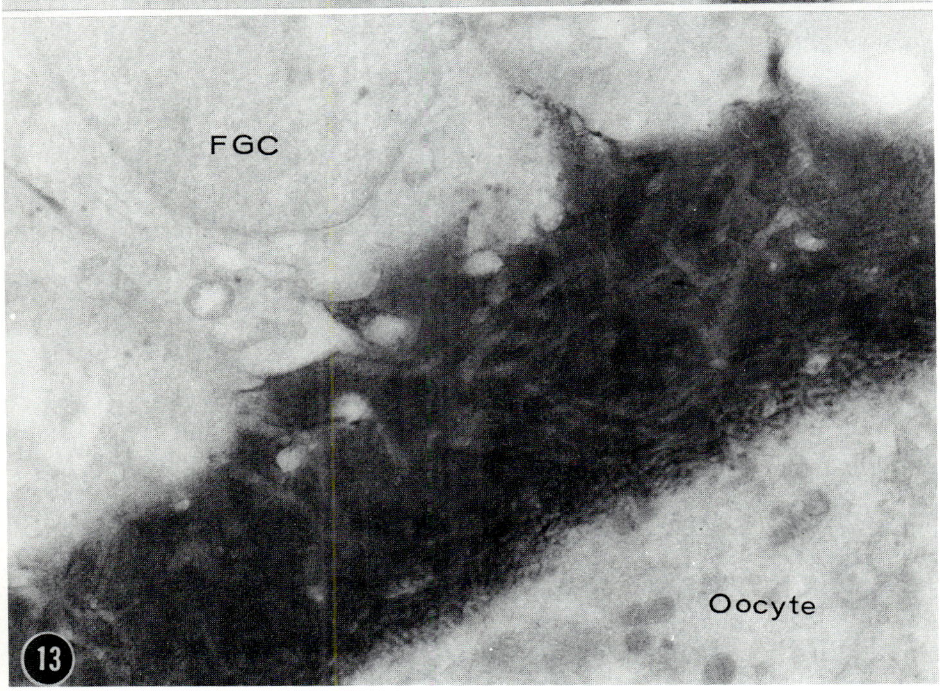

(presumably gap junctions) (*arrows*) are clearly discerned by this technique (Fig. 11). Fig. 10, × 64,800 Fig. 11, × 40 50 Fig. 12, × 7200 and Fig. 13, × 9000

DISCUSSION

The chemical composition and properties, and possible physiological role of follicular fluid, have been extensively investigated (Edwards, 1974; Chang et al., 1976). It contains several proteins, amino acids, sugars, specific enzymes (collagenase, hyaluronidase, transaminase, alkaline and acid phosphatases) (Zachariae and Jensen, 1958; Rondell, 1970; Caucig et al., 1971), mucopolysaccharides (Zachariae and Jensen, 1958), gonadotropin (LH) (Bjersing et al., 1972), steroid hormones (estrogens, androgens, progestagens) (Short, 1962a,b), and salts. The proteins present in the follicular fluid are primarily low molecular weight plasma proteins that are derived from the vascular system by transudation across the follicle basement lamina (von Kaulla et al., 1958; Mancini et al., 1963; Yatvin and Leathem, 1964; Glass, 1970; Shalgi et al., 1973). Large protein molecules, such as low-density lipoprotein, IgM, IgG, and fibrinogen, are found in lower concentration than serum (Shalgi et al., 1973; Edwards, 1974; Chang et al., 1976) or are excluded.

The antral space presumably contains two different types of mucopolysaccharides. A viscous type, containing hyaluronic acid, is distributed primarily in intercellular spaces of granulosa cells. A less viscous type, rich in chondroitin sulfate, is present in the follicular fluid (Zachariae, 1957). The presence of mucopolysaccharides in the follicular fluid has also been revealed by histochemical techniques (Jacoby, 1962; Motta et al., 1971), and by cytochemistry (Kang, 1974). Autoradiographic techniques suggest that granulosa cells secrete mucopolysaccharides into the antrum (Odeblad, 1954; Zachariae, 1957; Hadek, 1963; Motta, 1965a,b).

The chromic acid phosphotungstic acid technique consistently stains mucopolysaccharides (or glycoproteins) in the Golgi saccules and vesicles, multivesicular bodies of various nonsecretory and secretory cell (Rambourg et al., 1969), and in the cortical granules of oocytes (Fléchon, 1970). These results indicate that the zona and structures within the antrum stain intensely positive for acid mucopolysaccharides.

Conventional electron microscopy using Epon-embedded material suggests that the follicular fluid is an amorphous floccular material present in the widened intercellular spaces between granulosa cells. As revealed by ultrathin sections stained with chromic acid-phosphotungstic acid, a previous study indicates that the liquor folliculi is a fine reticulated structure of mucopolysaccharides (Kang, 1974). In thick sections treated similarly, it is clear that the antral space contains, in part, a complex network of coarse fibrous mucopolysaccharides with narrow spaces containing soluble nonstructural components of the fluid. The mucopolysaccharide network is the major constructional framework of the antrum and provides the interstices for movement and retention of blood-borne macromolecules (e.g., tracers). The fibrous network appears to be a product of the granulosa cells and is continuous with the cell surface coat and the zona. The consistency of the antral colloid varies with the developmental stage of the follicle. In the cycling rat the fluid is a gel; however, it is much less gelled in the PMSG-primed rats. The fluid in the porcine follicles is not gelled (Chang et al., 1976).

The zona pellucida has long been recognized as a filamentous mucopolysaccharide-rich structure. Young follicles possess zona containing nonfibrous mucopolysaccharide-rich material that is transformed into a coarse fibrous structure of mature follicles. The Golgi apparatus in the oocyte appears to sequester the zona-type mucopolysaccharide which is deposited in the extracellular space. The fibrous zona does not impede movement of macromolecule-like peroxidase (34 to 40 Å diameter) even though it may be an effective barrier to cells and large macromolecules.

ACKNOWLEDGMENT

This work was supported by grant NICHD 9140 from the National Institutes of Health and by Howard University Cancer Core grant 5PO-CA14718-01.

REFERENCES

Anderson, E. (1967): Observations on the uptake of horseradish peroxidase by developing oocytes of the rabbit. *J. Cell Biol.*, 35:160 (*abstract*).

Anderson, E. (1972): The localization of acid phosphatase and the uptake of horseradish peroxidase in the oocyte and follicle cells of mammals. In: *Oogenesis*, edited by J. D. Biggers and A. W. Schuetz, pp. 86-177. University Park Press, Baltimore.

Anderson, W. A. (1972): Permeability of ovarian vessels and follicles of juvenile rat. *Microvasc. Res.*, 4:373-384.

Bjersing, L. M., Hay, F., Kann, G., Moore, R. M., Naftolin, F., Scaramuzzi, R. J., Short, R. V., and YoungLai, E. V. (1972): Changes in gonadotrophins, ovarian steroids and follicular morphology in sheep at oestrus. *J. Endocrinol.*, 52:465-479.

Caucig, H., Friedrich, Hager, R., and Colob, E. (1971): Enzymes in follicular and cyst fluid in human ovaries. *Acta Endocrinol. (Kbh.)*, 66(Suppl. 152):52.

Chang, S. C., Ryan, R. J., Lee, C. Y., Ellefson, R., and Jones, J. D. (1976): The porcine ovarian follicle. I. Selected chemical analysis of follicular fluid at different developmental stages. *Biol. Reprod.*, 15:321-328.

Chang, S. C., Anderson, W. A., Lewis, J., Ryan, R. J., and Kang, Y. H. (1977): The porcine ovarian follicle. II. Electron microscopic study of surface features of granulosa cells at different stages of development. *Biol. Reprod.*, 16:349-357.

Edwards, R. G. (1974): Follicular fluid. *J. Reprod. Fertil.*, 37:189-219.

Fléchon, J. E. (1970): Nature glycoprotieque des granules corticaus de l'oeuf de lapine mis en evidence par l'utilisation comparee de techniques cytochimiques ultrastructurales. *J. Microscopie*, 9:221-242.

Glass, L. E. (1970): Translocation of macromolecules. In: *Cell Differentiation*, edited by O. A. Schjeida and L. E. Glass, pp. 42-118. Van Nostrand-Reinhold, New York.

Graham, R. C., and Karnovsky, M. J. (1966): The early stages of absorption of injected horseradish peroxidase in the proximal tubules of mouse kidney by a new technique. *J. Histochem. Cytochem.*, 14:291-302.

Gwatkin, R. (1967): Passage of Mengo virus through the zona pellucida of the mouse morula. *J. Reprod. Fertil.*, 13:577-578.

Hadek, R. (1963): Electron microscopic study on the primary liquor folliculi secretion in mouse ovary. *J. Ultrastruct. Res.*, 9:445-458.

Hastings II,R.A. Enders, A. C., and Schlafke, S. (1972): Permeability of the zona pellucida to protein tracers. *Biol. Reprod.*, 7:288-296.

Jacoby, F. (1962): Ovarian histochemistry. In: *Ovary*, Vol. 1, edited by S. Zuckerman, pp. 189-234. Academic Press, New York.

Kang, Y. H. (1974): Development of the zona pellucida in the rat oocyte. *Am. J. Anat.*, 139:535-566.

Karnovsky, M. J. (1965): A formaldehyde-glutaraldehyde fixative of high osmolality for use in electron microscopy. *Cell Biol.*, 27:137-138.

Leduc, E. H., and Bernhard, W. (1967): Recent modification of the glycol methacrylate embedding procedure. *J. Ultrastruct. Res.*, 19:196-199.

Lee, C. Y. (1975): Human chorionic gonadotrophin binding and stimulation of adenyl cyclase of porcine granulosa cells during follicular development. Presented at the 26th Annual Meeting of Tissue Culture Association.

Lee, C. Y. (1976): The porcine ovarian follicle. III. Development of chorionic gonadotropin receptors associated with increase in adenylate cyclase activity during follicle

maturation. *Endocrinology*, 99:42-48.
Luft, L. H. (1961): Improvement in epoxy resin embedding methods. *J. Biophys. Biochem. Cytol.*, 9:409-414.
Mancini, R. E., Vilar, O., Heinrich, J. J., Davidson, O. W., and Alvarez, B. (1963): Transference of circulating serum proteins to the follicle of the rat. *J. Histochem. Cytochem.*, 11:80-88.
Motta, P. (1965*a*): Sulla formazione del liquor folliculi nell' ovajo dells coniglia. *Biol. Lat.*, 18:252-271.
Motta, P. (1965*b*): Sur l'ultrastructure des "corpus d Call et Exner" dans l'ovarie du lapin. *Z. Zellforsch.*, 68:308-319.
Motta, P., Taheva, S., and Nesu, E. (1971): Etude ultrastructurall et histochemique des rapports entre les cellules folliculaires et l'ovocytoependant le development du follicule ovarien chez les mammiferes. *Acta Anat.*, 80:537-562.
Odeblad, E. (1954): Studies on the physiology of the follicular fluid. *Acta Endocrinol. (Kbh.)*, 15:313-316.
Rambourg, A. (1967): Detection des glycoproteines en microscopie electronique: coloration de la surface cellulaine et de l'appareil de Golgi par un melange acid chromique-phosphotungstique. *C. R. Acad. Sci.*, 265:1426-1428.
Rambourg, A., Hernandez, W., and Leblond, C. P. (1969): Detection of complex carbohydrates in the Golgi apparatus of rat cells. *J. Cell Biol.*, 40:395-415.
Rondell, P. (1970): Biophysical aspects of ovulation. *Biol. Reprod. Suppl.*, 2:64-89.
Shalgi, R., Kraicer, P., Rimon, A., Pinto, M., and Soferman, N. (1973): Proteins of human follicular fluid; the blood-follicle barrier. *Fertil. Steril.*, 24:429-434.
Short, R. V. (1962*a*): Steroids present in the follicular fluid of the cow. *J. Endocrinol.*, 23:401.
Short, R. V. (1962*b*): Steroids in follicular fluid and corpus luteum of the mare. A "two-cell type" theory of ovarian steroid synthesis. *J. Endocrinol. (Kbh.)*, 24:59-63.
Venable, J. H., and Coggeshall, R. (1965): A simplified lead citrate stain for use in electron microscopy. *J. Cell Biol.*, 25:407-408.
Von Kaulla, K. N., Aikawa, J. K., and Pettigrew, J. D. (1958): Concentration in the human ovarian follicular fluid of radioactive tracers and drugs circulating in the blood. *Nature*, 182:1238-1239.
Yatvin, M. B., and Leathem, J. H. (1964): Origin of ovarian cyst fluid: studies on experimentally induced cysts in rat. *Endocrinology*, 75:733-740.
Zachariae, F. (1957): Studies on the mechanism of ovulation. Autoradiographic observations on the uptake of radioactive sulfate (^{35}S) into the ovarian follicular mucopolysaccharides. *Acta Endocrinol. (Kbh.)*, 26:215-223.
Zachariae, F., and Jensen, C. E. (1958): Studies on the mechanism of ovulation; histochemical and physico-chemical investigations on genuine follicular fluid. *Acta Endocrinol. (Kbh.)*, 27:343-355.

Ovarian Follicular Development and Function,
edited by A. R. Midgley and W. A. Sadler.
Raven Press, New York © 1979.

Incidence and Characterization of Gap Junctions in the Graafian Follicle

A. Amsterdam and H. R. Lindner

Department of Hormone Research, Weizmann Institute of Science, Rehovot, Israel

We have identified gap junctions in the ovarian follicle by examining thin sections and freeze cleavage replicas (Amsterdam et al., 1974*a*,*b*, 1976) in the following locations: (1) between theca cells, (2) between granulosa cells, and (3) between cytoplasmic processes of corona cells and the oolemma (thin-sectioned material only; Amsterdam et al., 1976).

Dr. E. Anderson (*this volume*) emphasized the importance of demonstrating hexagonal lattice structure when reporting on the occurrence of gap junctions. This can conveniently be achieved by a technique we recently described (Amsterdam et al., 1974*a*,*b*, 1976), namely by submitting electron micrographs of freeze-cleavage replicas to optical diffraction analysis (Amsterdam et al., 1974*a*,*b*, 1976). In this way the organization of intramembrane particles within the junction can be quantitatively defined. Large junctions between granulosa or thecal cells were shown to consist of a mosaic of lattices identical in center-to-center spacing of particles but differing in orientation. This pattern is compatible with their formation from coalescence, on growth, of multiple centers of aggregation similar to the "plaques" described by Dr. Anderson (*this volume*) in early antral follicles.

Prior to ovulation, gap junctions appear to be dismantled, perhaps by internalization as suggested by the occurrence of many "annular junctions."

We suggested (Amsterdam et al., 1975, 1976) that gap junctions between granulosa cells, and between cumulus cells and the oocyte, may serve to propagate an LH-initiated signal toward the interior of the follicle, since only theca cells and the mural granulosa cells, but not the periantral and cumulus granulosa cells, could be shown to possess LH receptors by autoradiographic analysis (Amsterdam et al., 1975).

ACKNOWLEDGMENT

This work was supported by the Ford Foundation and the Population Council, New York. H. R. L. was the recipient of a travel grant from Serono Laboratories Ltd.

REFERENCES

Amsterdam, A., Josephs, R., Liebermann, E., and Lindner, H. R. (1974*a*): Organisation of intramembrane particles of granulosa cell gap junctions in rat ovarian follicles. *J. Cell Biol.*, 63 (2 Pt. 2):8a.

Amsterdam, A., Josephs, R., Liebermann, E., and Lindner, H. R. (1974*b*): Membrane junctions between granulosa cells of Graafian follicles from rat ovary. Israel J. Med. Sci., 10:1578 (*abstract*).

Amsterdam, A., Josephs, R., Liebermann, E., and Lindner, H. R. (1976): Organisation of intramembrane particles in freeze-cleaved gap junctions of rat Graafian follicles: optical diffraction analysis. *J. Cell Sci.*, 21:93

Amsterdam, A., Koch, Y., Liebermann, E., and Lindner, H. R. (1975): Distribution of binding sites for human chorionic gonadotrophin in the preovulatory follicle of the rat. *J. Cell Biol.*, 67:894.

Ovarian Follicular Development and Function,
edited by A. R. Midgley and W. A. Sadler.
Raven Press, New York © 1979.

Discussion Summary: Session on Follicular Morphology

Everett Anderson

Following the presentation of the morphologic aspects of the differentiating ovarian follicle by a variety of methodologies, the spirited discussion concerned itself with each compartment of the follicle and terminated with questions on follicular atresia. The general tone of the discussion was the clarification, search, and presentation of new ideas concerning the structural and physiological characteristics of the developing follicle.

There was some unanimity for the acceptance of the concept that the membrane specialization, in the form of a gap junction between the oocyte and granulosa cells, may mediate metabolic cooperation. However, the ontogeny of a similar specialization between granulosa cells produced an interesting discussion, largely brought about by the relatively recent presentation of the fluid mosaic model of the plasma membrane as presented by Singer and Nicholson (1972). The model shows that the plasma membrane is bilayered and contains a large and heterogeneous population of protein molecules. These molecules are thought to migrate laterally in the membrane. With the development of the freeze-fracture technique, it is now possible to examine the internal organization of cell membranes in general. The polemics centered around the fact that the differentiation of the surface of granulosa cells may be (1) related to the formation of gap junctions for cell-to-cell communication and (2) the acquisition of gonadotropic receptors. During the discussion, evidence was presented that in the rat granulosa cells are electrically coupled and that radioiodinated gonadotropin hormone preparations reveal the presence of receptors on or in the cell membrane. The autoradiographic technique left unanswered whether the specific receptors were in any way related to any specific particles as visualized by the freeze-fracture methodology. Moreover, the question of whether granulosa cell gap junctions or their precursor particles are direct targets of tropic hormones is still outstanding. Nevertheless, the discussion proceeded to the biological significance of the internalization (phagocytosis) of certain gap junctions during the differentiation of the follicle. Since the follicle is a highly developing system where cells are undergoing unequal shape changes it was suggested that internalization of gap junctions may represent a means of removing them from the cell surface for continued differentiation. Further discussions were concerned with the following questions: If there is a relationship between gap junctions and receptors, whether the internalization is a mechanism for clearing the surface of "used" receptors, and if there is a turnover and reformation of the gap junction. It was agreed that a good quantitation (morphometric analysis) of gap junctions is needed for both the normal developing and atretic follicle.

The discussion proceeded to the extracellular matrices (antrum, zona pellucida, and membrana limitans or basal lamina) of the ovarian follicle. While the origin of these matrices is unknown, cytochemistry reveals them to be composed of glycoproteins and complex mucopolysaccharides. It was pointed out that the membrana limitans (basement lamina) adjacent to the outer granulosa cells and the theca interna must be secreted by the outer granulosa cells. If this is the case, the question was raised as to whether or not these granulosa cells were different from the remaining population, for it is known that all granulosa cells within a given follicle do not make contact with the membrana limitans. By use of high-voltage electron microscopy, the aforementioned matrices were found to be composed of coarse networks of fibers. The hypothesis was presented that the extracellular matrices or compartments of follicles control, to some extent, the permeability characteristics of the follicles.

Precious little was said about the theca interna and externa, and there is therefore urgent need to clarify the origin and precise function of these important layers during folliculogenesis.

During the discussion questions were raised concerning the part gap junctions play, if any, in the phenomenon of atresia. This aspect of the discussion was not satisfying largely due to our lack of knowledge of the molecular events of this important phenomenon.

It is well known that organisms vary in the production of healthy follicles for ovulation. Little consideration was given during the discussion to the comparative approach to follicular development and function. In an effort to understand the assumed common physiological and biochemical denominator of folliculogenesis, it is imperative and fruitful to approach the subject from a comparative point of view with as many methodologies as possible.

OVULATION

Ovarian Follicular Development and Function,
edited by A. R. Midgley and W. A. Sadler.
Raven Press, New York © 1979.

Studies of the Enzymatic Basis and Hormonal Control of Ovulation

Sidney Strickland and William H. Beers

The Rockefeller University, New York, New York 10021

The study of ovulation was begun over 300 years ago by Regnier de Graaf (1), who provided a detailed and accurate description of the ovarian follicle and the events that accompany the release of the ovum. Since that time, many theories have been proposed to account for the deterioration of the follicle wall at the time of ovulation, but none have been substantiated (2). One attractive hypothesis, originally proposed by Schochet in 1916 (3), is that specific proteolytic enzymes are responsible for the disruption of the wall and release of the ovum. In recent years, evidence consistent with this proposal has been accumulating. Central to this issue are the observations of Espey that several serine proteases can weaken the follicle wall in a manner similar to that observed at ovulation (4,5).

A firm basis for the validity of this mechanism, however, would require the demonstration of an ovarian enzyme that has the following properties. First, it should be produced within the follicle and achieve maximum levels near the time of follicular rupture. Second, the level of activity should be controlled by hormones known to induce ovulation, e.g., gonadotropins. Third, it should be capable of physically weakening the follicle wall.

Recently, we have shown that the production of plasminogen activator by granulosa cells satisfies these requirements (6). The extracellular level of the enzyme is closely correlated with ovulation; it is modulated by gonadotropins, cyclic nucleotides, and prostaglandins, and, by its action on plasminogen, which is present in follicular fluid, it generates the protease plasmin, which can weaken the follicle wall (7). This chapter describes some of the experiments on which these conclusions are based, and presents a specific model for the biochemical basis and hormonal control of ovulation.

CORRELATION OF INCREASED PLASMINOGEN ACTIVATOR PRODUCTION WITH OVULATION

In a previously published paper, we have shown that granulosa-cell-associated plasminogen activator activity is markedly increased in preovulatory follicles, both in immature rats hormonally primed to ovulate, and in mature cycling adults (6). These studies were performed using a fibrin agar overlay procedure that allows direct visualization of fibrinolysis. Although this method of analysis yields results that are quantitative, data collection is tedious and time consuming. Consequently, in more recent studies (8) we have employed a different assay procedure that is more convenient. In this method, granulosa cells are harvested from follicles and cultured in vessels that are coated with ^{125}I-fibrin. Plasminogen-dependent fibrinolysis can then be detected by the appearance of ^{125}I-labeled fibrinopeptides in the growth medium. The results described below were obtained using this assay and confirm that granulosa cell plasminogen activator levels and ovulation are closely correlated.

When immature rats are induced to ovulate by administered gonadotropins, granulosa cells can be collected and assayed from animals at a known time prior to ovulation. As seen in Fig. 1, the amount of fibrinolytic activity increases as the time of ovulation approaches. This activity is completely plasminogen dependent, indicating that it is due to plasminogen activator. Furthermore, elevated levels of the enzyme are associated only with cells obtained from preovulatory follicles (Fig. 2). In experiments conducted with mature cycling adults, elevated levels of granulosa cell plasminogen activator can only be demonstrated late on the night of proestrus, shortly before ovulation (Fig. 2). These two sets of experiments indicate that increased production of plasminogen activator and ovulation are closely coupled and suggest that the level of enzyme is controlled by circulating gonadotropins. It was therefore of interest to investigate the nature of the endocrine control of plasminogen activator synthesis and secretion.

IN VITRO STUDIES

Granulosa cells from antral (but not fully developed) follicles produce barely detectable amounts of plasminogen activator. However, following suitable stimulation *in vitro*, e.g., exposure to gonadotropins, the amount of extracellular activity increases 20- to

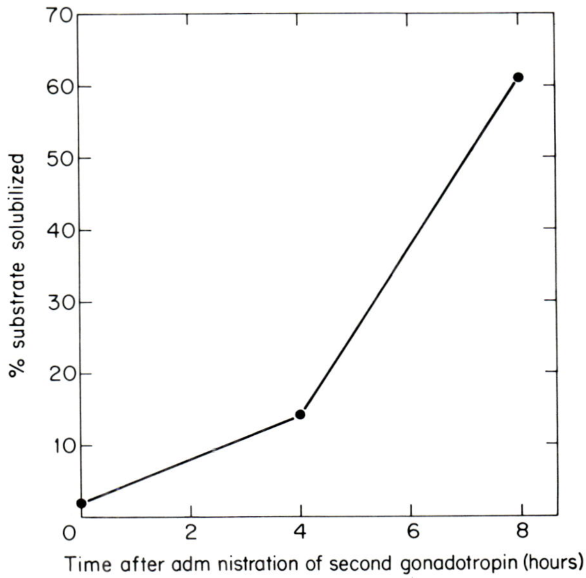

FIG. 1. Temporal correlation between the time of ovulation and the production of plasminogen activator by granulosa cells. Immature rats received PMSG for 48 hr and then LH. Ovulation occurs 8 to 10 hr after the administration of LH. At the indicated times, granulosa cells were collected and then cultured and assayed on ^{125}I-fibrin-coated vessels. The results are presented as the percentage of the radioactive substrate solubilized. The solubilization of fibrin by granulosa cells is plasminogen dependent, indicating that it is due to plasminogen activator. For exact culture and assay conditions see ref. 8. Results equivalent to those shown above are obtained if FSH or HCG are substituted for LH as the second gonadotropin.

40-fold within a few hours. Figure 3 demonstrates the dose-dependent characteristics of the response when the cells are incubated with ovine LH[1] *in vitro*. Half maximal response is achieved with approximately 3 μg/ml of LH in the presence or absence of the cyclic nucleotide phosphodiesterase inhibitor, theophylline. The magnitude of the response is potentiated by theophylline, however, which increases the level of activity by as much as twofold. This effect suggests that the response to gonadotropins may be mediated by cyclic nucleotides and is consistent with the finding that cAMP itself is capable of increasing the level of plasminogen activator in granulosa cells (6).

---

[1]The abbreviations used are: PMSG, pregnant mare serum gonadotropin; LH, luteinizing hormone; FSH, follicle-stimulating hormone; TSH, thyroid-stimulating hormone; HCG, human chorionic gonadotropin; cAMP, cyclic adenosine monophosphate; cGMP, cyclic guanosine monophosphate.

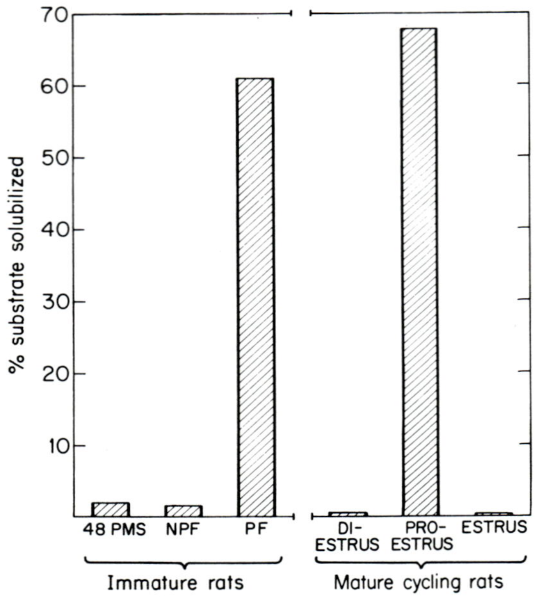

FIG. 2. Correlation of elevated plasminogen activator with granulosa cells from preovulatory follicles. Immature rats: Animals received PMSG for 48 hr and then LH for 0 or 8 hr. 48 PMSG: Cells collected from animals after 48 hr exposure to PMSG. NPF: Cells collected from nonpreovulatory follicles after 8 hr exposure to LH. PF: Cells collected from preovulatory follicles after 8 hr exposure to LH. Mature cycling rats: Cells were collected from animals in the indicated stage of the estrous cycle. The cells were cultured and assayed as in Fig. 1.

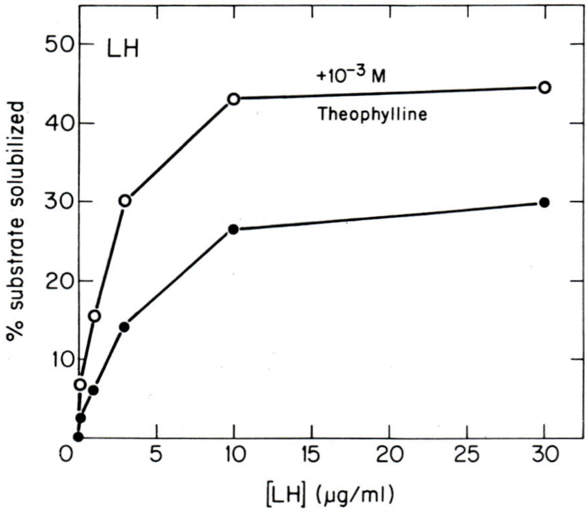

FIG. 3. Dose dependence of plasminogen activator production by granulosa cells in response to ovine LH *in vitro*. Animals received PMSG for 48 hr. The cells were collected, cultured in the presence of the indicated amount of LH with or without 10^{-3} M theophylline, and assayed for plasminogen activator production.

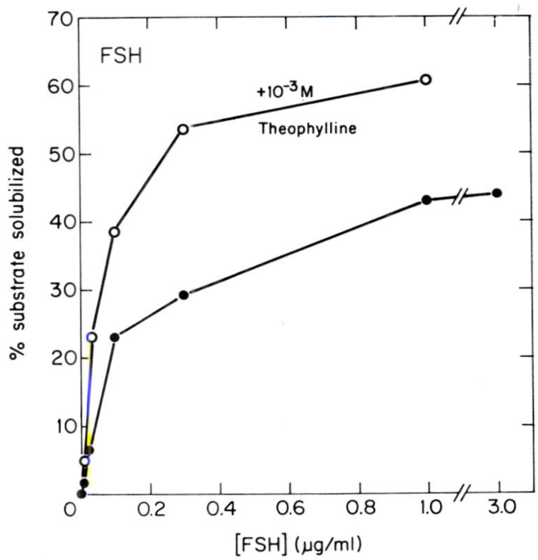

FIG. 4. Dose dependence of plasminogen activator production by granulosa cells in response to ovine FSH *in vitro*. This experiment was conducted as in Fig. 3 except that FSH was used in place of LH.

The method of analysis used in the above study allows the measurement of extracellular plasminogen activator. To determine the effect of LH on intracellular enzyme levels, experiments were performed to analyze the levels of plasminogen activator in cell lysates. These data indicate that the level of intracellular enzyme also increases with increasing concentrations of LH and suggest that the cells respond to gonadotropins by increasing *de novo* synthesis of enzyme. This conclusion is strengthened by the observations that actinomycin D and cycloheximide inhibit the response of the cells, indicating that RNA and protein synthesis are necessary for the appearance of elevated levels of enzyme.

Experiments designed to determine the specificity of the *in vitro* response to gonadotropins have revealed that LH, the presumptive ovulatory hormone, is not the most potent effector of increased plasminogen activator synthesis. Of the several gonadotropins tested, FSH is by far the most effective, and the characteristics of the response of the cells to it are shown in Fig. 4. The effect of FSH is similar to that of LH and the level of activity observed is elevated by theophylline. FSH, however, is active at much lower concentrations. Half maximal stimulation is seen at approximately 0.1 μg/ml with FSH as compared with 3 μg/ml with LH. Thus, on a weight basis the ovine FSH preparations are minimally 30-fold more active than equivalent LH preparations.

Other anterior pituitary hormones have also been tested for their ability to stimulate granulosa cells (Table 1). Of these, FSH was by far the most potent. TSH preparations were as active as LH, but the prolactin and growth hormone were inactive. PMSG and HCG, which exhibit high FSH and LH activity, respectively, were also studied. The cells respond to PMSG at much lower concentrations than HCG, a result that is consistent with the specificity observed using ovine hormones.

TABLE 1. *Potency of various polypeptide hormones in stimulating plasminogen activator production by granulosa cells*

Hormone	Approximate concentration for half-maximal response (μg/ml)
o-FSH	0.1
o-LH	3
o-TSH	3
o-PRL	>>10
o-GH	>>10
r-FSH	0.1
r-LH	>10

Animals received PMSG for 48 hr. The dose dependence for each hormone was determined as in Fig. 3.
o-, ovine hormones.
r-, rat hormones.

While the above studies were of interest regarding the specificity of the rat granulosa cell response, their significance could not be judged until similar experiments were conducted using homologous rat hormones. As seen in Table 1, rat FSH is more than 100 times more potent than rat LH, suggesting that the specificity observed previously has relevance to the physiological situation. Finally, in more recent experiments, using highly purified preparations of human FSH and LH, the same pattern is seen.

The enhancement of the effect of gonadotropins produced by theophylline suggests that cAMP is a mediator of increased plasminogen activator synthesis. Consequently, cAMP was tested for its ability to stimulate the cells *in vitro*. Plasminogen activator levels do increase on exposure of the cells to cAMP, and dibutyryl cAMP. As seen in Table 2, the effect of

TABLE 2. *In vitro response of granulosa cells to dibutyryl cAMP, theophylline, and prostaglandin-E_1*

Addition to medium	% Substrate solubilized
None	2.2 ± 1.0
10^{-3} M theophylline	5.0 ± 1.0
10^{-3} M DBC	18.6 ± 0.6
10^{-3} M DBC + 10^{-3} M theophylline	33.5 ± 3.0
10^{-6} M PGE_1	39.9 ± 2.2
10^{-6} M PGE_1 + 10^{-3} M theophylline	67.7 ± 3.7
10 μg/ml o-FSH	46.6 ± 4.2
10 μg/ml o-FSH + 10^{-3} M theophylline	65.9 ± 5.8

Animals received PMSG for 48 hr. The cells were collected, cultured in the presence of the indicated compounds, and assayed as in Fig. 1.

DBC, $N^6,O^{2'}$-dibutyryl cAMP.

PGE_1, prostaglandin E_1.

dibutyryl cAMP is potentiated by theophylline. Other cyclic nucleotide phosphodiesterase inhibitors, such as methylisobutyl xanthine also enhance the response to dibutyryl cAMP. In contrast to the effect of cyclic adenine nucleotides, granulosa cells are not stimulated by cGMP or its derivatives.

Since prostaglandins have been suggested to possess gonadotropin-like activities and to play a role in ovulation, several of these hormones were tested for their ability to induce plasminogen activator in granulosa cells. Prostaglandins E_1 and E_2 elicited a response at physiological concentrations, while prostaglandins $F_{1}\alpha$ and $F_{2}\alpha$ were completely inactive. The response to E-prostaglandins was enhanced by theophylline, again suggesting a role for cAMP (Table 2).

Although the response to gonadotropins also appears to involve cAMP, two lines of evidence suggest that prostaglandins and gonadotropins act independently, although possibly through a common final pathway, in their ability to stimulate the cells. First, indomethacin, an inhibitor of prostaglandin biosynthesis, does not inhibit the effect of gonadotropins. Second, prostaglandins will elevate the level of plasminogen activator produced by cells exposed to saturating levels of FSH.

On structural grounds, as well as on the basis of current theories regarding the actions of cyclic nucleotides and prostaglandins, the possibility existed that the inactive compounds tested (cGMP and its derivatives and prostaglandins $F_1\alpha$ and $F_2\alpha$) might function as inhibitors of the active agents that have been studied. However, under several experimental conditions, these drugs were without effect on the amplitude of the response or the dose-response characteristics of the various effectors (gonadotropins, cAMP, E-prostaglandins). Several steroids were also studied for their ability to stimulate the cells to produce plasminogen activator and as inhibitors of the response to gonadotropins. All of the steroids tested were without effect on this aspect of granulosa cell metabolism. The compounds used were: progesterone, 17α hydroxyprogesterone, 20α dihydroprogesterone, 17β estradiol, testosterone, and dexamethasone. Finally, cyanoketone, which blocks progesterone biosynthesis by inhibiting 3β hydroxy-steroid dehydrogenase had no effect on the *in vitro* response to gonadotropins.

DISCUSSION

The experiments described above demonstrate that the production of plasminogen activator by granulosa cells is closely correlated with ovulation. Furthermore, we have been able to show that this function is hormonally controlled and amenable to analysis under *in vitro* conditions. The *in vitro* studies suggest that the response to gonadotropins and prostaglandins is mediated by cAMP and involves an increase in the rate of synthesis and secretion of plasminogen activator. Since plasminogen is present in follicular fluid, an elevation of the level of extracellular plasminogen activator would yield increasing intrafollicular concentrations of plasmin, a serine protease that can act on basement membrane (9) and possibly other substrates to weaken the follicle wall. This sequence of events is outlined in the scheme shown in Fig. 5.

While this mechanism probably represents a simplified version of the events leading to the disruption of the follicle wall at ovulation, it should provide the basis for investigations into the specific biochemical events that are involved. It is worth noting that several events occurring prior to ovulation, in addition to the weakening of the wall, could be explained by the presence of plasmin. These include the observations that, in the rat, the granulosa cells become dissociated from one another and are shed into the antrum (10), and that the ovum and its investments are dissociated from the underlying granulosa layers and become free floating in the follicular fluid (11).

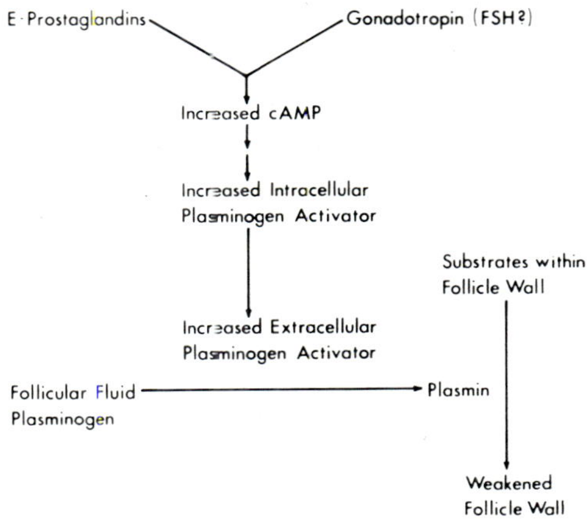

FIG. 5. Working hypothesis for the hormonal control and enzymatic basis of ovulation.

The *in vitro* stimulation of granulosa cells allows a well-defined analysis of the specificity of the response to gonadotropins. These studies indicate that FSH is much more effective than LH in causing the cells to produce plasminogen activator. In fact, since the LH preparations used are likely to contain traces of FSH, the possibility exists that the observed response to LH is due to contamination with FSH. The observation that TSH, which has no known function in the ovary was as active as LH, also suggests that the stimulation by these two hormones might be due to the presence of FSH in the preparations.

If FSH is the only gonadotropin which, at physiological concentrations, will stimulate the granulosa cells to produce plasminogen activator, then the scheme outlined in Fig. 5 suggests that FSH is the hormone ultimately responsible for the disruption of the follicle wall. While LH is generally regarded to be the ovulatory hormone, an examination of the literature indicates that the identity of the ovulatory hormone is not firmly established. In fact FSH, apparently free of LH activity, can induce ovulation in the rat (12-14). Moreover, both LH and FSH reach peak levels in the blood at the same time prior to ovulation (15) and granulosa cells in preovulatory follicles have receptors for both hormones (16). Thus, it would seem likely that both FSH and LH are involved in ovulation, but that they may be responsible for different biochemical events. Our results suggest that FSH controls the level of plasminogen activator and that this enzyme initiates the events that weaken the follicle wall. LH may be involved in other functions important at the time of ovulation, such as preparing the cells for luteinization if fertilization occurs or initiating the second meiotic division of the oocyte.

It is apparent that for an understanding of the endocrinology and biochemistry of ovulation at this level, analyses must be conducted under well-defined experimental conditions with highly purified hormone preparations. We feel that further examination of the production of plasminogen activator by granulosa cells should prove useful in this regard and point the way toward experimental systems suitable for studies of other aspects of follicular development, oocyte maturation, and ovulation.

ACKNOWLEDGMENTS

This work was supported by grants from the Rockefeller Foundation (RF 70095), the National Institutes of Health (CA 08290-11), the American Cancer Society (PDT-1G), and the Jane Coffin Childs Memorial Fund for Medical Research.

REFERENCES

1. de Graaf, R. (1672): "A New Treatise Concerning the Generative Organs of Women," De Mulierum Organis Generationi Inservientibus Tractatus Novus; English translation: (1974). *J. Reprod. Fert.*, Suppl. 17:77.
2. Espey, L. L. (1974): "Ovarian Proteolytic Enzymes and Ovulation." *Biol. Reprod.*, 10:216-235.
3. Schochet, S. S. (1916): "A Suggestion as to the Process of Ovulation and Ovarian Cyst Formation." *Anat. Rec.*, 10:447-457.
4. Espey, L. L. (1970): "Effect of Various Substances on Tensile Strength of Sow Ovarian Follicles." *Am. J. Physiol.*, 219:230-233.
5. Espey, L. L. (1967): "Tenacity of Porcine Graafian Follicle as it Approaches Ovulation. *Am. J. Physiol.*, 212:1397-1401.
6. Beers, W. H., Strickland, S., and Reich, E. (1975): "Ovarian Plasminogen Activator: Relationship to Ovulation and Hormonal Regulation." *Cell*, 6:387-394.
7. Beers, W. H. (1975): "Follicular Plasminogen and Plasminogen Activator and the Effect of Plasmin on Ovarian Follicle Wall." *Cell*, 6:379-386.
8. Strickland, S., and Beers, W. H. (1976): "Studies on the Role of Plasminogen Activator in Ovulation." *J. Biol. Chem.*, 251:5694-5702.

9. Bray, B. A., Hsu, K. C., Wigger, H. J., and LeRoy, E. C. (1975): "Association of Fibrinogen and Microfibrils with Trophoblast Basement Membrane." *Connect. Tissue Res.*, 3:55-71.
10. Parr, E. L. (1974): "Histological Examination of the Rat Ovarian Follicle Wall Prior to Ovulation." *Biol. Reprod.*, 11:483-503.
11. Brambell, F. W. R. (1956): "Ovarian Changes." In: *Marshall's Physiology of Reproduction*, 3rd edition, Vol. I, part 1 (Parkes, A. S., ed.), p. 445. Longman, Green and Co., London.
12. Goldman, B. D., and Mahesh, V. B. (1969): "A Possible Role of Acute FSH-Release in Ovulation in the Hamster, as Demonstrated by Utilization of Antibodies to LH and FSH." *Endocrinol.*, 84:236-243.
13. Harrington, F. E., Bex, F. J., Elton, R. L., and Roach, J. B. (1970): "The Ovulatory Effects of Follicle Stimulating Hormone Treated with Chymotrypsin in Chlorpromazine Blocked Rats." *Acta Endocrinol.*, 65:222-228.
14. Tsafriri, A., Lieberman, M. E., Koch, Y., Bauminger, S., Chobsieng, P., Zor, U., and Lindner, H. R. (1976): "Capacity of Immunologically Purified FSH to Stimulate Cyclic AMP Accumulation and Steroidogenesis in Graafian Follicles and to Induce Ovum Maturation and Ovulation in the Rat." *Endocrinology*, 98:655-661.
15. Gay, V. L., Midgley, A. R., Jr., and Niswender, G. D. (1970): "Patterns of Gonadotrophin Secretion Associated with Ovulation." *Fed. Proc.*, 29:1880-1887.
16. Midgley, A. R., Jr. (1973): "Autoradiographic Analysis of Gonadotropin Binding to Rat Ovarian Tissue Sections." In: *Receptors for Reproductive Hormones* (O'Malley, B., and Means, A., eds.), pp. 365-378. Plenum Press, New York.

Ovarian Follicular Development and Function,
edited by A. R. Midgley and W. A. Sadler.
Raven Press, New York © 1979.

Regarding the Role of Plasminogen Activator in Ovulation

Lawrence L. Espey and Jon M. R. Rawson

Trinity University and the University of Texas
Health Science Center, San Antonio, Texas 78284

Strickland and Beers (3,4) have presented an exciting hypothesis in their statement that gonadotropins induce ovulation by stimulating the synthesis of a plasminogen activator in mature ovarian follicles. If this idea is correct, it represents a valuable contribution to our present knowledge about the biochemical processes in mammalian ovulation.

This hypothesis describes the nature of the ovulatory enzyme and the manner by which proteolysis is confined to the follicular tissue without causing extensive damage to the rest of the ovary. Presumably, as plasminogen activator is secreted in the follicle it activates the conversion of local plasminogen into plasmin. Some of the newly formed plasmin is probably inactivated by alpha$_2$-macroglobulin and other serum antiproteases (10), but a sufficient amount apparently remains active long enough to degrade follicular connective tissue and weaken the follicle wall to the point of rupture.

It is possible that the follicular "collagenolytic" enzyme, which we have been studying during the past few years, is actually plasmin (7,8). This activity requires a period of time for activation, as does plasminogen, and it is inhibited also by alpha$_2$-macroglobulin. Also, it may be quite relevant that serum antiplasmin has been identified as the same as the anticollagenase, alpha$_2$-macroglobulin (11). Furthermore, it has been suggested "that, *in vivo*, humoral factors such as kallikrein and plasmin could play a role in the activation of procollagenase and that interconnections may exist between collagenolysis and both fibrinolysis and the various actions of kallikrein" (6).

A number of questions still need to be answered before the role of plasmin in ovulation is confirmed.

Some of these questions follow:

1. If plasmin is the ovulatory enzyme, then an ovulatory follicle contains a considerable amount of fibrinolytic activity; and, if so, then how does a blood clot form so readily at the point of rupture? Would not one expect the ovary to bleed profusely under such conditions?

2. What is the chemical nature of the plasminogen activator in the follicle? Is it urokinase? Could it possibly be the strong reducing agent, ascorbic acid?

3. Is the granulosa the principal source of plasminogen activator? It should not be ignored that there are at least four other distinct types of cells in the wall of a mature ovarian follicle. These include epithelial cells on the surface of the follicle, fibroblasts in the theca externa and tunica albuginea, secretory cells in the theca interna, and endothelial cells in the massive capillary network in the follicle. Plasminogen activator is not only common in epithelial cells, but it is also found in fibroblasts and endothelial cells (2). Therefore, it would seem premature to conclude that the granulosa is the only source of activator, especially in view of the fact that the present information does not clearly establish the granulosa as the origin of the activator. This preliminary deduction by Strickland and Beers is based on their electron micrographic analysis of cell pellets in which 88% of the viable intact cells in the preparations had features characteristic of granulosa cells. However, the activator could be produced, at least in part, by the various cell types that compose the remaining 12% of the pellets.

4. It is presumptuous to conclude that follicular fluid is the principal source of plasminogen. As Strickland and Beers have pointed out, serum has the same concentration of plasminogen as does follicular fluid, and, therefore, the extensive capillary network in the follicle wall also represents a potential source of plasminogen.

5. Is FSH indeed the gonadotropin responsible for the production of plasminogen activator? If so, then what function does LH have in the ovulatory process? Is it limited to the regulation of steroidogenesis?

In the present report, the tropic hormones are compared on a crude weight basis, rather than by specific activity of the various preparations. Unless more information is available about the purity of the hormones, it is premature to conclude that FSH is the hormone that stimulates the production of plasminogen activator.

Schwartz et al. (15) have demonstrated that anti-ovine LH will effectively block ovulation in the rat when administered at the appropriate time during the estrous cycle, while antiovine FSH will not. Other studies have shown that FSH in the presence of anti-LH serum will not induce ovulation in the rat (*personal communication with Dr. Neena Schwartz*). These studies indicate that LH is essential to the ovulatory process.

Why is TSH (i.e., a thyrotropin) as effective as LH in producing plasminogen activator in gonadal tissue?

6. Could the primary function of plasminogen activator be to reconstruct tissue, rather than cause ovulatory degradation? Indirect support for this possibility comes from studies that have shown that (a) fibrinolytic activity usually results from an increase in plasminogen activator that is released by cells responding to tissue injury (1,9), and ovulation involves tissue damage; (b) fibrinolytic activity is associated with polymorphonuclear leukocytes (13), and these cells accumulate in ovulatory follicles (*unpublished observations*); (c) the stimulation of fibrinolytic activity may be due to the products of an acute inflammatory response (13), and ovulatory follicles undergo inflammatory changes (14).

7. Could the exposure of the experimental samples of Strickland and Beers to 1 N HCl for 2 hr (3) remove the tissue too far from physiological conditions? It may be that this acid treatment has other effects in addition to the removal of inhibitors from the samples. For example, it has been demonstrated that acid alters the plasminogen molecule, resulting in changes in both the molecular weight and the substrate specificity (12,16).

8. What is the interrelationship among gonadotropins, steroids, cAMP, prostaglandins, and plasminogen activator? It would be interesting to know the mechanism by which prostaglandins E_1 and E_2 stimulate the production of plasminogen activator. Just as important is the question of why indomethacin, which inhibits prostaglandin synthesis and ovulation, does not inhibit the production of plasminogen activator. [Also, there is the question of why cyanoketone, which blocks ovulation in the PMSG-treated rat (5), has no effect on the action of plasminogen activator.]

In concluding, it will be interesting to examine the hypothesis that gonadotropic stimulation of plasminogen activator might influence collagenolysis in tissues, such as the ovarian follicle. It may be that the plasminogen activator-plasmin model can also account for other physiological and pathological changes in connective tissues. For example, during aging the increased incidence of connective tissue disease might be related to a sequence of changes that would begin at the time of senesence of the gonads, followed by a compensatory increase in gonadotropin production, and, ultimately an increase in plasmin levels sufficient to cause inflammation of sensitive connective tissues. Thus, the approach of Strickland and Beers may lead to a number of rewarding investigations in the decade ahead.

REFERENCES

1. Astrup, T. (1966): *Fed. Proc.*, 25:42-51.
2. Astrup, T. (1975): In: *Proteases and Biological Control*, edited by E. Reich, D. B. Rifkin, and E. Shaw, pp. 343-355. Cold Spring Harbor Laboratory, Cold Spring Harbor, N. Y.
3. Beers, W. H. (1975): *Cell*, 6:379-386
4. Beers, W. H., Strickland, S., and Reich, E. (1975): *Cell*, 6:387-394.
5. Bullock, D. W., and Dappauf, B. H. (1973): *Endocrinology*, 92:1625-1628.
6. Eeckhout, Y., and Vaes, G. (1974): *Arch. Int. Physiol. Biochim.*, 82:786.
7. Espey, L. L. (1974): *Biol. Reprod.*, 10:216-235.
8. Espey, L. L., and Coons, P. J. (1976): *Biol. Reprod.*, 14:233-245.
9. Guest, M. M. (1966): *Fed. Proc.*, 25:73-76.
10. Heimburger, N. (1975): In: *Proteases and Biological Control*, edited by E. Reich, D. B. Rifkin, and E. Shaw, pp. 367-386. Cold Spring Harbor Laboratory, Cold Spring Harbor, N. Y.
11. Iwamoto, M., and Abiko, Y. (1970): *Biochem. Biophys. Acta*, 214:402-410.
12. Kline, D. L. (1966): *Fed. Proc.*, 25:31-33.
13. Kwaan (1966): *Fed. Proc.*, 25:52-56.
14. Parr, E. L. (1974): *Biol. Reprod.*, 11:483-503.
15. Schwartz, N. B., Krone, K., Talley, W. L., and Ely, C. A. (1973): *Endocrinology*, 92:1165-1174.
16. Slotta, K. H., Michl, H., and Santos, B. G. (1962): *Biochem. Biophys. Acta*, 58:459-466.

Ovarian Follicular Development and Function,
edited by A. R. Midgley and W. A. Sadler.
Raven Press, New York © 1979.

Prevention of Ovulation in Rabbits by Antihistamine

*Eric Knox, **Suzanne Lowry, and *Lee Beck

*Department of Pediatrics and ** Obsterics and Gynecology, University of Alabama, University Station, Birmingham, Alabama

It has been suggested that a number of different biochemical mediators--including histamine--are necessary for ovulation to occur (1). Direct evidence supporting an obligatory association between histamine and ovulation is not available. Previous studies have failed to establish histamine as a necessary component in the ovulatory mechanism (2). The experiment reported here demonstrates that antihistamine can block ovulation in the rabbit and suggests for the first time that histamine is necessary for normal ovulation to occur.

MATERIALS AND METHODS

New Zealand white rabbits weighing 2 to 6 kg were used in these experiments. Rabbits were maintained on a 14/10-hr light/dark regimen at a controlled temperature (20°C).

The antihistamine, chlorprophenpyridamine maleate (chlor-trimeton) was dissolved in sterile saline to a final concentration of 10 mg/ml. In the rabbits receiving chlor-trimeton, a dosage of 10 mg/ml of the stock preparation was injected i.m. at 4-hr intervals. This dosage schedule has been previously shown to suppress skin reactions to intradermal injections of histamine (3). Human chorionic gonadotropin (HCG) was used to induce ovulation (100 IU i.v. via the lateral ear vein).

Estrous female rabbits (N = 13) were injected with chlor-trimeton (10 mg/kg i.m.) every 4 hr beginning at time zero and extending to 96 hr. Ovulation was induced in the experimental group of rabbits (N = 10) at 48 hr by injection of HCG. An equal number of control rabbits were pretreated with saline without antihistamine and had HCG administered at the same time as the experimental group.

Blood samples were obtained from all rabbits prior to and at 0, 24, 48, 72, 96, and 144 hr. All serum samples were stored frozen at -70° until assayed. Total immune reactive serum estrogens and progestins were measured in the serum samples by specific radioimmunoassay by methods previously described (4). Serum FSH and LH were kindly measured by Dr. Claude Desjardins using methods previously described (5-7).

All rabbits were sacrificed on the 10th day of the experiment and both ovaries were removed for morphological and histological examination. Standard histological methods were used in the preparation and examination of the tissues.

RESULTS

The mean number of ovarian corpus lutea observed in rabbits treated with chlor-trimeton (0.9 ± 0.72) is significantly less ($p < 0.001$) than the mean number of corpus lutea observed in the control rabbits (10.12 ± 2.27). Histological examination of the ovaries of the rabbits treated with chlor-trimeton revealed multiple large hemorrhagic follicles containing retained ova. No evidence of luteinized granulosa cells could be found in the ovarian follicles of treated rabbits. Histological examination of the ovaries of the control rabbits revealed multiple normal-appearing luteinized follicles. Similar results were obtained when mating was used as the ovulatory stimulus.

The serum FSH and LH values before and after treatment with chlor-trimeton but before ovulation induction are shown in Table 1. It is apparent that treatment

TABLE 1. *Serum LH and FSH of rabbits following 48 hr of treatment with antihistamine and before ovulation induction with HCG*[a]

	Animals before antihistamine or saline treatment (N = 6)	Animals after histamine or saline treatment and before HCG administration	
		Histamine treatment 10 mg/kg every 4 hr × 48 hr (N = 3)	Saline control (N = 3)
LH (ng/ml)	1.273 ± 0.276	1.514 ± 0.509	1.496 ± 0.315
FSH (ng/ml)	1.327 ± 0.790	1.526 ± 0.840	2.346 ± 1.527

[a]Values given are mean ± standard deviation in nanograms per milliliter.

with chlor-trimeton has no effect on peripheral gonadotropin levels. The slight increase in gonadotropins observed in both the antihistamine-treated and saline control groups is consistent with handling and bleeding of the animals.

Figure 1 shows the results of total estrogens and immunoreactive progestin determinations throughout the course of the experiment. The serum levels of estrogens in rabbits treated with chlor-trimeton are similar to those in the control rabbits at all time intervals tested. No significant amount of serum progestin could be detected in rabbits treated with chlor-trimeton. However, in the rabbits that did not receive antihistamine, significant levels of immunoreactive progestins appeared in the blood following the induction at 96 (3.8 ± 1.9 ng/ml) and 144 hr (7.2 ± 2.4 ng/ml). The difference in the mean serum progestin levels at 96 and 144 hr between the control and antihistamine-treated rabbits is statistically significant ($p < 0.01$).

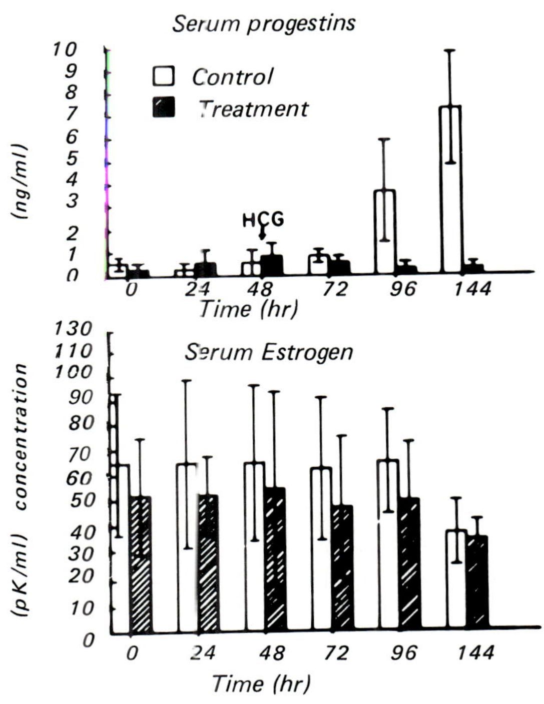

FIG. 1. Serum total estrogens and immunoreactive progestins in rabbits treated with HCG following pretreatment with either antihistamine or saline.

DISCUSSION

This study clearly suggests that treatment of estrous rabbits with antihistamine for 48 hr effectively blocks ovulation induced by either HCG or natural mating. No histological evidence of luteinization in the antihistamine-treated rabbits could be demonstrated.

The data presented suggest that antihistamine acts to block ovulation by an action at the ovarian level. Treatment with chlor-trimeton does not affect secretion of either FSH or LH (Table 1). In addition, the estrogen data (Fig. 1) demonstrate that the steroidogenic capacity of the preovulatory ovarian follicles is not affected by antihistamine. Serum estrogen levels between the control and treated animals were similar at all times during the experiment and are not statistically distinguished from each other. These results plus a previous study (8) in which exogenously administered gonadotropins were unable to stimulate ovarian hyperemia all suggest that the action of antihistamine is not centrally mediated. In addition, no luteolytic effect of antihistamine can be demonstrated. There is no evidence that significant progesterone production is ever initiated in the rabbits treated with antihistamine. Thus, it is suggested that chlor-trimeton is acting to block ovulation through a classical H_1 competitive receptor blockade at the ovarian level.

Previous studies concerned with the role histamine plays in ovulation are contradictory. On the one hand, histamine has been suggested to be a mediator of the ovarian hyperemia induced by administration of gonadotropins (8,9). There is a depletion of ovarian histamine following the administration of LH to female rats (10), and antihistamine effectively blocks the ovarian hyperemia produced in rats by treatment with HCG (2). On the other hand, HCG still causes hyperemia in ovaries chemically depleted of histamine and histamine blockade in the rat does not prevent ovulation (2). These contrasting findings neither confirm nor disprove the suggestion that histamine has a vital function in the process of ovulation. Findings of the present studies suggest a direct association between histamine and ovulation in the rabbit, although pharmacological effects of chlor-trimeton other than antihistamine blockade cannot be excluded. The failure of previous investigators to block ovulation in rats by antihistamine blockade may be due to species differences or a dose dependency of the response.

The hemorrhagic follicles of the rabbits treated with chlor-trimeton with their retained ova appear similar to those described in rabbits in which ovulation has been inhibited by administration of indomethacin (11,12). In these and in other studies (13), the effect of indomethacin seems limited to the events leading to follicular rupture and does not include the mediation of other LH effects (such as luteinization) on the ovary. The endocrine and histological findings of our study reveal no luteinization, and thus suggest that the mechanism of action of antihistamine in blocking ovulation is different from that of indomethacin. It can therefore be inferred that, although both histamine and prostaglandin play a role in ovulation, each mediator acts at a different step of the overall process. The interaction between histamine, prostaglandins, and other mediators, as well as the temporal sequence of that interaction, are necessary components to an overall understanding of normal ovulation and should provide fascinating questions for future research.

REFERENCES

1. Schwartz, N. B., and McCormack, C. E. (1972): In: *Annual Review of Physiology, Vol. 34,* edited by J. Comroe, Jr., A. Griese, and R. Sonnenschein, p. 436. Annual Reviews, Inc., Palo Alto.
2. Lipner, H. (1971): *Proc. Soc. Exp. Biol.,* 136:111.
3. Knicker, W. T., and Cochrane, C. G. (1968): *J. Exp. Med.,* 127:119.
4. Boots, L. R., Younger, J. B., Beck, L. R., and Buchina, E. S. (1973): *Ala. J. Med. Soc.,* 10:445.
5. Niswender, G. D., Midgley, A. R., Jr., Monroe, S. E., and Reichert, L. E., Jr. (1968): *Proc. Soc. Exp. Biol. Med.,* 128:807.
6. Ewing, L. L., Stratton, L. G., and Desjardins, J. (1973): *Reprod. Fertil.,* 35:245.
7. Goodman, A. L. (1975): Ph.D. dissertation, Emory University.
8. Knox, G. E. (1974): *Am. J. Obstet. Gynecol.,* 118:992.
9. Wurtman, J. (1964): *Endocrinology,* 75:927.
10. Szego, C. M., and Gitin, E. S. (1964): *Nature,* 201:682.
11. Grinwich, D. L., Kennedy, T. G., and Armstrong, D. J. (1972): *Prostaglandins,* 1:89.
12. O'Grady, J. P., Caldwell, B. V., Auletta, F. T., and Speroff, L. (1972): *Prostaglandins,* 1:97.
13. Linder, H. R., Zor, U., Bauminger, S., Tsafuri, A., Lamprecht, S. A., Koch, Y., Antibe, S., Schwartz, A. (1965): In: *Prostaglandin Synthetase Inhibitors,* Vol. 1, edited by H. J. Robinson and J. R. Vane, p. 271. Raven Press, New York.

Ovarian Follicular Development and Function,
edited by A. R. Midgley and W. A. Sadler.
Raven Press, New York © 1979.

Discussion Summary: Session on Ovulation

S. Strickland

The first part of the discussion concerned questions raised in the paper by Espey and Rawson regarding the role of plasminogen activator in ovulation:

1. If an ovulatory follicle contains fibrinolytic activity at the time of ovulation, how is a blood clot able to form? There are two factors that make this possible. First, the production of plasminogen activator by granulosa cells reaches a maximum at the time of ovulation, but decreases rapidly thereafter. For example, approximately 2 hr after ovulation, enzyme production is only 33% of the maximum level. Second, the follicle contains serum protease inhibitors that will quickly neutralize proteolytic activity that is not being continually produced. The combined effect of these two factors rapidly reduces the fibrinolytic activity in the follicle after ovulation, and allows the formation of a stable blood clot.

2. What is the chemical nature of the ovarian plasminogen activator and could it be ascorbic acid? The molecular weight of the granulosa cell enzyme has been investigated by sodium dodecyl sulfate gel electrophoresis, and the predominant form is a protein of molecular weight 75,000 (1). The catalytic specificity of this enzyme is similar to other mammalian plasminogen activators, including urokinase (1). Ascorbic acid does not cleave peptide bonds, and hence cannot activate plasminogen.

3. What is the principal source of follicular plasminogen activator? Granulosa cells have been reported to be a major source of follicular plasminogen activator, but not the *only* source. Published reports have acknowledged that other ovarian cell types might also produce this enzyme (2). However, the experimental basis for believing the enzyme is primarily produced by granulosa cells should be reviewed: the fibrin-agar overlay allows the positive identification of single, fibrinolytically active cells. The position of these cells in the petri dish can be determined so that after fixing and processing, the *active* cells can be

analyzed by electron microscopy. In every case to date, the active cells have had the morphological characteristics of granulosa cells. This demonstrates that the granulosa cells are at least one source of plasminogen activator. This observation, coupled with the fact that the prepared cell cultures are approximately 90% granulosa cells, strongly suggests that these cells are the principal source.

4. What is the plasminogen source for the granulosa cell plasminogen activator? There are no capillaries in the granulosa cell layer, since these cells exist in the interior of the follicle that is avascular. The interior of the follicle is bathed in follicular fluid, a filtrate of blood. The blood-follicular barrier is permeable to many serum proteins, including plasminogen. Thus, the only local source of plasminogen is the follicular fluid.

5. What are the respective roles of FSH, LH, and TSH in ovulation? Both LH and FSH probably have a role in ovulation, since there is a concomitant surge of both gonadotropins before follicle rupture. If FSH is indeed the hormone regulating plasminogen activator synthesis, then LH possibly functions in controlling steroidogenesis, meiosis, etc. Many different preparations of LH and FSH have now been tested on granulosa cells. These include ovine, bovine, rat, porcine, and equine samples as well as highly purified human samples. In every case, FSH has been much more effective than LH in stimulating the cells *in vitro*. The TSH preparation used was active probably because of contamination with FSH. It should be noted that FSH, purified by treatment with antibodies to the β-subunit of LH, will induce ovulation in the rat (3).

6. Is the proteolytic activity observed a consequence of tissue reconstruction or inflammation? The activity is not associated with cells normally involved in these processes, e.g., polymorphonuclear leukocytes or macrophages (2).

7. Does acid treatment of the samples affect the results? The same results are obtained if the samples are not acid treated. This treatment simply makes the assay more sensitive.

8. What is the interrelationship between the compounds effective *in vitro* and why don't indomethacin and cyanoketone inhibit the response? The available evidence suggests that prostaglandins and gonadotropins act independently at the cell surface to stimulate the cells. However, both classes of compounds appear to work via cyclic AMP. All steroids tested have been inactive.

There is no reason to expect indomethacin to inhibit

plasminogen activator production by granulosa cells. It has been demonstrated that prostaglandins will stimulate the cells *in vitro* to produce the enzyme, but the source of prostaglandins in the intact animal is not known. The sequence of events presumably would be distant synthesis of prostaglandins, diffusion to the granulosa cell layer, and then stimulation of the cells. It is entirely consistent that indomethacin can block ovulation but not inhibit the response of granulosa cells *in vitro*. The same reasoning applies to cyanoketone.

Other areas discussed are summarized below:

1. The question of the identity of the follicular substrate for plasmin was raised. A likely substrate for this enzyme is basement membrane since (a) plasmin will digest basement membrane and (b) the granulosa cells are surrounded by a layer of basement membrane that must be degraded at the time of ovulation.

2. It was noted that the fact that the granulosa cells respond better to FSH than to LH could reflect more lability in the LH response system.

3. Cathepsin(s) were proposed as the source of the follicular protease. This enzyme can be ruled out for the following reasons: (a) Cathepsin does not activate plasminogen, and all of the follicular activity is plasminogen dependent, (b) the granulosa cell enzyme has maximum activity at pH 7 to 8, and (c) the granulosa cell enzyme is inhibited by DFP whereas cathepsin is not.

4. Attempts are being made to determine if only the granulosa cells in the region of the follicle dome produce the enzyme. This would explain why the follicle usually ruptures in such a way that the ovum is released near the oviductal ampulla.

5. It was noted that the production of plasminogen activator by granulosa cells in response to gonadotropins (a) occurs at very low gonadotropin concentrations, (b) can be assayed at the single cell level, and (c) is relatively rapid. These characteristics make this system attractive as a potential bioassay for gonadotropin.

6. The role of histamine in the granulosa cell response was discussed. Histamine has been examined as a possible regulatory substance for plasminogen activator production in the ovary. This molecule has no effect. The effect of antihistamines has not been examined.

7. Finally, it was emphasized that the production of plasminogen activator by granulosa cells is one well-defined biochemical event among undoubtedly many reactions that collectively constitute the process of ovulation.

1. Strickland, S., and Beers, W. H. (1976): "Studies on the Role of Plasminogen Activator in Ovulation." *J. Biol. Chem.*, 251:5694-5702.
2. Beers, W. H., Strickland, S., and Reich, E. (1975): "Ovarian Plasminogen Activator: Relationship to Ovulation and Hormonal Regulation," *Cell*, 6:387-394.
3. Tsafriri, A., Lieberman, M. E., Koch, Y., Bauminger, S., Chobsieng, P., Zor, U., and Lindner, H. R. (1976): "Capacity of Immunologically Purified FSH to Stimulate Cyclic AMP Accumulation and Steroidogenesis in Graafian Follicles and to Induce Ovum Maturation and Ovulation in the Rat." *Endocrinol.*, 98:655-661.
4. Beers, W.H., and Strickland, S. (1978): A cell culture assay for follicle stimulating hormone. J. Biol. Chem., 253:3877-3881.

FOLLICULAR STEROIDOGENESIS

Ovarian Follicular Development and Function,
edited by A. R. Midgley and W. A. Sadler.
Raven Press, New York © 1979.

REGULATION OF FOLLICULAR ESTROGEN BIOSYNTHESIS

David T. Armstrong, Alan K. Goff, and
*Jennifer H. Dorrington

Departments of Physiology and Obstetrics and
Gynaecology, University of Western Ontario,
London, Canada, and *Banting and Best Department
of Medical Research, University of Toronto,
Toronto, Canada

The follicle is the principal source of estrogen secretion in the follicular phase of the ovarian cycle in mammals. The physiological factors involved in the regulation of estrogen secretion have been extensively investigated, and the pituitary gonadotropic hormones, luteinizing hormone (LH) and follicle-stimulating hormone (FSH), are of paramount importance. The early investigations of Greep et al. (1) and Lostroh and Johnson (2) demonstrated that these two gonadotropins acted synergistically to stimulate estrogen secretion as judged by increased uterine weights. LH elicited histological signs of stimulation of interstitial and thecal cells in hypophysectomized immature rats, whereas FSH stimulated follicle growth without demonstrable effects on thecal and interstitial cells.

The biochemical sites at which the pituitary gonadotropins stimulate steroidogenesis have been studied more recently. In testes and ovaries of numerous species, there is compelling evidence that LH acts at the level of conversion of cholesterol to pregnenolone--a site common to the biosynthesis of all gonadal steroids (3-5). Early reports (6,7) that FSH acted at the level of aromatase, the enzyme that catalyzes the conversion of androgens to estrogens, were not widely accepted because of the high dosages of impure FSH preparations used, and numerous subsequent reports that the ability of FSH preparations to stimulate steroidogenesis could be attributed to LH contamination (8-10).

Our recent finding that highly purified preparations of FSH, but not of LH, stimulated aromatization of androgens by testicular Sertoli cells (11), led us to reexamine the hormonal regulation of follicular estrogen secretion by granulosa cells.

MATERIALS AND METHODS

In Vitro Studies with Isolated Granulosa Cells

Granulosa cells were isolated from follicles of hypophysectomized or intact immature rats, as described previously (12). In most experiments, rats were injected with 1 mg estradiol-17β in 0.1 ml sesame oil daily, for 3 to 5 days before isolating the cells. This treatment has been found to stimulate granulosa cell proliferation markedly in both hypophysectomized and intact rats, increasing both the yield and uniformity of the cells. In one experiment, rats were injected subcutaneously with 12 IU pregnant mare's serum gonadotropin (PMSG) to induce follicular growth prior to isolation of granulosa cells 2 days later.

Granulosa cells were either plated in Falcon plastic culture dishes, and grown for 3 days in modified Eagle's minimum essential medium, as described previously (12), for measurement of estrogen and androgen biosynthesis, or incubated for 10 min in Krebs-Ringer bicarbonate buffer containing glucose (1 mg/ml) and methylisobutyl xanthine (MIX, 0.5 mM), for measurement of cyclic adenosine-3',5'-monophosphate (cAMP) production. Methods used for assay of steroids and cAMP were essentially as described previously (11,13).

In Vivo Studies with Hypophysectomized Rats

Sprague-Dawley rats were hypophysectomized at 21 to 22 days of age. Beginning 2 to 3 days later, they were administered daily subcutaneous injections of LH, FSH, or FSH in conjunction with LH or with various androgens, as described under *Results and Discussion*. At autopsy, after 3 days of treatment, uteri and ovaries were weighed and fixed for histological examination, and ovaries were assayed for their immunoreactive estradiol-17β or androgen contents. Experimental details and analytical procedures have been reported previously (14).

RESULTS AND DISCUSSION

In Vitro Studies with Isolated Granulosa Cells

In preliminary experiments with granulosa cells isolated from ovaries of immature hypophysectomized rats measurable amounts of immunoreactive estradiol 17β were produced when cells were cultured in medium containing testosterone and FSH. It was subsequently found that the yield and viability of granulosa cells could be increased by pretreatment of hypophysectomized rats with estradiol-17β. Figure 1 summarizes the effects of purified gonadotropins on estrogen production by granulosa cells isolated from ovaries of hypophysectomized, estrogen-treated rats. Cells

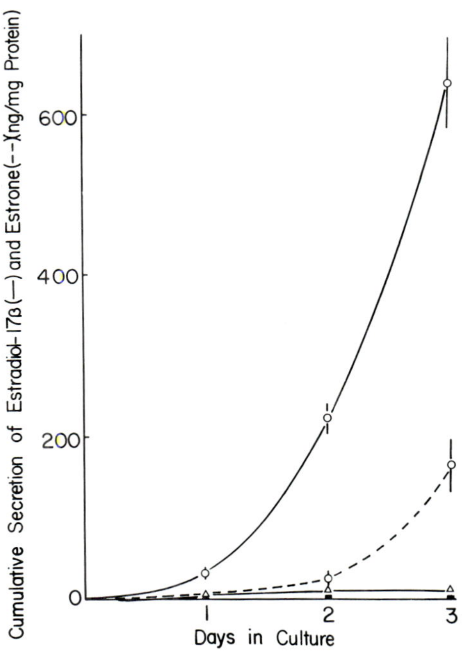

FIG. 1. Effects of purified preparations of FSH (FSH-HP, 0.25 μg/ml) (*circles*) and LH (LH-HP, 0.25 μg/ml) (*triangles*) on estrogen secretion by granulosa cells isolated from ovaries of 30-day-old rats hypophysectomized at 21 days of age and pretreated for 5 days with 1 mg estradiol-17β/day, when cultured in medium containing 0.5 μM testosterone. Controls (*squares*), cultured without testosterone, or with testosterone but without gonadotropins produced negligible amounts of estrogens.

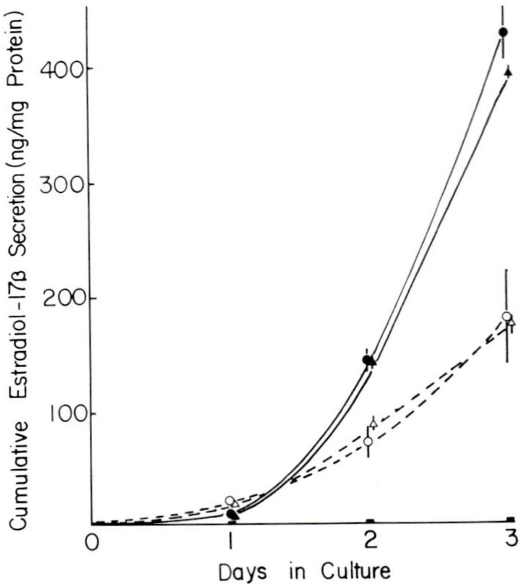

FIG. 2. Effects of FSH-HP (0.25 μg/ml) alone (*circles*), or together with LH-HP (0.25 μg/ml) (*triangles*) on estradiol-17β secretion by granulosa cells from 28-day-old untreated (*broken line*) or estradiol-17β pretreated (1 mg/day for 5 days) (*solid line*) rats with intact pituitary glands, when cultured in medium containing 0.5 μM testosterone. Controls (*squares*) cultured without gonadotropins, or with LH alone, produced negligible amounts of estradiol-17β.

cultured in medium containing testosterone (0.5 μM), FSH, or LH alone produced negligible amounts of immunoreactive estradiol-17β and estrone. FSH was highly effective in stimulating synthesis of both estradiol-17β and estrone in cells grown in the presence of testosterone (0.5 μM), production of estradiol-17β being four- to ninefold greater than that of estrone. LH produced only a borderline stimulation of estradiol-17β synthesis in this type of granulosa cell cultured in the presence of testosterone.

Similar results have been obtained with granulosa cells from immature rats with pituitary glands intact (Fig. 2). As observed with cells from hypophysectomized rats, pretreatment of intact rats with estradiol-17β increased the yield of granulosa cells, and the responsiveness of these cells to FSH, when cultured with testosterone was greater and more consistent than that of cells from intact, untreated rats. LH was essentially without effect on estrogen production by these granulosa cells cultured with testosterone, whether added by itself or with FSH.

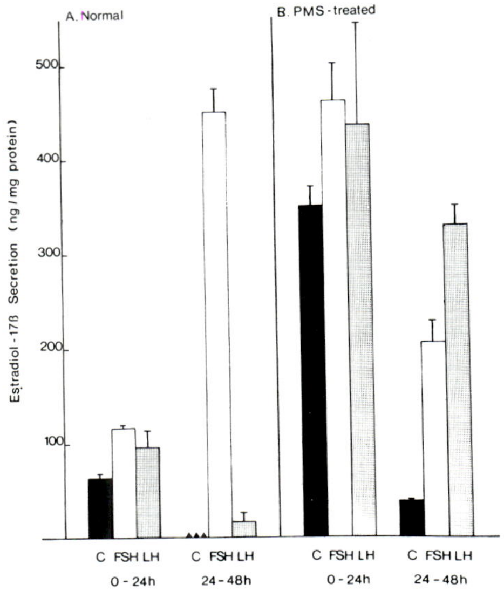

FIG. 3. Effects of FSH-LER 1932 (0.05 μg/ml) and LH-HP (0.05 μg/ml) on estradiol-17β production during first and second day in culture, of granulosa cells obtained from 26-day-old untreated rats (A) or from 26-day-old rats which had received 12 IU PMSG at 24 days of age (B). All media contained 0.5 μM testosterone.

Granulosa cells from rats pretreated with PMSG, when cultured with testosterone alone, produced much larger amounts of estradiol-17β during the first day in culture than those from hypophysectomized or intact untreated or estrogen-treated rats (Fig. 3). During the second day in culture, estradiol-17β production declined markedly in the absence of gonadotropins. Both FSH and LH were effective in preventing this decline in estradiol-17β production by cells from PMSG-treated rats, whereas only FSH prevented the decline (and markedly stimulated) estradiol-17β production by cells from intact, untreated rats (Fig. 3). This stimulatory action of LH only on cells from PMSG-treated rats is consistent with previous reports of the ability of FSH preparations to increase LH binding by granulosa cells (15,16). It cannot be concluded from the present studies whether FSH and LH responsiveness are functions of the same cells, or whether they are functions of different subpopulations of granulosa cells from the PMSG-treated rats.

Granulosa cells cultured without testosterone failed to produce detectable amounts of immunoreactive androgens (testosterone and 5α-dihydrotestosterone, DHT)

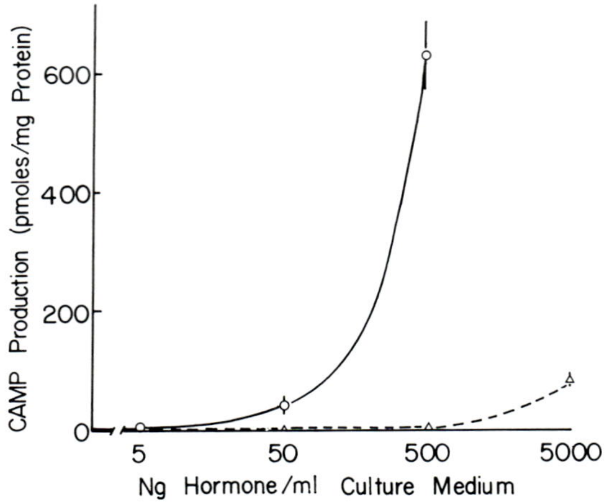

FIG. 4. Effects of varying concentrations of FSH-HP (*circles*) and NIH-LH-B8 (*triangles*) on cAMP production during 10-min incubation of granulosa cells isolated from 28-day-old intact rats pretreated for 3 days with 1 mg estradiol-17β/day.

whether cultured in the absence or presence of FSH or LH, supporting the conclusion of previous investigators that granulosa cells have limited, if any, ability to produce C_{19} steroids (17,18). Thus, it seems likely that the failure of estrogen synthesis by FSH-stimulated granulosa cells cultured without testosterone is due to lack of substrate for aromatization.

We are exploring the mechanism of action of FSH in stimulating aromatization. The observations of Erickson and Ryan (19) that dibutyryl cAMP stimulated estradiol-17β production by cultured rabbit granulosa cells raised the possibility that cAMP may be an intracellular mediator of the action of FSH. Accordingly, we investigated the ability of FSH to stimulate cAMP production by isolated granulosa cells. The results, summarized in Fig. 4, demonstrate the effectiveness of purified FSH in stimulating cAMP production by granulosa cells from intact, estrogen-treated immature rats. Although stimulation by LH also was observed, this occurred only at high concentrations of the hormone, at which contamination with FSH could not be excluded.

The results of these studies are consistent with the hypothesis that FSH stimulates granulosa cell aromatase activity via activation of a hormone-sensitive adenylate cyclase enzyme. Evidence obtained with isolated testicular Sertoli cells has led to similar conclusions concerning the mechanism of action of FSH (20).

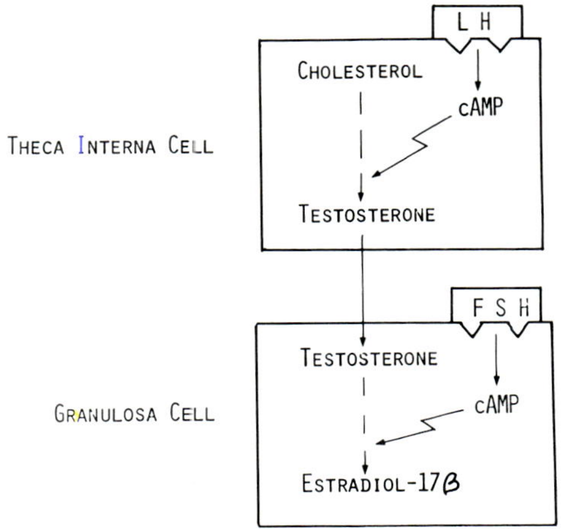

FIG. 5. Cellular and biochemical sites of action of LH and FSH in regulation of estrogen biosynthesis in follicles: a working hypothesis.

In attempting to place our findings in physiological perspective, it seems clear that this one ovarian cell type (the granulosa cell) can carry out the final enzymatic reaction in estrogen biosynthesis (aromatization) when stimulated by FSH, but that it cannot synthesize the substrate(s) necessary for this reaction. Nothing can be concluded concerning the role of other ovarian cells from these studies. However, because of previous evidence that thecal cells of the follicle contain LH receptors (16), and that isolated thecal tissue can produce androgens after stimulation by endogenous (21) or exogenous (22) LH, it is tempting to conclude that follicular estrogen biosynthesis occurs through the co-operation of two separate cell types, with the theca interna cells (under LH stimulation) producing the androgens, which are then transported to the granulosa cells, where they are converted (under FSH stimulation) to estrogens. This hypothesis, illustrated diagrammatically in Fig. 5, thus may provide an explanation, in biochemical terms, for the involvement of two ovarian cell types previously shown by Falck (23) to be required for estrogen secretion *in vivo*.

In Vivo Studies with Hypophysectomized Rats

Having formulated the hypothesis proposed in Fig. 5 on the basis of *in vitro* studies, its applicability *in vivo* was assessed using hypophysectomized immature

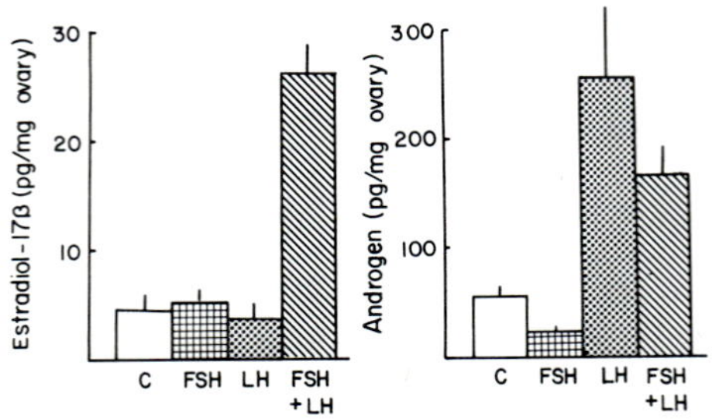

FIG. 6. Effects of FSH-HP (2.5 μg/day) and LH-HP (1 μg/day) administered separately and together for 3 days to hypophysectomized immature rats, on ovarian contents of immunoreactive estradiol-17β and androgens (testosterone + DHT).

rats, as previously employed by Greep et al. (1) and by Lostroh and Johnson (2). The effects of purified LH and FSH preparations administered alone and in combination on the uterine and ovarian weights of hypophysectomized rats are summarized in Table 1. These findings confirm the reports of previous investigators (1,2) that low doses of either of these gonadotropins

TABLE 1. *Effect of FSH and LH alone and in combination on uterine and ovarian weights of hypophysectomized immature rats*

Treatment	Ovarian weight (mg ± SE)	Uterine weight (mg ± SE)
None	6.8 ± 0.4	25.7 ± 1.2
FSH-HP (2.5 μg/day, 3 days)	10.6 ± 1.2	25.8 ± 1.2
LH-HP (1.0 μg/day, 3 days)	8.6 ± 1.0	32.8 ± 2.6
FSH-HP + LH-HP	14.2 ± 0.8	48.4 ± 1.2

alone are ineffective in stimulating uterine growth, indicative of estrogen secretion, in contrast to their effectiveness when administered concomitantly. The amounts of immunoreactive androgens and estradiol-17β present in the ovaries of the same rats are summarized in Fig. 6. It is evident that LH increased ovarian androgen contents when administered either by itself or together with FSH. On the other hand, neither gonadotropin, administered alone, influenced ovarian estradiol-17β contents, but levels of this steroid were markedly increased when FSH and LH were administered together.

The ability of FSH to increase ovarian estrogen content and uterine weight only under conditions in which

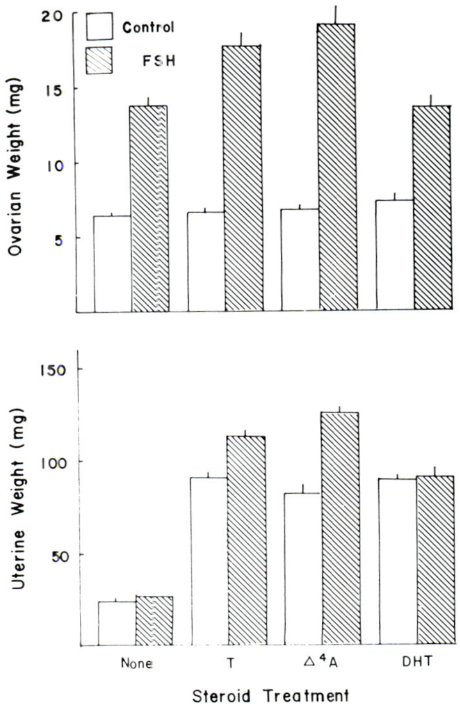

FIG. 7. Interactions between FSH-HP (2.5 μg/day for 3 days) and androgens (T, testosterone; Δ^4A, androstenedione; DHT, 5α-dihydrotestosterone; each 5 mg/day for 3 days) in stimulating uterine and ovarian weight increase in hypophysectomized immature rats.

ovarian androgen content was elevated (i.e., after LH treatment) prompted us to attempt to mimic the effects of LH by subcutaneous administration of high doses of exogenous androgens. In preliminary studies with rats with intact pituitaries, daily doses of 5 mg testosterone (or other androgens) were found to elevate ovarian androgen contents severalfold (24). Therefore, 5 mg/day testosterone, androstenedione, or DHT were administered alone, or together with purified FSH, and the same uterine and ovarian indices of increased estrogen secretion were examined. All three androgens stimulated uterine growth, including myometrial and endometrial stromal hypertrophy in the absence of FSH (Fig. 7). These are apparently true androgen effects, rather than the result of conversion to estrogens, since in the absence of FSH, the response to the nonaromatizable androgen, DHT, appeared neither qualitatively nor quantitatively different from responses to equivalent doses of testosterone or androstenedione.

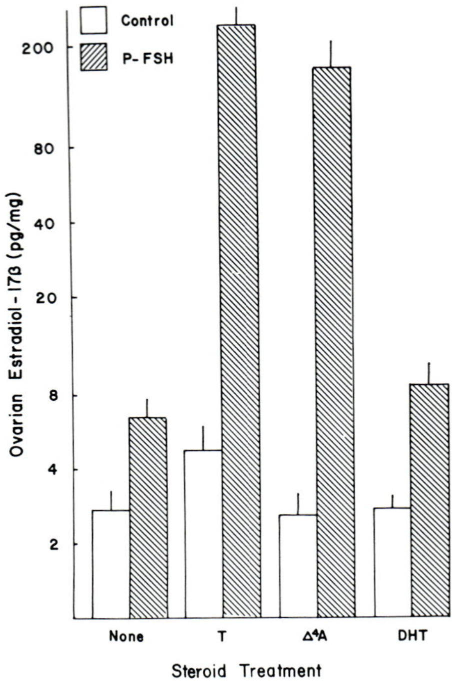

FIG. 8. Interactions between FSH-HP and various androgens in increasing ovarian contents of immunoreactive estradiol-17β in hypophysectomized rats. Note logarithmic scale on ordinate (see legend to Fig. 7).

A very different picture emerged when the three androgens were administered together with FSH; testosterone and androstenedione were much more effective in stimulating uterine weight, and in particular in stimulating hypertrophy of uterine mucosal epithelial cells, when administered together with FSH. In contrast, FSH did not enhance the ability of DHT to alter uterine weight or histology.

The ability of FSH to increase ovarian weight was also enhanced by concomitant administration of testosterone or androstenedione, but not of DHT (Fig. 7). Similarly, ovarian estradiol-17β contents were markedly increased by FSH when administered together with testosterone or androstenedione, but not when administered together with DHT (Fig. 8). These combined data suggest that FSH stimulates estrogen secretion in hypophysectomized rats by stimulating the aromatization of C_{19}-androgens (testosterone and androstenedione) to C_{18} estrogens (estradiol-17β), and, by inference, that the role of LH is to increase the production of the C_{19} androgens for aromatization.

The greater effectiveness of FSH in stimulating ovarian weight when administered with testosterone or androstenedione, than when given alone or with DHT, suggests a role of ovarian estradiol-17β in the stimulation of follicular growth. This interpretation receives additional support from previous reports that estrogen treatment increases ovarian weight primarily due to granulosa cell proliferation when administered alone or in combination with FSH to hypophysectomized rats (25-27), and that antiestradiol serum decreases the ovarian weight response to exogenous gonadotropins (28).

The results of these *in vivo* studies support the conclusions, reached previously on the basis of *in vitro* experiments, that FSH and LH regulate ovarian estrogen secretion by actions at biochemically distinct sites: LH stimulates the synthesis of androgens, which are then converted to estrogens under specific stimulation by FSH. Our *in vivo* studies provide no information concerning the cell types upon which the separate actions of FSH and LH are exerted. However, the results of *in vitro* studies (reviewed above), of *in vivo* studies reporting separate cytological responses to FSH and LH (1), and of experiments demonstrating separate cellular localization of FSH versus LH receptors (16), considered together, offer compelling support for the hypothesis summarized in Fig. 5. Whether granulosa cells are the only follicular site of aromatization cannot be concluded from the present studies. It is possible that the theca interna may act as an additional source of estrogen formation, which may assume greater quantitative importance in some species or physiologic states than in others.

ACKNOWLEDGMENTS

We thank the Hormone Distribution Office, National Institutes of Health, Bethesda, Md., for donating luteinizing hormone (NIH-LH-B8); Dr. Harold Papkoff, University of California, San Francisco, Calif., for providing purified preparations of ovine FSH (FSH-HP, batch No. G4-150C, with FSH activity of approximately 50 times that of NIH-FSH-S1 and LH activity of less than 0.01 times that of NIH-LH-S1) and LH (LH-HP, batch No. G3-222B, with LH activity of approximately 2.75 times that of NIH-LH-S14 and FSH activity 0.014 times that of NIH-FSH-S11); and Dr. Leo E. Reichert, Emory University, Atlanta, Ga., for providing purified ovine

FSH (FSH-LER, batch No. 1932, with FSH activity of approximately 44 times that of NIH-FSH-S1). The technical assistance of Ms. Heather McKeracher, Dr. Jemima Laryea, and Mr. Ram Srivastava is gratefully acknowledged. This research was supported by grants MT-3392 and MA-5772 from the Medical Research Council (Canada), and grant 74228 from the World Health Organization. Dr. Armstrong is an Associate of the Medical Research Council (Canada).

REFERENCES

1. Greep, R. O., Van Dyke, H. B., and Chow, B. F. (1942): *Endocrinology,* 30:635.
2. Lostroh, A. J., and Johnson, R. E. (1966): *Endocrinology,* 79:991.
3. Solod, E. A., Armstrong, D. T., and Greep, R. O. (1966): *Steroids,* 7:607.
4. Major, P. W., Armstrong, D. T., and Greep, R. O. (1967): *Endocrinology,* 81:19.
5. Hall, P. F., and Young, D. G. (1968): *Endocrinology,* 82:559.
6. Hollander, N., and Hollander, V. (1958): *J. Biol. Chem.,* 233:1097.
7. Kaiser, J. (1964): *Acta Endocrinol.,* 47:676.
8. Savard, K., Marsh, J. M., and Rice, B. F. (1965): *Recent Prog. Horm. Res.,* 21:285.
9. Mills, T. M., Davies, P. J. A., and Savard, K. (1971): *Endocrinology,* 88:857.
10. Dorrington, J. H., and Kilpatrick, R. (1966): *J. Endocrinol.,* 35:53.
11. Dorrington, J. H., and Armstrong, D. T. (1975): *Proc. Natl. Acad. Sci. U.S.A.,* 72:2677.
12. Dorrington, J. H., Moon, Y. S., and Armstrong, D. T. (1975): *Endocrinology,* 97:1328.
13. Goff, A. K., and Major, P. W. (1975): *J. Endocrinol.,* 65:73.
14. Armstrong, D. T., and Papkoff, H. (1976): *Endocrinology* 99:1144
15. Channing, C. P., and Kammerman, S. (1974): *Biol. Reprod.,* 10:179.
16. Richards, J. S., and Midgley, A. R., Jr. (1976): *Biol. Reprod.,* 14:82.
17. Short, R. V. (1962): *J. Endocrinol.,* 24:59.
18. Bjersing, L., and Carstensen, H. (1967): *J. Reprod. Fertil.,* 14:101.
19. Erickson, G. F., and Ryan, K. J. (1975): *Endocrinology,* 97:108.
20. Dorrington, J. H., Fritz, I. B., and Armstrong, D. T. (1976): *Mol. Cell. Endocrinol. (in press).*

21. Makris, A., and Ryan, K. J. (1975): *Endocrinology*, 96:694.
22. Fortune, J. E., and Armstrong, D. T. (1976): In: Proceedings of NIH Workshop on Ovarian Follicle, Santa Cruz, Calif. Raven Press, New York.
23. Falck, B. (1959): *Acta Physiol. Scand.,* 47 (Suppl. 163):1.
24. Armstrong, D. T., Moon, Y. S., and Leung, P. C. K. (1976): *Biol. Reprod. (in press).*
25. Williams, P. C. (1940): *Nature*, 145:388.
26. Pencharz, R. I. (1940): *Science*, 91:554.
27. Goldenberg, R. L., Vaitukaitis, J. L., and Ross, G. T. (1972): *Endocrinology*, 90:1492.
28. Reiter, E. O., Goldenberg, R. L., Vaitukaitis, J. L., and Ross, G. T. (1972): *Endocrinology*, 91:1518.

Ovarian Follicular Development and Function,
edited by A. R. Midgley and W. A. Sadler.
Raven Press, New York © 1979.

Rabbit Follicular Steroidogenesis

E. V. YoungLai

Department of Obstetrics and Gynecology,
Reproductive Biology Program, McMaster University
Medical Center, Hamilton, Ontario, Canada L8S 4J9

The concept of a functional relationship between the theca interna and granulosa cells of the graafian follicle was first introduced by Falck (1959), and later modified by Short (1962). Since then several laboratories have been investigating and elaborating on this concept using various experimental techniques and animal species. In the mare, rhesus monkey, and human the theca interna seems to be capable of producing estradiol both *in vivo* and *in vitro* (Ryan et al., 1968; YoungLai and Short, 1970; Channing and Coudert, 1976). In the rabbit and guinea pig (YoungLai, 1973*a*; Makris and Ryan, 1975), although individual cell types can produce estrogens it appears that a combination of both is necessary for maximal estrogen production. Rabbit theca cells obtained after mating were not able to convert labeled pregnenolone and progesterone to estrogens, whereas whole sliced follicles were (YoungLai, 1973*a*). Also, Mills et al. (1971) found that rabbit follicles were extremely sensitive to LH with good incorporation of ^{14}C-acetate to estrogens. On the other hand, labeled androstenedione and testosterone were converted to estrogens by separated granulosa and theca cells (YoungLai, 1973*b*) and cultured rabbit granulosa cells converted androstenedione to estrogens in the presence of LH/FSH and dibutyryl cyclic AMP (Erickson and Ryan, 1975). In the rabbit it would be important to determine steroidogenesis by isolated and recombined theca and granulosa cells.

Our laboratory has also been investigating the control of testosterone production by the isolated rabbit ovarian follicle (YoungLai, 1975). We have demonstrated that this steroid is consistently produced in large quantities by the intact follicle after stimulation by LH and that the response can be elicited by cyclic nucleotides as well as a mere 1-sec exposure to the gonadotropic hormone. Moreover, the rapidity of decline of testosterone production after removal of cyclic AMP compared to the slow decline after LH stimulation suggests that cyclic AMP produces a labile factor that is formed continuously in the presence of LH, for about 3 hr.

Further studies with gonadotropic hormones (Table 1) have indicated that the isolated follicle can also respond to FSH (10 μg/ml) with the production of testosterone and estradiol that is significantly different from controls. The estradiol response to LH (5 μg/ml) stimulation was not different from controls. If the LH activity in FSH is neutralized by an antiserum to LH, only estradiol production is increased, depending on the conditions. We have previously shown that rabbit follicles contain testosterone, and it could be surmised from Dr. Armstrong's work and the present data that this endogenous testosterone is aromatized as a response to FSH.

TABLE 1. *Effect of FSH and LH on follicular steroidogenesis*

	Testosterone (ng/follicle)	E_2 (ng/follicle)
Control (10)	1.97 ± 0.2	0.65 ± 0.1
LH (12)	15.4 ± 2.4[a]	1.08 ± 0.4
FSH (7)	19.3 ± 2.6[a]	3.31 ± 1.2[b]
FSH + āLH (10)	5.98 ± 0.9[a]	1.83 ± 0.5[b]
āLH + FSH (10)	1.64 ± 0.4	3.65 ± 0.9[b]

Single intact follicles were incubated for 2 hr with medium alone before adding medium containing test substance and incubating for another 24 hr. Controls had medium only for both time periods.

[a] $p < 0.01$.
[b] $p < 0.05$.

REFERENCES

Channing, C. P., and Coudert, S. P. (1976): *Endocrinology*, 98:590.
Erickson, G. F., and Ryan, K. J. (1975): *Endocrinology*, 97:108.
Falck, B. (1959): *Acta Physiol. Scand.*, 47 (Suppl.):163.
Makris, A., and Ryan, K. J. (1975): *Endocrinology*, 96:694.
Mills, T. M., Davies, P. J., and Savard, K. (1971): *Endocrinology*, 88:857.
Ryan, K. J., Petro, F., and Kaiser, J. (1968): *J. Clin. Endocrinol.*, 28:355.
Short, R. V. (1962): *J. Endocrinol.*, 24:59.

YoungLai, E. V., and Short, R. V. (1970): *J. Endocrinol.*, 47:321.
YoungLai, E. V. (1973*a*): *Acta Endocrinol.*, 74:775.
YoungLai, E. V. (1973*b*): *Steroids Lipids Res.*, 4:11.
YoungLai, E. V. (1975): *J. Reprod. Fertil.*, 45:575.

Ovarian Follicular Development and Function,
edited by A. R. Midgley and W. A. Sadler.
Raven Press, New York © 1979.

Steroidogenesis in the Rabbit Ovary

Thomas M. Mills

Department of Endocrinology, Medical College of Georgia, Augusta, Georgia 30902

The *in vitro* synthetic capacity of large ovarian follicles has been studied in unmated and mated rabbits (3). Large follicles were dissected from the ovaries and then incubated for 3 hr in Krebs-Ringer bicarbonate buffer with the resulting incubate analyzed for steroid content. Steroid synthesis was quantified by measuring the incorporation of ^{14}C-acetate into steroids using standard isolation and purification techniques. More recently, steroids have been quantified by radioimmunoassay (RIA). The results of these studies show that the rabbit follicle is capable of synthesizing *in vitro* a wide range of steroids including progesterone, 20α-dihydroprogesterone, testosterone, androstenedione, 17β-estradiol, and estrone. Addition of luteinizing hormone (LH) to the incubation medium significantly increases the synthesis of all steroids measured. Using the rate of incorporation of acetate ^{14}C into labeled estradiol and estrone to quantify steroid synthesis, follicles respond to LH concentrations as low as 0.003 μg/ml medium (4). A lower sensitivity to LH is evident from the RIA results presented in Fig. 1. The follicular content of progesterone, 17OH-progesterone, testosterone, and 17β-estradiol show progressive elevation over the range of 0.05 μg LH/ml to 0.5 μg LH/ml; the response appears to plateau at LH concentrations from 0.5 to 50 μg/ml. Figure 1 also demonstrates that follicles isolated from unmated rabbits are highly responsive to stimulation by cycle adenosine 3',5'-monophosphate (cAMP). Concentrations of this cyclic nucleotide from 0.2 to 20 μmole cAMP/ml increase the synthesis of all four steroids. Thirty and 40 μmole cAMP/ml appear to diminish steroidogenesis.

To determine whether LH and cAMP stimulate steroidogenesis by the same mechanism, follicles were incubated with 5 μg LH/ml and 20 μmole cAMP/ml (2).

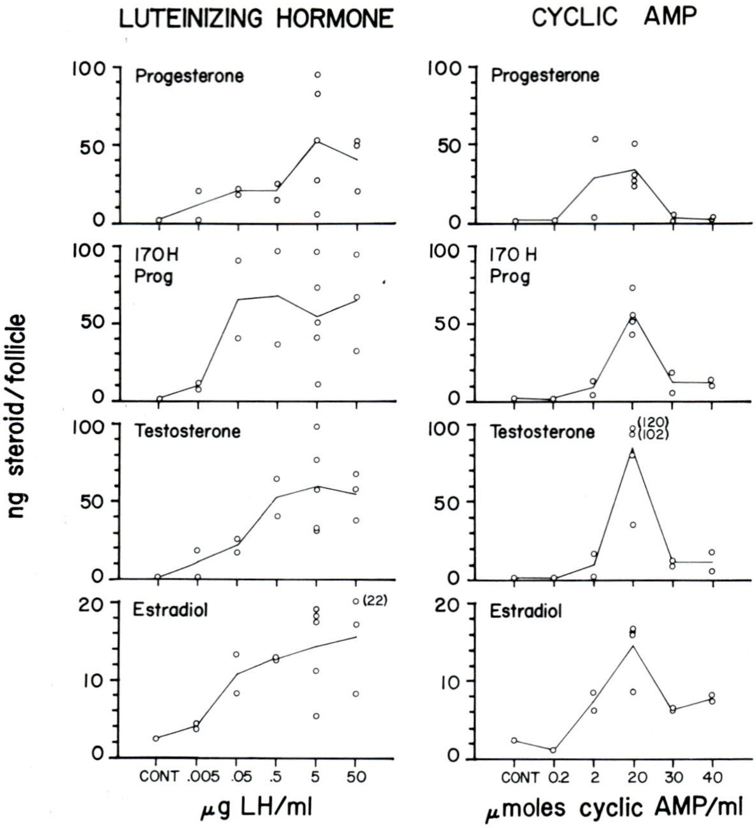

FIG. 1. The stimulation of steroid synthesis in isolated ovarian follicles incubated with LH or cyclic AMP. Each point represents a separate assay of the steroid, the solid line connects the average value for each group. Control values (CONT) in the LH experiments are the average of 11 separate determinations, while control values in the cyclic AMP experiments represent the mean of nine determinations. (Reprinted with permission.)

The results of these incubations (Fig. 2) confirm that both of these agents independently cause a marked increase in all four of the measured steroids. When follicles are incubated simultaneously with both LH and cAMP, the total of the four steroids synthesized is no different than the totals after incubation with either agent alone. Previously published results have demonstrated that follicular cAMP synthesis is stimulated by LH *in vitro* (1). Taken together, these two studies strongly support the role of cAMP as a second messenger in the action of LH on steroidogenesis in the rabbit follicle.

In the rabbit, coital stimulation results in ovulation and in a surge in the ovarian secretion of several steroids into the general circulation (5). Figure 3

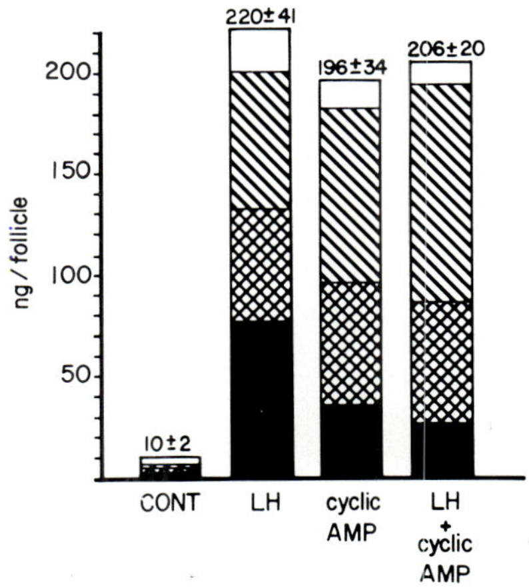

FIG. 2. The effect of incubation of isolated follicles in the presence of 5 μg LH/ml, 20 μmole cyclic AMP/ml or a combination of LH and cyclic AMP on the synthesis of steroids. The solid portion of each bar is ng progesterone/follicle, cross-hatched portion is ng 17-hydroxyprogesterone/follicle, hatched portion is ng testosterone/follicle, open portion is ng 17β-estradiol/follicle. Each steroid was determined separately in three experiments and averages are presented here. The numbers at the top of each bar are mean ± SE of the sum of all four steroids. The total steroid synthesized in the presence of LH, cyclic AMP, or the combination of the two agents are significantly different from control incubations ($p < 0.05$) but are not significantly different from each other. (Reprinted with permission.)

depicts this transitory postcoital rise in the blood levels of progesterone, 17OH-progesterone, testosterone, and 17β-estradiol. The serum levels of all four steroids have declined at 12 hr after mating, a time at which most follicles have ovulated. When large follicles are isolated from the ovaries of rabbits mated 0, 2, 6, and 12 hr before sacrifice, an identical pattern of postcoital surge and decline is evident in the *in vitro* pattern of steroidogenesis (ref. 3; Fig. 4). Follicles from estrous rabbits (0 hr) show low steroid synthetic rates but in the presence of LH (5 μg/ml), these follicles convert about 10 times more ^{14}C into steroids than do control follicles. At 2 hr after mating, ^{14}C incorporation rates in both the control and LH groups are high; presumably the follicles are responding maximally to endogenous gonadotropin at this

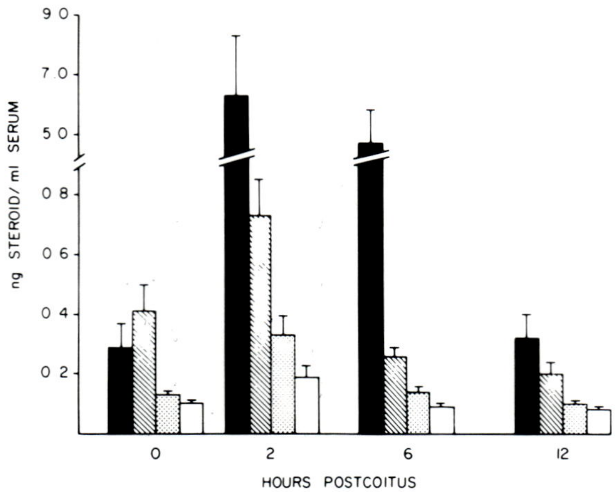

FIG. 3. The concentration of selected steroids in peripheral blood of rabbits immediately after (0 hr) and 2, 6, and 12 hr after mating. Solid bars, progesterone; hatched bars, 17α-hydroxy progesterone; stippled bars, testosterone; open bars, 17β-estradiol. Each bar represents the mean of 8 to 13 separate determinations; brackets represent one SEM. (Reprinted with permission.)

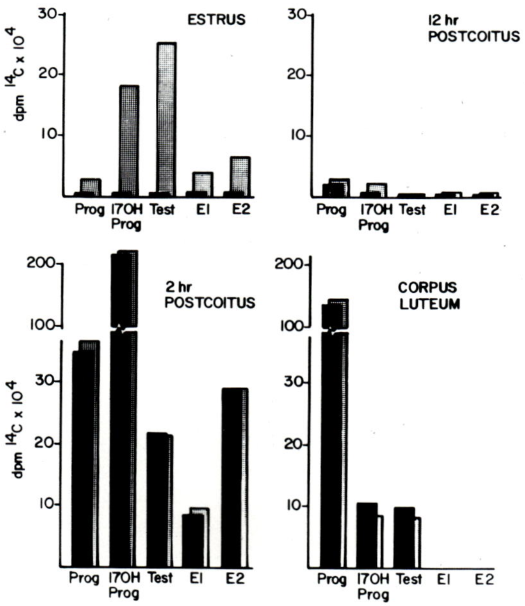

FIG. 4. The *in vitro* incorporation of ^{14}C from acetate-1-^{14}C into selected steroids by follicles and corpora lutea isolated from rabbit ovaries before and after mating. Solid bars, 0.025 μg BSA/ml; cross-hatched bars, 0.025 μg LH ml. Abbreviations include: Prog, progesterone; 17OH Prog, 17OH progesterone; Test, testosterone; E1, estrone; E2, 17β-estradiol. (Reprinted with permission.)

time. Ovulated follicles (12 hr after mating) synthesize little steroid and, furthermore, these follicles are refractory to LH stimulation. A distinct qualitative shift in follicular steroidogenesis is also evident in ovulated follicles with less estrogen and more progesterone synthesis in 12 hr postcoitus follicles than in estrous (0 hr) follicles (Fig. 4).

REFERENCES

1. Marsh, J., Mills, T., and LeMaire, W. (1972): Cyclic AMP synthesis in rabbit graafian follicles and the effect of luteinizing hormone. *Biochim. Biophys. Acta*, 273:389.
2. Mills, T. (1975): Effect of luteinizing hormone and cyclic adenosine 3',5'-monophosphate on steroidogenesis in the ovarian follicle of the rabbit. *Endocrinology*, 96:440.
3. Mills, T., and Savard, K. (1973): Steroidogenesis in ovarian follicles isolated from rabbits before and after mating. *Endocrinology*, 92:788.
4. Mills, T., Davies, P., and Savard, K. (1971): Stimulation of estrogen synthesis in rabbit follicles by luteinizing hormone. *Endocrinology*, 88:857.
5. Waterston, J., and Mills, T. (1976): Peripheral blood steroid concentrations in the preovulatory rabbit. *J. Steroid Biochem.*, 7:15.

Ovarian Follicular Development and Function,
edited by A. R. Midgley and W. A. Sadler.
Raven Press, New York © 1979.

Androgen Production by Isolated Components of Rat Ovarian Follicles

Joanne E. Fortune and David T. Armstrong

Departments of Obstetrics and Gynaecology and Physiology, University of Western Ontario, London, Canada

Ovaries or isolated granulosa cells from hypophysectomized immature rats secrete estradiol-17β when cultured in the presence of follicle-stimulating hormone (FSH) and testosterone (1,2). This observation prompted us to investigate the cellular source and gonadotropic control of the testosterone precursor needed for estradiol-17β synthesis by granulosa cells. Previous reports have indicated that luteinizing hormone (LH) stimulates ovarian androgen secretion (3-6), and other studies have implicated the theca as the major site of follicular androgen production (7). The present study combined these two approaches in an attempt to demonstrate simultaneously both the cell type(s) responsible for androgen production and how this production is controlled.

MATERIALS AND METHODS

Isolation of Theca and Granulosa

Prepubertal rats were injected with 4 IU pregnant mare's serum gonadotropin (Equinex; Ayerst Laboratories, Inc.) on the morning of day 28 of age and killed 48 hr later (the equivalent of the morning of proestrus). Ovaries were removed and the largest ovarian follicles (0.9 to 1.1 mm in diameter) were isolated and trimmed of interstitial tissue. Theca and granulosa were isolated in the following manner. Each follicle was cut in half and the inside of the follicle was scraped with a glass needle or hair loop to clean the theca of granulosa cells. Scraped pieces of theca (two follicles per dish) were cultured in plastic Petri dishes in 1 ml of the appropriate medium.

Oocytes and surrounding cumulus cells were discarded. Granulosa cells from each set of two follicles were collected, centrifuged for 5 to 10 min at 1,000 rpm in an International Clinical Centrifuge, resuspended in 2 ml of the appropriate medium, and grown in monolayer culture in plastic tissue culture flasks. Granulosa cells had attached to the culture flask within 3 hr.

Culture Conditions

Theca and granulosa were cultured in medium 199 (Grand Island Biological Co.) containing 10% fetal calf serum, 50 units penicillin/ml, and 50 μg streptomycin/ml. Tissue in some culture dishes was treated with the following highly purified gonadotropins (supplied by Dr. Harold Papkoff): LH (lot No. G3-222B), 0.1 μg/ml; FSH (lot No. G4-150C), 0.1 μg/ml. Medium was collected, frozen for subsequent steroid analysis, and replaced at 3, 6, and 12 hr of culture. In some experiments thecal preparations were treated with various concentrations of LH and medium collected after 6 hr.

Androgen Radioimmunoassay

Samples were assayed for immunoreactive androgens by direct radioimmunoassay of 5- to 100-μl aliquots of culture medium, without extraction. The accuracy of direct assay of culture medium was verified by comparison with values obtained for aliquots of the same samples assayed after ether extraction. The correlation coefficient was 0.93 ($p < 0.001$; $Y = 0.95\ X + 0.01$) when 25 μl of medium was assayed and 0.96 ($p < 0.001$; $Y = 0.76\ X + 0.03$) for an assay of 100 μl. Antiserum, prepared by immunizing rabbits with testosterone conjugated to bovine serum albumin, was obtained from Dr. B. V. Caldwell (8). When the antiserum was tested for specificity in our laboratory, it did not cross-react with estradiol-17β or progesterone, but the following cross-reactivities were obtained for other androgens: 17β-OH-5α-androstan-3-one (DHT), 53%; 5α-androstan-3α,17β-diol, 30%; and 5α-androstan-3β,17β-diol, 13%. Thus, we will refer to the steroids measured by the assay as "androgens" rather than "testosterone."

RESULTS

Androgen Production by Theca

Thecal preparations secreted androgen in the absence of exogenous gonadotropins (Fig. 1). LH or LH plus FSH significantly increased ($p < 0.001$) the rate of thecal androgen production as compared to control tissue. FSH alone had no significant effect ($p > 0.10$) and there was no significant interaction between LH and FSH ($p > 0.10$).

Effect of Different Concentrations of LH on Thecal Androgen Production

Concentrations of LH ranging from 0.0001 to 1.0 μg/ml were applied to isolated pieces of theca and medium was collected after 6 hr of culture. The dose-response curve obtained is shown in Fig. 2. The lowest effective dose of LH was between 0.001 and 0.01 μg/ml and a maximal response was elicited by doses of 0.01 μg/ml or higher.

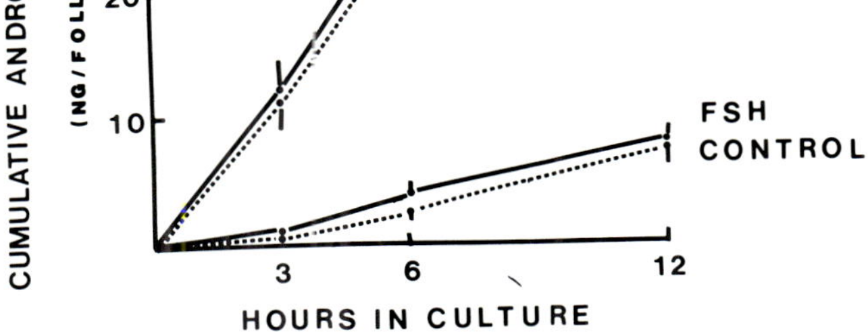

FIG. 1. Cumulative androgen secretion by thecal preparations (two follicles/dish; n = 4) cultured for 12 hr with or without LH (0.1 μg/ml) and/or FSH (0.1 μg/ml).

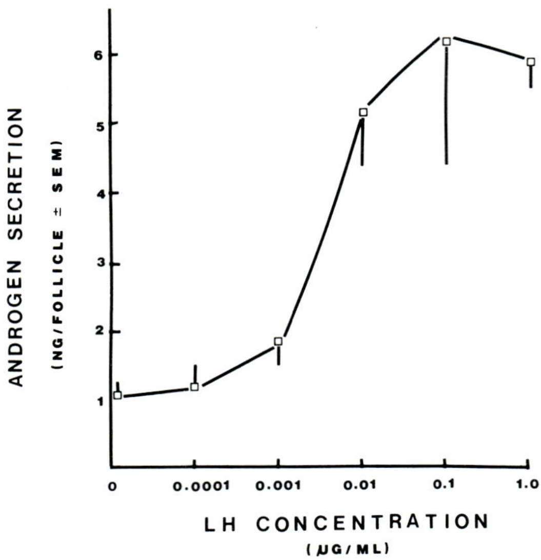

FIG. 2. Androgen production by thecal preparations (two follicles/dish; n = 6) cultured in the presence of various concentrations of LH for 6 hr.

Androgen Production by Granulosa Cells

Granulosa cells produced negligible amounts of androgen under the conditions of these experiments. Androgen values were always below the sensitivity of the assay (12.5 pg) when 100 µl of medium was assayed and binding to the antibody was similar to that observed in standard tubes containing no testosterone. No differences among the four treatment groups were noted.

DISCUSSION

Under the conditions provided in these experiments granulosa cells were incapable of androgen production. This finding as well as the previous report that granulosa cells synthesize estradiol-17β only if supplied with exogenous testosterone indicates that the granulosa cells do not contribute significantly to follicular androgen production. The present results strongly suggest that the theca is responsible for most or all androgen secretion by proestrous rat follicles and that androgen production is under the control of LH. Since the thecal preparations used were not entirely free of granulosa cell contamination, the possibility cannot be excluded that the granulosa cells in some way influence androgen production by theca or that a cooperation between the two cell types is required for androgen secretion. These findings suggest that follicular estradiol-17β is produced through the cooperation of both types of follicle cells, i.e., the androgen secreted by the theca cells is aromatized by granulosa cells to estradiol-17β.

ACKNOWLEDGMENTS

We are grateful to Dr. H. Papkoff (University of California, Berkeley) for his generous gifts of purified gonadotropins, to Dr. B. V. Caldwell (Yale University) for supplying the antitestosterone serum, and to Dr. Y. S. Moon and Ms. J. Calaresu for skilled technical advice and assistance.

REFERENCES

1. Moon, Y. S., Dorrington, J. H., and Armstrong, D. T. (1975): Stimulatory action of follicle-stimulating hormone on estradiol-17β secretion by hypophysectomized rat ovaries in organ culture. *Endocrinology*, 97:244.
2. Dorrington, J. H., Moon, Y. S., and Armstrong, D. T. (1975): Estradiol-17β biosynthesis in cultured granulosa cells from hypophysectomized immature rats; stimulation by follicle-stimulating hormone. *Endocrinology*, 97:1328.
3. Armstrong, D. T., Dorrington, J. H., and Robinson, J. (1976): Effects of indomethacin and aminoglutethimide phosphate on luteinizing hormone-induced alterations of cyclic adenosine monophosphate, prostaglandin F and steroid levels in preovaluatory rat ovaries. *Can. J. Biochem.*, 54:796.
4. YoungLai, E. V. (1975): Steroid production by the isolated rabbit ovarian follicle. III. Actinomycin D-insensitive stimulation of steroidogenesis by LH. *Endocrinology*, 96:468.
5. Lieberman, M. E., Barnea, A., Bauminger, S., Tsafriri, A., Collins, W. P., and Lindner, H. R. (1975): LH effect on the pattern of steroidogenesis in cultured graafian follicles of the rat: Dependence on macromolecular synthesis. *Endocrinology*, 96:1533.
6. Louvet, J.-P., Harman, S. M., Schreiber, J. R., and Ross, G. T. (1975): Evidence of a role for androgens in follicular maturation. *Endocrinology*, 97:366.

7. Makris, A., and Ryan, K. J. (1975): Progesterone, androstenedione, testosterone, estrone, and estradiol synthesis in hamster ovarian follicle cells. *Endocrinology*, 96:694.
8. Auletta, F. J., Caldwell, B. V., and Hamilton, G. L. (1974): Androgens: Testosterone and dihydrotestosterone. In: *Methods of Hormone Radioimmunoassay* (B. M. Jaffe and H. R. Behrman, eds.), Academic Press, New York, p. 359.

Ovarian Follicular Development and Function,
edited by A. R. Midgley and W. A. Sadler.
Raven Press, New York © 1979.

Steroidogenesis by Mitochondria Prepared from Porcine Ovarian Tissues

Michael J. Dimino, Frances E. DeLisle, and James R. Downing

Department of Research, Sinai Hospital of Detroit, Detroit, Michigan

As concluded by the studies of Dr. Armstrong et al., granulosa cells can synthesize estrogen from C_{19} steroids but have limited ability to synthesize the C_{19} substrate. Our laboratory has been concerned with the syntheses of pregnenolone and progesterone by ovarian mitochondria, and we would like to present some preliminary data that relate to these conclusions. For our experiments, we prepared mitochondria from granulosa, theca, and corpora lutea of porcine ovaries and measured their ability to convert [4-^{14}C]-cholesterol to [4-^{14}C]-pregnenolone and [4-^{14}C]-progesterone.

Granulosa cells were obtained by aspirating follicular fluid from medium and large follicles according to the method of Channing and Ledwitz-Rigby (1). The collapsed follicles were removed, inverted, gently scraped, and rinsed to remove any remaining granulosa cells. However, we could not be certain that our granulosa and thecal preparations were free of contamination or that our isolation procedure did not result in some loss of thecal steroidogenic cells. The cell preparations were homogenized and centrifuged at 1,000 $\times g$ for 5 min to remove heavy debris and nuclei. The mitochondrial pellets then were collected at 7,500 $\times g$ for 10 min. Corpora luteal mitochondria were prepared as previously reported (2). In our studies, succinate was used to support [4-^{14}C]-cholesterol conversion, since this substrate requires functionally intact mitochondria in order for reducing equivalents to be made available for steroidogenesis.

As seen in Table 1, mitochondrial preparations of granulosa and theca of medium follicles (3 to 6 mm in diameter) showed some ability to convert [4-^{14}C]-cholesterol to [4-^{14}C]-pregnenolone and [4-^{14}C]-progesterone. Steroidogenic activity was markedly improved in

TABLE 1. *Steroidogenesis by mitochondria of porcine ovarian tissues*[a]

Mitochondria obtained from	Percent of [4-^{14}C]-cholesterol converted to	
	[4-^{14}C]-pregnenolone	[4-^{14}C]-progesterone
Medium follicles[b]		
Granulosa	1.9	1.1
Theca	1.4	1.5
Large follicles[c]		
Granulosa	5.1	1.2
Theca	5.0	3.4
Corpora lutea	2.6	8.8

[a]Steroidogenesis was measured by the method of Robinson and Stevenson (6), modified by using 0.8 mg mitochondrial protein/assay and a 2-hr incubation. Substrates used were succinate (source of reducing equivalents) and [4^{14}C]-cholesterol (25,000 cpm/assay).

[b]Medium follicles: 3 to 6 mm in diameter.

[c]Large follicles: 7 to 12 mm in diameter.

mitochondria obtained from tissues of large follicles (7 to 12 mm), with thecal mitochondria having somewhat more ability to synthesize [4-^{14}C]-progesterone. In contrast to mitochondrial preparations from follicular tissues, mitochondria obtained from corpora lutea had a much greater ability to convert [4-^{14}C]-cholesterol to [4-^{14}C]-progesterone.

Although it is likely that our mitochondrial preparations contain some microsomal contamination, we have already shown that highly purified luteal mitochondria prepared by zonal centrifugation do contain the enzyme system required for progesterone synthesis (2). It has been reported that other subcellular fractions of luteal cells can synthesize progesterone (3). However, in this respect, we are exploring the possibility that progesterone synthesis in these nonmitochondrial preparations may actually be due to contamination by mitochondrial fragments.

The lower [4-^{14}C]-cholesterol conversion values found for mitochondria of medium follicles is in agreement with studies of other investigators (4). [4-^{14}C]-pregnenolone synthesis increased appreciably in granulosa and thecal mitochondria of large follicles. In contrast to studies of Ryan and Petro (5), which show a higher progesterone synthesis rate in granulosa than in theca of human ovaries, we found higher [4-^{14}C]-progesterone formation in thecal mitochondria than in granulosa mitochondria of large follicles. We have not yet determined whether the difference we see is significant or if this progesterone synthesizing activity is actually due to microsomal contamination. In anticipation of the various possibilities, we have developed two working hypotheses. If thecal mitochondria (or other subcellular compartments) synthesize progesterone more readily than their granulosa equivalents, then it is possible that granulosa cells cannot synthesize C_{19} steroids, at least in part, because of their inability to synthesize steroids which precede testosterone.

On the other hand, it is possible that neither thecal nor granulosa mitochondria contain the enzymes required for progesterone synthesis. Rather, progesterone synthesis occurring in our mitochondrial preparations could be due to enzyme activity associated with microsomal contamination. If this is the case, then differences in the ability of luteal and follicular mitochondria to synthesize progesterone may help explain differences in the amounts of progesterone and estrogen secreted by luteal and follicular tissue. We are now investigating these possibilities in our laboratory and feel that validation of either hypothesis would offer insight into the differentiation of ovarian cells, as well as the regulation of ovarian steroidogenesis.

REFERENCES

1. Channing, C. P., and Ledwitz-Rigby, F. (1975): *Methods Enzymol.*, 34:183.
2. Dimino, M. J., and Campbell, M. D. (1976): *Proc. Soc. Exp. Biol. Med.*, 152:54.
3. Cheatum, S. G., and Warren, J. C. (1966): *Biochim. Biophys. Acta*, 122:1.
4. Bjersing, L., and Carstensen, H. (1967): *J. Reprod. Fertil.*, 14:101.
5. Ryan, K. J., and Petro, Z. (1966): *J. Clin. Endocrinol.*, 26:46.
6. Robinson, J., and Stevenson, P. M. (1971): *Eur. J. Biochem.*, 24:18.

Ovarian Follicular Development and Function,
edited by A. R. Midgley and W. A. Sadler.
Raven Press, New York © 1979.

Synergistic Action of Androgens and FSH on Progesterone Secretion by Rat Granulosa Cells in Culture

*Jennifer H. Dorrington and †David T. Armstrong

†Departments of Physiology and Obstetrics and Gynaecology, University of Western Ontario, London, Canada, and *Banting and Best Department of Medical Research, University of Toronto, Toronto, Canada

There are two peaks of progesterone biosynthesis during the estrous cycle of the rat (Barraclough et al., 1971; Smith et al., 1975). The preovulatory increase is coincident with the surge of luteinizing hormone (LH) and follicle-stimulating hormone (FSH) on the afternoon of proestrus. The plasma levels of progesterone return to basal values on the morning of estrus and remain low for at least a day before rising again as luteinization proceeds and the corpus luteum is formed (Smith et al., 1975).

Follicles removed from rats on the morning of proestrus and maintained in culture responded to both LH and FSH with an increase in the rate of release of progesterone (Lindner et al., 1974). The effect of gonadotropins was significant after 1 hr. This response seems to be too rapid to be a reflection of luteinization, but rather may account for the increase in preovulatory progesterone occurring at the time of the proestrous increase in gonadotropins. These observations suggest that the follicle may be responsible for the preovulatory peak of progesterone biosynthesis.

Progesterone secretion was elevated 12 hr after the addition of LH to porcine granulosa cells (Kolena and Channing, 1972) and within 2 days after the addition of LH/FSH (Pergonal) to rabbit granulosa cells in culture (Erickson and Ryan, 1975).

In the present communication, we report a stimulatory effect of FSH and androgens on progesterone biosynthesis by rat granulosa cells in culture. While this work was in progress, Schomberg et al. (1976) presented evidence that androgens stimulate progesterone synthesis by cultured porcine granulosa cells.

MATERIALS AND METHODS

Luteinizing hormone (LH-HP, batch No. G4-233, with LH activity of approximately 2.06 NIH-LH-S1 units/mg and FSH activity of approximately 0.01 NIH-FSH-S1 units/mg) and follicle-stimulating hormone (FSH-HP, batch No. G4-150C, with FSH activity of 50 NIH-FSH-S1 units/mg and LH activity of 0.01 NIH-LH-S1 units/mg) were gifts from Dr. H. Papkoff. Follicle-stimulating hormone (FSH-LER, batch No. 1932 with FSH activity of 44 NIH-FSH-S1 units/mg and LH activity of less than 0.01 NIH-LH-S1 units/mg) was a gift from Dr. L. E. Reichert. Testosterone, 17β-OH-5α-androstan-3-one (DHT) and estradiol-17β were obtained from Sigma Chemical Co.

Female Wistar rats (20 days of age) were obtained from Canadian Breeding Laboratories, Montreal. Estradiol-17β (1 mg in 0.1 ml sesame oil) was administered subcutaneously each day for 3 to 4 days prior to autopsy at 25 days of age.

Granulosa cells were obtained by puncturing individual follicles with a fine needle (Dorrington et al., 1975*a*). The cells were suspended in culture medium consisting of Eagle's minimum essential medium modified as described previously (Dorrington et al., 1975*b*) but containing reduced levels of antibiotics (Fungizone, 0.6μg/ml; penicillin, 50 units/ml; streptomycin, 50 μg/ml). Aliquots (0.5 ml) of the cell suspension were dispensed into Linbro culture wells (24 wells/plate), test substances were added, and the cells were cultured at 37°C for 72 hr in a water-saturated atmosphere of 95% air and 5% CO_2. After each 24-hr period of culture, the medium was removed and frozen, and replaced with new medium containing the same test substances. The cells were washed with Dulbecco medium after 72 hr, removed in 0.2 ml 1% sodium dodecyl sulfate containing 1 mM EDTA, and the protein content determined by the method of Lowry et al. (1951).

The progesterone content of the medium was measured by radioimmunoassay employing an antiserum with a high specificity for progesterone, prepared by immunizing rabbits with 11α-hydroxyprogesterone hemisuccinate conjugated to bovine serum albumin. The characteristics of this antiserum, obtained from Dr. H. R. Behrman, Merck Institute for Therapeutic Research, Rahway, N.J., have been described (Orczyk et al., 1974).

RESULTS AND DISCUSSION

In 25-day-old rats pretreated with estradiol-17β for 4 days, the mean ovarian weight was 11.2 mg, whereas in untreated animals of the same age the mean weight was 7.8 mg. Histological examination revealed that the increase in weight was due to increased numbers of granulosa cells, as described by others previously (Goldenberg et al., 1972). Pretreatment of rats with estradiol-17β was adopted therefore on a routine basis in order to increase the yield of granulosa cells for biochemical studies.

The amounts of progesterone secreted into the medium by granulosa cells from estrogen-treated immature rats over a period of 3 days in culture are shown in Fig. 1. Cells cultured in medium alone, or in medium containing testosterone (0.5 μM) alone, secreted negligible amounts of progesterone. Partially purified FSH (0.25 μg FSH-LER/ml) stimulated progesterone secretion to a small extent during the first day in culture, after which time no further increases were observed. In the presence of both testosterone (0.5 μM) and FSH (0.25 μg FSH-LER/ml) the progesterone level in the medium after 1 day in culture was 15-fold greater

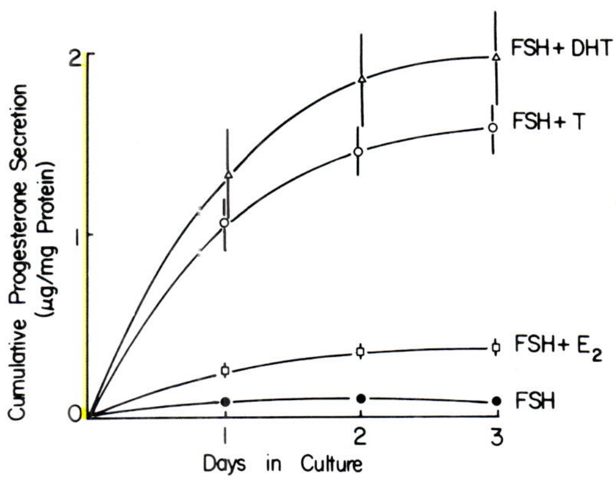

FIG. 1. Progesterone secretion by cultured granulosa cells from 25-day-old rats pretreated with 1 mg estradiol-17β s.c. daily, for 3 days before autopsy. FSH: 0.25 μg FSH-LER/ml; DHT: 0.5 μM 17β-OH-5α-androstan-3-one; T: 0.5 μM testosterone; E_2: 0.5 μM estradiol-17β. Secretion by cells cultured in media containing no gonadotropin or steroid, or containing DHT, T, or E_2 alone was undetectable.

than that produced by FSH alone. Secretion continued under these conditions during the next 2 days, but the rate was lower than that observed during the first day of culture. This type of granulosa cell did not respond to LH (0.25 μg LH-HP/ml) alone, or to testosterone (0.5 μM) and LH (0.25 μg LH-HP/ml) added together.

The results demonstrate synergism between a polypeptide hormone (FSH) and a steroid hormone (testosterone) in the stimulation of progesterone secretion by cultured granulosa cells from estrogen-treated immature rats. When the cells were incubated in the presence of testosterone (0.5 μM), the progesterone released from the cells depended upon the concentration of FSH, as shown in Fig. 2. Increase in progesterone secretion were detectable in the presence of 10 ng FSH-HP/ml and approached maximal levels at 250 ng FSH-HP/ml.

We have shown that FSH stimulates the conversion of testosterone to estradiol-17β by similarly prepared granulosa cells (Dorrington et al., 1975*a*; Armstrong et al., 1976*b*). There is evidence to suggest that estrogens may be involved in the regulation of progesterone secretion by corpora lutea of pregnant rats (Takayama and Greenwald, 1973). However, it seemed unlikely that the ability of testosterone to augment the effect of FSH on progesterone levels was due to its conversion to estrogens, since exogenous estradiol-17β was not as effective as testosterone in enhancing the FSH effect (Fig. 1). This conclusion received further support from the observation that DHT, a

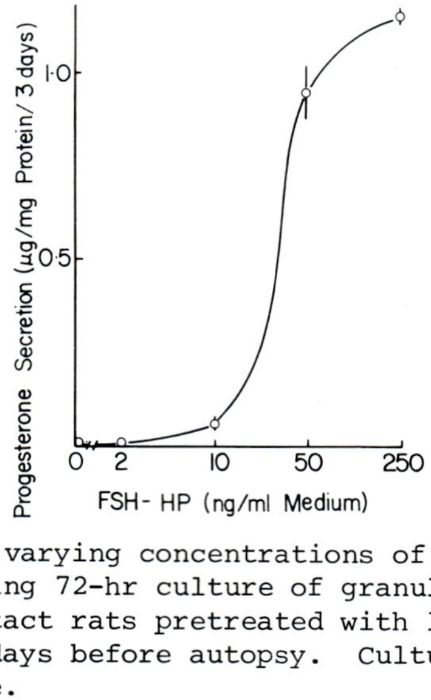

FIG. 2. Effect of varying concentrations of FSH-HP on progesterone secretion during 72-hr culture of granulosa cells isolated from 25-day-old intact rats pretreated with 1 mg estradiol-17β s.c. daily, for 4 days before autopsy. Culture media contained 0.5 μM testosterone.

potent androgen that cannot be converted to testosterone (Wilson, 1972) and hence to estrogens, was even more effective than testosterone in stimulating progesterone synthesis in the presence of FSH (Fig. 1).

FSH stimulates cyclic AMP accumulation in isolated granulosa cells (Kolena and Channing, 1972; Armstrong et al., 1976), and it is possible that the cyclic AMP generated stimulates steroidogenesis by increasing the rate of conversion of cholesterol to pregnenolone, since it is well established that 3'5'-AMP has this ability in other steroid synthesizing cells (Marsh, 1976). It is likely that any agent that has the capacity to increase 3'5'-AMP levels would mimic the effect of FSH. Furthermore, granulosa cells from preovulatory follicles, which have acquired LH receptors, may respond to LH and testosterone with an increase in progesterone biosynthesis to an extent comparable to that elicited by FSH and testosterone. Studies on granulosa cells known to have LH receptors as well as FSH receptors are under way.

In addition to being target cells for FSH and LH action, granulosa cells are also target cells for androgens as shown by the presence of androgen receptors in granulosa cell cytosol (Schreiber et al., 1976), and by the present demonstration that androgens influence granulosa cell metabolism. The mechanism by which testosterone enhances the effect of FSH on progesterone secretion is not clear, and may involve stimulation of biosynthesis of progesterone or inhibition of its metabolism. A marked elevation of ovarian androgen levels occurs in prepubertal rats within minutes after the injection of LH (Armstrong et al., 1976*a*), and it is tempting to speculate that this androgen of thecal origin (Fortune and Armstrong, 1976) or of interstitial cell origin (Louvet et al., 1975) interacts with the increased levels of gonadotropins to evoke the marked preovulatory increase in progesterone that occurs in rats on the afternoon of proestrus.

ACKNOWLEDGMENTS

We thank Dr. H. Papkoff, University of California, for donations of purified FSH and LH, and Dr. L. E. Reichert for purified FSH. The technical assistance of Ms. Heather McKeracher and Mr. Gerald Barbe is gratefully acknowledged. This research was supported by grants MT-3392 and MA-5772 from the Medical Research Council (Canada), and grant 74228 from the World Health Organization. Dr. Armstrong is an Associate of the Medical Research Council (Canada).

REFERENCES

1. Armstrong, D. T., Dorrington, J. H., and Robinson, J. (1976): Effects of indomethacin and aminoglutethimide phosphate *in vivo* on luteinizing-hormone-induced alterations of cyclic adenosine monophosphate, prostaglandin F and steroid levels in preovulatory rat ovaries. *Can. J. Biochem.,* 54:796-802.
2. Armstrong, D. T., Goff, A., and Dorrington, J. H. (1978): Regulation of follicular estrogen biosynthesis. *This volume.*
3. Barraclough, C. A., Collu, R., Massa, R., and Martini, L. (1971): Temporal interrelationships between plasma LH, ovarian secretion rates and peripheral plasma progestin concentrations in the rat: Effects of Nembutal and exogenous gonadotropins. *Endocrinology,* 88:1437-1447.
4. Dorrington, J. H., Moon, Y. S., and Armstrong, D. T. (1975*a*): Estradiol-17β biosynthesis in cultured granulosa cells from hypophysectomized immature rats; stimulation by follicle-stimulating hormone. *Endocrinology,* 97:1328-1331.
5. Dorrington, J. H., Roller, N. F., and Fritz, I. B. (1975*b*): Effects of follicle-stimulating hormone on cultures of Sertoli cell preparations. *Mol. Cell. Endocrinol.,* 3:57-70.
6. Erickson, G. F., and Ryan, K. J. (1975): The effect of LH/FSH, dibutyryl cyclic AMP, and prostaglandins on the production of estrogens by rabbit granulosa cells *in vitro*. *Endocrinology,* 97:108-113.
7. Fortune, J. E., and Armstrong, D. T. (1978): Androgen production by isolated components of rat ovarian follicles. *This volume.*
8. Goldenberg, R. L., Vaitukaitis, J. L., and Ross, G. T. (1972): Estrogen and follicle stimulating hormone interactions on follicle growth in rats. *Endocrinology,* 90:1492-1498.
9. Kolena, J., and Channing, C. P. (1972): Stimulatory effects of LH, FSH, and prostaglandins upon cyclic 3',5'-AMP levels in porcine granulosa cells. *Endocrinology,* 90:1543-1550.
10. Lindner, H. R., Tsafriri, A., Lieberman, M. E., Zor, U., Koch, Y., Bauminger, S., and Barnea, A. (1974): Gonadotropin action on cultured graafian follicles: Induction of maturation division of mammalian oocyte and differentiation of the luteal cell. *Rec. Progr. Hormone Res.,* 30:79-138.
11. Louvet, J.-P., Harman, S. M., Schreiber, J. R., and Ross, G. T. (1975): Evidence for a role of androgens in follicular maturation. *Endocrinology,* 97:366-372.
12. Lowry, O. H., Rosebrough, N. J., Farr, A. L., and Randall, R. J. (1951): Protein measurement with the Folin phenol reagent. *J. Biol. Chem.,* 193:265-275.
13. Marsh, J. M. (1976): The role of cyclic AMP in gonadal steroidogenesis. *Biol. Reprod.,* 14:30-53.

14. Orczyk, G. P., Hichens, M., North, G., and Behrman, H. R. (1974): Progesterone. In: *Methods in Hormone Radioimmunoassay* (eds. B. M. Jaffe, and H. R. Behrman). Academic Press, New York, p. 347-358.
15. Schomberg, D. W., Stouffer, R. L., and Tyrey, L. (1976): Modulation of progestin secretion in ovarian cells by 17β-hydroxy-5α-androstane-3-one (dihydrotestosterone): A direct demonstration in monolayer culture. *Biochem. Biophys. Res. Commun.*, 68:77-81.
16. Schreiber, J. R., Reid, R., and Ross, G. T. (1976): A receptor-like testosterone-binding protein in ovaries from estrogen-stimulated hypophysectomized immature female rats. *Endocrinology*, 98:1206-1213.
17. Smith, M. S., Freeman, M. E., and Neill, J. D. (1975): The control of progesterone secretion during the estrous cycle and early pseudopregnancy in the rat: Prolactin, gonadotropin and steroid levels associated with rescue of the corpus luteum of pseudopregnancy. *Endocrinology*, 96:219-226.
18. Takayama, M., and Greenwald, G. S. (1973): Direct luteotropic action of estrogen in the hypophysectomized-hysterectomized rat. *Endocrinology*, 92:1405-1413.
19. Wilson, J. D. (1972): Recent studies on the mechanism of action of testosterone. *New Eng. J. Med.*, 287:1284-1291.

Ovarian Follicular Development and Function,
edited by A. R. Midgley and W. A. Sadler.
Raven Press, New York © 1979.

Androgen/FSH Synergism in Stimulation of Progestin Synthesis in Immature Granulosa Cells In Vitro

A. Nimrod and H. R. Lindner

Department of Hormone Research, Weizmann Institute of Science, Rehovot, Israel

Granulosa cells from immature rat follicles possess FSH receptors (1-3) and respond to FSH, but not to LH, in culture with increased progesterone synthesis (4; Erickson, G. F., Nimrod, A., and Ryan, M., unpublished observations). These cells are also endowed with an androgen receptor (5), the function of which is not quite clear. Androgen has been reported to counteract the mitogenic action of FSH on granulosa cells of immature follicles (6).

We examined the interaction of androgen and FSH in primary cultures of granulosa cells from two sources: (i) diethylstilbestrol-treated (3) hypophysectomized immature rats (Hx-DES) and (ii) preovulatory follicles (PO cells). Progesterone (P) and 20α-dihydroprogesterone (20α-OH-P) were determined separately in the medium by specific radioimmunoassay (7). Ten to 15 times more 20α-OH-P than P was produced by both cultures (8). Only the combined output of P + 20α-OH-P will be shown here Table 1); more detailed data are being published elsewhere (8).

Hx-DES cells produced lesser amounts of progestins than PO cells, and even after stimulation with FSH (2-fold) the rate of secretion per cell was below the basal level in untreated PO cells. Hx-DES cells responded, however, to dibutyryl cyclic AMP (1.0 mM) with a 10 to 20-fold increase in progestin secretion, indicating that these cells had considerable steroidogenic capability. Combined treatment of Hx-DES cells with androgen (androstenedione, testosterone or dihydrotestosterone; 1.0 μg/ml) and FSH (NIAMD-I-3, courtesy of Dr. A. F. Parlow; 0.1 μg/ml) increased progestin secretion 5 to 10-fold over the level attained upon FSH-treatment alone during the first 48 h of culture (Table 1); in PO cells this synergism was present, but

TABLE 1. *Effect of various androgens and oestrone on progestin accumulation in the medium of Hx-DES cultures, and their synergistic action with FSH. Progesterone (P) and 20α-dihydroprogesterone (20α-OH-P) concentrations in the medium after the first 2 days of culture were assayed in duplicate by radioimmunoassay, using three culture dishes per treatment.*

Steroid added (1 μg/ml)	Progestin (P + 20α-OH-P) accumulation (10^{-15} g/cell; mean ± SEM)	
	Without FSH	FSH (0.1 μg/ml)
None (control)	31 ± 1	53 ± 7
Androstenedione	80 ± 5***	303 ± 10***
5α-Dihydrotestosterone	69 ± 9*	254 ± 42**
Testosterone	124 ± 22**	450 ± 72**
Oestrone	33 ± 2	51 ± 6

Significantly different from corresponding control (head of column) by Student's "t"-test at the level of: $^*p < 0.05$, $^{**}p < 0.01$, $^{***}p < 0.001$.

less pronounced (8). Androstenedione on its own (1.0 μg/ml) had only a small stimulatory effect (1.5 to 3-fold) in Hx-DES cultures. Oestrone and oestradiol-17β (1.0 μg/ml) were without effect. When the culture period was extended beyond 48 h, Hx-DES cells ceased to produce progestins in basal medium and failed to respond to FSH alone. However, such cells maintained their progestin production and responded to FSH with a 10 to 40-fold increase in progestin production during the 48-96 h culture period, provided androgen (androstenedione, 1.0 μg/ml) was present throughout (0-96 h) in the culture medium.

It was concluded that the synergistic action of androgen is not due to its aromatisation, since 5α-DHT (not a substrate of aromatase) was effective while oestrogens were not. Nor could the effect be attributed to a progestin-sparing action through saturation of 5α-reductase, since 5α-DHT is not a substrate of this enzyme, yet is active, and androstenedione potentiated the action of FSH even at very low concentrations (30 ng/ml).

The immature granulosa cells examined are endowed with both FSH-receptors and steroidogenic potential as evident from their response to cyclic AMP. Perhaps maturation of the mechanism responsible for coupling the FSH receptor to adenylate cyclase requires theca-derived androgen.

Supported by the Ford Foundation and the Population Council, New York. H.R.L. was the recipient of a travel grant from Serono Laboratories Ltd.

REFERENCES

1. Midgley, A. R., Jr. (1973): *Advances in Exper. Medicine and Biology, 36*, 365.
2. Presl, J., Prospisil, J., Figarova, V., and Krabec, Z. (1974): *Endocrinologica experimentalis, 8*, 2.
3. Nimrod, A., Erickson, G. F., and Ryan, M. J. (1976): *Endocrinol., 98*, 56.
4. Channing, C. P. (1974): *Endocrinol., 94*, 1215.
5. Schreiber, J. R., and Ross, G. T. (1975): Program of the 57th Meeting, U. S. Endocrine Society, Abstract 120, p. 110.
6. Louvet, J. P., Harman, M., Schreiber, J. R., and Ross, G. T. (1975): *Endocrinol., 97*, 366.
7. Bauminger, S., Kohen, F., and Lindner, H. R. (1974): *J. Steroid Biochem., 5*, 739.
8. Nimrod, A., and Lindner, H. R. (1976): *Molecular and Cellular Endocrinology, 5*, 315.

Ovarian Follicular Development and Function,
edited by A. R. Midgley and W. A. Sadler.
Raven Press, New York © 1979.

Influence of Serum, Estrogen, and Gonadotropins on Growth and Progesterone Secretion by Cultures of Granulosa Cells from Small Porcine Follicles

K. H. Thanki and C. P. Channing

Department of Physiology, School of Medicine,
University of Maryland at Baltimore,
Baltimore, Maryland 21201

A number of differences between granulosa cells obtained from small (1- to 2-mm) and large (6- to 12-mm) porcine follicles have been observed. Compared to small follicle granulosa cells, those from large follicles bind more HCG/LH (1,2) and accumulate more cAMP and secrete more progesterone in response to LH or HCG (3,4). In culture, granulosa cells from large follicles undergo spontaneous luteinization as defined in terms of morphology and progesterone secretion (5,6). The present study was undertaken to determine whether these maturational changes could be induced in granulosa cells obtained from small follicles by hormone treatment in culture.

MATERIALS AND METHODS

Highly purified pituitary gonadotropins used in the study were the gifts of NIAMDD. Human follicle stimulating hormone (HFSH; LER 1366) had 3473 IU FSH and 202 IU LH per milligram. Human luteinizing hormone (HLH; LER 960) had 4620 IU LH and 1.07 IU FSH per milligram. Estradiol 17β (E_2) and progesterone were purchased from Steraloids. Other commercially obtained hormones were cortisol (Sigma), thyroxin (Nutritional Biochemical Corp.), and insulin (Calbiochem). Tissue culture media were purchased from Grand Island Biological Co. (GIBCo). Tissue culture medium 199 with Earles salts and 25 mm Hepes buffer was used as the basic medium. Steroids were dissolved in ethyl alcohol: the final concentration of alcohol in the cultures was 1 μl/ml or less.

Porcine serum was purchased from GIBCo and contained 20 pg/ml estrogen and 1 ng/ml progesterone as measured by radioimmunoassay described in detail elsewhere (7). The porcine serum contained <1 ng/ml LH in terms of NIH-LH-S1 as measured by a radioreceptor assay.

Granulosa Cell Culture Procedure

Pig ovaries were collected from a local slaughterhouse and the granulosa cells from follicles measuring less than 2 mm were collected by the method of Channing (4,7). The time interval between the collection of the ovaries and the initiation of the cultures was between 2 and 3 hr. The cultures were grown in Falcon Multiwells (Cat. No. 3008) with 1 ml medium per well. The cells were routinely inoculated at a density of 4×10^5 viable cells per well, and the cultures were incubated at 37°C in 5% CO_2 in air. The cultures were normally started in medium 199 + 0.4% bovine serum albumin (GIBCo) + 0.2% lactalbumin hydrolysate (199D) and 4% porcine serum, unless indicated otherwise. The cultures were allowed to become established in the absence of exogenous hormones for the first 2 days of the culture period. The hormones were added every other day to the culture medium starting at day 2. Culture media were changed every 2 days and frozen at -20°C until assayed for progesterone. Each experiment was repeated at least three times on separate batches of granulosa cells. Progesterone radioimmunoassay was carried out as described previously (7) using an antiserum provided by Dr. Gordon Niswender. Viable cell counts were taken with trypan blue before the cultures were started, and the number of cells growing in culture were determined by using the crystal violet method of Paul (8).

RESULTS

Effect of Serum on Cell Growth

Granulosa cells in cultures initiated in the absence of serum attached poorly to glass (Fig. 1). The plating efficiency on day 2 improved with the addition of serum. The number of cells in culture with 2% serum or less decreased over the 8-day period. Ten percent serum caused a doubling of cell number between days 2 and 4. Minimum growth or deterioration of cells was observed in 4% serum and therefore this medium was used

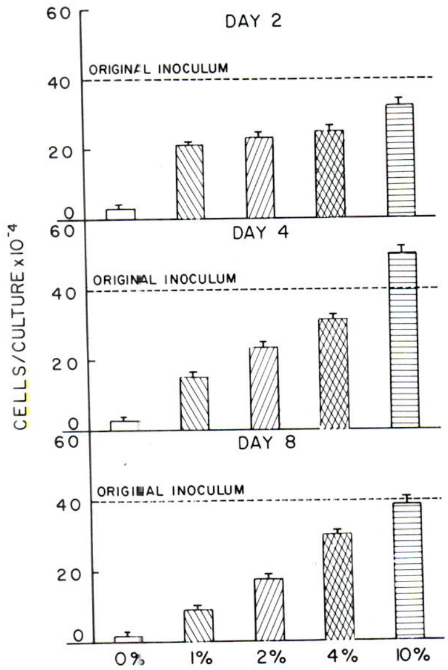

FIG. 1. Growth of granulosa cells obtained from small follicles (< 2 mm) and cultured in medium 199 + 0.2% lactalbumin hydrolysate + 0.4% BSA and various concentrations of *serum*. The cultures were sacrificed on days 2, 4, and 8 and the cell nuclei were counted. The values shown are the means ± SEM counts of replicate cultures ($n = 6$). The viable cells in the initial inoculum were $40 \pm 1 \times 10^4$ cells per culture.

for investigating hormonal control of granulosa cell maturation.

Effect of Hormones on Cell Growth, Morphology, and Progesterone Secretion

Hormones were first added to the culture on day 2, and their effects were examined on days 4, 6, and 8. Estradiol 17β (1 μg/ml) and FSH (0.1 μg/ml) stimulated cell growth (Fig. 2), increasing the cell number 40% and 30%, respectively. LH (0.1 μg/ml) was without any effect on cell number.

Estradiol (1 μg/ml) consistently inhibited the secretion of progesterone ($p < 0.01$, Fig. 3) by granulosa cells in culture. FSH and LH, on the other hand, stimulated progesterone secretion between days 4 and 6 and days 6 and 8 (Fig. 4). In the presence of cortisol (2.9×10^{-8} M) plus thyroxin (10^{-7} M) and insulin

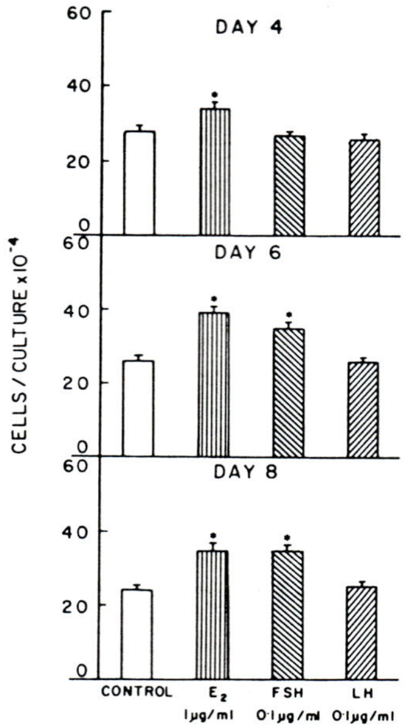

FIG. 2. Effect of estradiol 17β (E_2), HFSH, and HLH on the growth of granulosa cells obtained from small follicles and cultured in medium 199D + 4% serum. The cell number in control cultures remained constant over the 8-day period, while in the presence E_2 and FSH cell numbers increased. The values shown are (means ± SEM) representative of five such experiments (n = 6). The asterisk above the bar indicates $p < 0.05$ compared to control.

(1.7×10^{-7} M) FSH stimulated progesterone secretion between days 2 and 4.

Granulosa cells in the presence of FSH and LH acquired extensive granular cytoplasm containing a large number of lipid droplets (Fig. 5). The presence of lipid droplets was demonstrated by phase contrast microscopy and oil Red O staining. Granulosa cells exposed to estradiol or progesterone did not undergo such changes in morphology.

DISCUSSION

Serum is known to contain survival and growth factors (9-12). These factors may be responsible for the profound effects of serum on the growth of granulosa cells in culture. The cultures with 4% serum maintained a steady cell number over the 8-day period and was therefore used in subsequent studies.

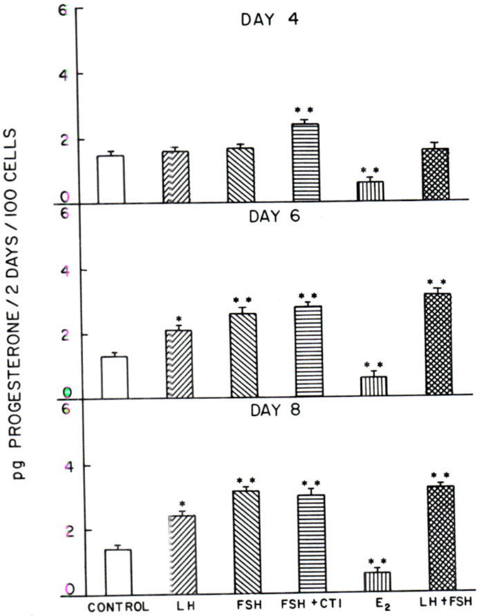

FIG. 3. Effect of various hormones on secretion of progesterone by granulosa cells obtained from small follicles. The cultures were initiated in medium 199D + 4% serum and the hormones were first added on day 2. The media were changed every other day and assayed for progesterone. FSH and LH increased while E_2 decreased progesterone secretion compared to control culture. The values shown are means ± SEM (n = 6). The single asterisk above the bar indicates $p < 0.05$ and the double asterisk indicates $p < 0.01$ compared to the control.

Estrogen implants in hypophysectomized immature female rats have been reported to cause hyperplasia of granulosa cells in preantral follicles (13,14). McNatty and Sawers (15) have observed stimulation of mitoses in human granulosa cell cultures by estrogen in the presence of FSH. We report for the first time that estrogen alone can stimulate granulosa cell growth in culture. This estrogen effect may not be observed in cultures with 10% serum, since cells in such a system may already be stimulated "maximally." Since the dose of estrogen (1 μg/ml) used is equal to the concentrations reached in the follicular fluid (15,16), our observation may have physiological significance.

FSH and LH induce morphological changes in the cytoplasm of granulosa cells associated with luteinization. This effect of gonadotropin does not seem to be

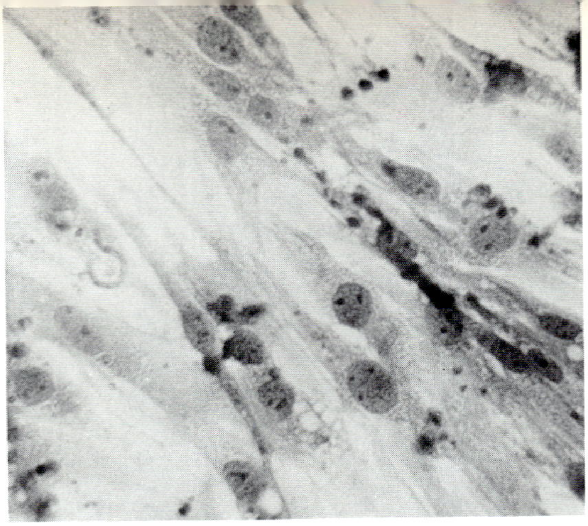

FIG. 4. Eight-day culture of granulosa cells (obtained from small follicles) grown in medium 199D + 4% serum. Note the spindle-shape morphology. Hematoxylin and eosin. ×460

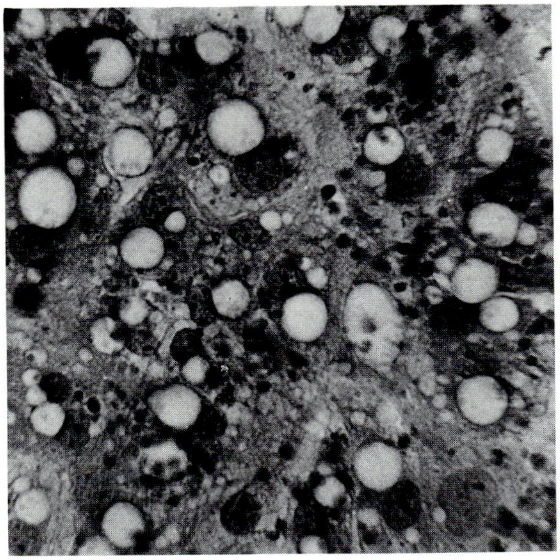

FIG. 5. Eight-day culture of granulosa cells (obtained from small follicles) grown in medium 199D + 4% serum + 0.1 μg/ml HFSH. Note the epithelial morphology. The light circles within the cytoplasm represent oil droplets (as indicated by oil Red O staining and phase contrast microscopy). Hematoxylin and eosin. ×460

mediated by estrogen or progesterone.

Estrogen has been shown to stimulate progesterone secretion in granulosa cells from preovulatory follicles (17) and in slices of luteinized ovaries (18). We report an inhibitory effect of estradiol (1 μg/ml) on progesterone secretion by granulosa cell cultures obtained from small (1- to 2-mm) follicles. Recently it has been reported (19) that estrogen inhibits progesterone secretion by granulosa cell cultures initiated from medium-sized follicles (3- to 5-mm). These studies suggest that just prior to ovulation granulosa cells undergo changes that modify their response to estrogen. It is possible that estrogen, at least in part, inhibits granulosa cells from secreting progesterone prematurely and allows them to divide until adequate numbers are present to form a corpus luteum.

ACKNOWLEDGMENTS

This research was supported by grants HD-08834 and HD-00435 from the National Institute of Child Health and Human Development.

REFERENCES

1. Channing, C. P., and Kammerman, S. (1973): Characteristics of Gonadotropin Receptors of Porcine Granulosa Cells during Follicle Maturation. *Endocrinology,* 92:531-540.
2. Channing, C. P., and Kammerman, S. (1974): Binding to Gonadotropins to Ovarian Cells. *Biol. Reproduction,* 10:179-198.
3. Channing, C. P. (1970): Effects of Stage of the Menstrual Cycle and Gonadotrophins on Luteinization of Rhesus Monkey Granulosa Cells in Culture. *Endocrinology*, 87:49-60.
4. Channing, C. P. (1970): Effect of Stage of Estrous Cycle and Gonadotrophins upon Luteinization of Porcine Granulosa Cells in Culture. *Endocrinology*, 87:156-164.
5. Channing, C. P., and Tsafriri, A. Regulation of Ovulatory Processes: Ovum Maturation, Follicular Rupture and Luteinization. In: *Advances in Fertility Regulation through Basic Research*, edited by W. A. Sadler and S. Segal. Plenum Press, N. Y. *(in press)*.
6. Kolena, J., and Channing, C. P. (1972): Stimulatory Effects of LH, FSH and Prostaglandins upon Cyclic 3,5-AMP Levels in Porcine Granulosa Cells. *Endocrinology*, 90:1543-1550.
7. Channing, C. P., and Ledwitz-Rigby, F. (1975): Methods for Assessing Hormone Mediated Differentiation of Ovarian Cells in Culture and in Short Term Incubations. *Methods in Enzymology*, Vol. 39, Part D, edited by J. G. Hardman and B. W. O'Malley, pp. 183-230. Academic Press, New York.
8. Paul, J. (1965): In: *Cell and Tissue Culture*, p. 327. E & S Livingstone Ltd., Edinburgh and London.
9. Pollock, R. E., Green, H., and Todaro, G. (1968): Growth Control in Cultured Cells: Selection of Sublines with Increased Sensitivity to Contact Inhibition and Decreased Tumor Producing Ability. *J. Proc. Nat. Acad. Sci. U.S.A.*, 122-133.

10. Temin, H. M. (1967): Studies on Carcinogenesis by Avian Sarcoma Viruses. III. The Differential Effect of Serum and Polyanions on Multiplication of Uninfected and Converted Cells. *J. Nat. Cancer Inst.*, 37:167-175.
11. Nishikawa, K., Armelin, H. A., Sato, G. (1975): Control of Ovarian Cell Growth in Culture by Serum and Pituitary Factors. *Proc. Nat. Acad. Sci.*, pp. 483-487.
12. Holley, R. W., and Kiernan, J. A. (1971): Growth Control of Cell Cultures, edited by G. E. Wolstenholme and J. Knight (CIBA FOUNDATION Symposium). Churchill Livingstone, Edinburgh and London.
13. Louvet, J., Harman, S. M., Schreiber, J. R., and Ross, G. T. (1975): Evidence for a Role of Androgens in Follicular Maturation. *Endocrinology*, 97:366-372.
14. Bradbury, J. (1961): Direct Action of Estrogen on the Ovary of the Immature Rat. *Endocrinology*, 68:115-120.
15. McNatty, K. P., and Sawers, K. P. (1975): Relationship between the Endocrine Environment Within the Graafian Follicle and the Subsequent Rate of Progestrone Secretion by Human Granulosa Cells in Vitro. *J. Endocrinology*, 66:391-400.
16. Edwards, R. J. (1974): Follicular Fluid, *J. Reprod. Fert.*, 37:189-219.
17. Goldenberg, R. L., Bridson, W. E., and Kohler, P. O. (1972): Estrogen Stimulation of Progesterone Synthesis by Porcine Granulosa Cells in Culture. *Biochem. Biophys. Res. Comm.*, 48:101-107.
18. Lee, C. Y., and Ryan, R. J. (1975): Estrogen Stimulation of Human Chorionic Gonadotrophin Binding by Luteinized Rat Ovarian Slices. *Endocrinology*, 95:1691-1693.
19. Schomberg, D. W., Stouffer, R. L., and Tyrey, L. (1976): Modulation of Progestin Secretion in Ovarian Cells by 17β-Hydroxy-5α-Androstan-3-one (Dihydro-testosterone). A Direct Demonstration in Monolayer Culture. *Biochem. Biophys. Res. Comm.*, 68:77-81.

Ovarian Follicular Development and Function,
edited by A. R. Midgley and W. A. Sadler.
Raven Press, New York © 1979.

Discussion Summary: Session on Follicular Steroidogenesis

D. T. Armstrong

The discussion following this series of papers centered around the question of whether the granulosa cells were the exclusive follicular site of aromatization, or whether this might be a property of theca cells and interstitial cells under some circumstances in some species. Species in which granulosa cells have definitively been shown to carry out aromatization *in vitro* include rat, rabbit, hamster, and mare. Thecal preparations from mare and human follicles have been shown to convert radioactive androgen to estrogen, *in vitro*, but this type of study does not permit conclusions to be drawn about quantitative significance.

A discussion of a recent publication of Channing and Coudert (*Endocrinology*, 98:590,1976) concerning the source of follicular estrogen in the rhesus monkey ensued. On the basis of *in vivo* experiments in which follicular fluid and granulosa cells had been aspirated from the single larger preovulatory follicle, with no significant reduction in estrogen secretion becoming apparent, these authors had concluded that the theca cells, rather than the granulosa cells, were the major source of the estrogen secreted during the preovulatory period in this species. The possibility was raised that granulosa cells from follicles other than the one destined to ovulate may have been contributing to the estrogen secreted during this period. The controls of Channing and Coudert, in which the entire follicle was removed, do not seem to exclude this possibility completely, both because of the considerably greater trauma to surrounding tissue associated with removal of the entire follicle, and because removal of this follicle may remove a major source of androgens (the theca cells), which might otherwise continue to be aromatized by granulosa cells of smaller follicles.

It was emphasized that the steroidogenic capabilities, as well as the gonadotropin responsiveness of granulosa cells vary with the state of development or differentiation of the follicles. The ability of granulosa cells from immature, and from hypophysectomized, estrogen-treated rats to respond to FSH but not to LH is undoubtedly due to lack of LH receptors at these stages, since it was shown that both LH and FSH were effective in stimulating estrogen production in granulosa cells from rats which had been pretreated with an FSH preparation (PMSG).

Factors involved in regulating progesterone production by granulosa cells were discussed at some length. Patterns of progesterone production during culture of granulosa cells from large follicles of sheep suggested considerable heterogeneity of the cell population, both concerning their responsiveness to gonadotropins, and the time required to reach maximal progesterone secretion. In studies with porcine granulosa cells, evidence was presented that estrogen inhibits progesterone secretion, while apparently stimulating cell division. It was suggested that specialized cell function (progesterone secretion) may not be compatible with active cell proliferation.

The source of the cholesterol from which granulosa cells produce progesterone received some consideration. Evidence was cited that follicular fluid contains a high-density lipoprotein that binds to granulosa cells, and the possibility was speculated upon that this may represent a mechanism for transport of cholesterol (and perhaps other steroid substrates?) into the granulosa cells.

Evidence was presented from two other laboratories (Lindner, Ross), similar in nature to that reported by Dorrington and Armstrong, indicating that androgens are capable of augmenting the production of progesterone by cultured rat granulosa cells, in response to gonadotropic stimulation. The mechanism of this interesting synergism was the subject of some speculation. The possibility was suggested that androgens play a role in inducing the FSH-responsive mechanisms (e.g., induction of FSH receptors?) in the immature rat. Another possibility, that androgens may inhibit further metabolism of progesterone by granulosa cells, rather than actively stimulating biosynthesis, seems to have been excluded. Further research is necessary to elucidate the mechanism of this action of androgens, as well as its physiological significance.

There was some discussion of the mechanism of the aromatization reaction and its regulation. It was pointed out that this several-step reaction occurs in a concerted fashion with the substrate probably remaining bound to the enzyme throughout the entire reaction; stimulatory agents such as FSH probably affect the overall reaction, rather than any one specific step in the sequence.

FOLLICULAR HORMONE RECEPTORS

Ovarian Follicular Development and Function,
edited by A. R. Midgley and W. A. Sadler.
Raven Press, New York © 1979.

Hormone Regulation of Hormone Receptors in Ovarian Follicular Development

JoAnne S. Richards

Reproductive Endocrinology Program, Department of Pathology, University of Michigan, Ann Arbor, Michigan 48109

The observation that exogenous estrogen increased the responsiveness of rat ovarian follicles to gonadotropins (Pencharz, 1940; Williams, 1945) has suggested that follicular production of estrogen and/or the response of follicular cells to estrogen may affect the rate of follicular growth and the differentiation of follicular cells (Richards and Midgley, 1976). In the following discussion, recent evidence is described indicating that granulosa cells are an estrogen target tissue possessing estrogen-specific binding components (hereafter referred to as receptors) and that the actions of estrogen may be required for follicle-stimulating hormone (FSH) to initiate the final stages of follicular differentiation including antrum formation, increased content of receptor for FSH, and the marked appearance of receptor for LH in granulosa cells.

Figure 1, which illustrates a presumptive pattern of follicular growth during four estrous cycles in the rat, has been derived from a kinetic analysis of follicular development in the mouse (Pederson, 1970) and by a morphometric analysis of follicular growth in the rat (Hirshfield, 1976). The relative stages of growth (represented on the vertical axis according to the classification of Pederson and Peters, 1968) include small follicles (Type 2) just entering the pool of committed, growing follicles (distinguished from the large pool of quiescent, nongrowing follicles); large preantral follicles (Types 5 and 6); and large antral, preovulatory follicles (Types 7 and 8). As designated by the solid lines, some follicles, apparently by a random, as yet unknown initiating event, begin to grow each day and continue to grow until atresia or ovulation occurs. Thus, those follicles that do ovulate on a given cycle, such as those represented by the heavy solid line, actually began to grow some 19 days earlier as part of a larger pool of growing follicles, represented by the stippled area. Most of these follicles continue to grow until, as large preantral follicles, they undergo atresia, as indicated by the cross-hatched area. Only a few follicles enter the final stages of follicular growth (Types 7 and 8). The selection of these follicles from the pool of large preantral follicles may be associated with the action of FSH, most specifically the surge of FSH in the cycle preceding ovulation, and with the ability of these follicles to produce and respond to estrogen such that they develop mechanisms by which to ovulate. From the diagram one would predict that a large number of follicles would arrive at the large preantral stage each day. Those reaching this stage slightly out-of-phase with the gonadotropic hormone surge may have developed mechanisms to respond

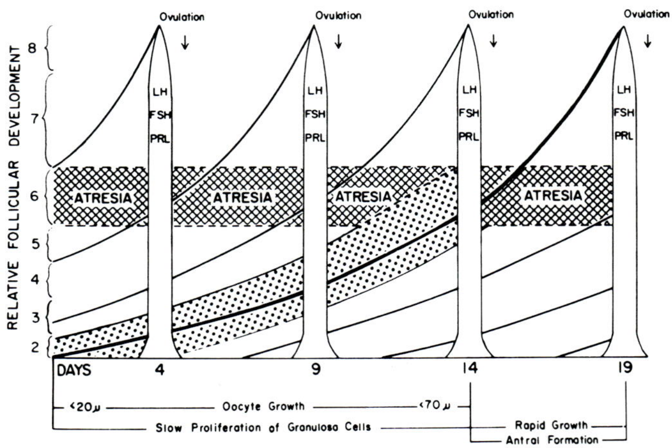

FIG. 1. Presumptive pattern of follicular growth during four estrous cycles in the rat. This diagram presupposes that a given number of follicles (*stippled area*) begins to grow each day and continues to grow until either atresia (*cross-hatched area*) or ovulation (*solid line*) occurs. On any given day, such as estrus, many stages of development are present. Note, however, that no large antral follicles are observed. In contrast, on proestrus one would predict from this diagram that large antral follicles but no small antral follicles would be present. Recent observations in the cycling rat by Hirshfield (1976) support this pattern of final growth and have indicated further that the FSH of one cycle stimulates the growth of those follicles destined to ovulate at the next cycle. (From Richards and Midgley, 1976.)

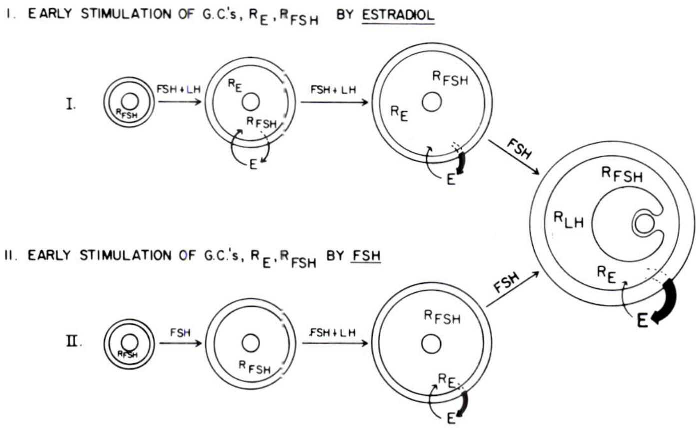

FIG. 2. Hormonal control of follicular development: two hypotheses. The first hypothesis suggests that estradiol (designated by *solid arrow*), synthesized in response to FSH and LH (acting on granulosa and theca cells, respectively), stimulates an increase in receptors for FSH (R_{FSH}) and estradiol (R_E). As more estradiol is produced and follicles gain more R_{FHS}, the responsiveness of granulosa cells to FSH is increased and final differentiation occurs. The second hypothesis indicates that FSH alone may determine the early growth of the follicle until at some point the follicle gains the ability to produce estradiol (by differentiation of theca and granulosa cells such that LH and FSH now synergistically support estradiol synthesis), and in the presence of estradiol, FSH can now stimulate the final growth phase including appearance of antra and receptor for LH (R_{LH}). (From Richards and Midgley, 1976.)

to a stimulatory surge of FSH and, therefore, degenerate if the surge fails to occur within a critical time. Thus, during the estrus cycle, the surges of FSH, LH, and PRL are superimposed on a continuum of committed, growing follicles all at different stages of development. The gonadotropic hormone surges appear, therefore, to have two primary functions: FSH dictates the entry of a select number of follicles into the final stage of growth while LH terminates their growth and initiates ovulation.

On the basis of these considerations it would appear that follicles are unequally responsive to gonadotropins and that the stage of follicular growth and the differentiation of the follicular cells (both granulosa and theca) determine which follicles will respond to the surges of gonadotropins and enter the final phase of follicular growth. Since granulosa cells of most follicles appear to possess receptor for FSH (Eshkol and Lunenfeld, 1972; Richards et al., 1976), while only large, preovulatory follicles appear to possess receptor for LH (Zeleznik et al., 1974; Richards et al., 1976), it would seem reasonable to postulate that changes in receptor content might determine which follicles are capable of responding to specific hormones. Furthermore, since estrogen changes the responsiveness of follicles to gonadotropins, it would seem reasonable to propose that estrogen might act to alter the content of receptor for FSH and/or LH in differentiating granulosa cells.

For example, an increase in the responsiveness of follicles to FSH might be associated with an increase in the number of FSH receptors per granulosa cell. Thus, FSH acting on granulosa cells and LH acting on theca cells of small follicles might be expected to increase the synthesis of estrogen. Estrogen, in turn, might cause granulosa cell proliferation and an increase in granulosa cell receptor for FSH (Fig. 2). As follicles grow they would be expected to produce more estrogen, gain more FSH receptor in response to the estrogen, until, as large preantral follicles, they become acutely responsive to FSH, and enter the final growth phase.

An alternative hypothesis (Fig. 2), based primarily on observations by Eshkol and Lunenfeld (1972), suggests that FSH alone may stimulate early follicular development. At some critical time during development, follicles would gain the ability to produce estrogen. The acute appearance of estrogen synthesis in some follicles, but not all, would then rapidly alter the responsiveness of these estrogen-producing follicles to FSH.

A third hypothesis (Fig. 3) proposes that a single hormone, estrogen or FSH, or a continuation of hormones, might act to stimulate different responses in follicles as they go through successive stages of differentiation. Thus, whether or not follicles begin to produce estrogen at some particular stage need not be critical. Rather, the responses of the granulosa cells to estrogen and/or FSH may change over time. Granulosa cells may go through successive stages of cell division, each followed by or associated with a period of cellular differentiation such that each leads to an altered pattern of differentiation. These have been arbitrarily designated Stages A, B, C, and D and are illustrated in Fig. 3. One is reminded of the sequential stages of development in the chick oviduct and in the rat uterus in response to continued estrogen treatment (O'Malley et al., 1972; Schimke et al., 1973). In the immature rat, the appearance of the uterine estrogen receptor precedes the ability of

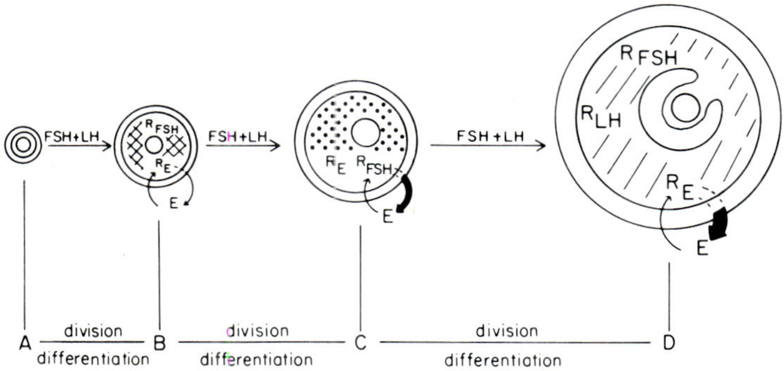

FIG. 3. Progressive differentiation of the follicles. This diagram illustrates the hypothesis that estradiol (and/or FSH) may affect granulosa cell function at many stages of development such that at each stage the response of the granulosa cell to estradiol (and/or FSH) changes. The stages of cell division and differentiation have arbitrarily been designated A, B, C, and D. (From Richards and Midgley, 1976.)

estrogen to stimulate induced protein (IP) synthesis, which, in turn, precedes increased protein, RNA, and DNA synthesis (Somjen et al., 1973; Katzenellenbogen and Greger, 1974). Thus, the response of the granulosa cell to estrogen and/or FSH may change during follicular growth.

The model system selected to study the interactions of estrogen and FSH on follicular morphology and the differentiation of granulosa cell function has been the immature, hypophysectomized rat. The effects of estradiol-17β and HFSH (LER-1577) on ovarian morphology and on the intraovarian distribution of binding sites for ^{125}I-HCG are illustrated in Fig. 4.

As shown in panel A, ovaries of intact immature rats (day 24 of age) contained many large preantral and several antral follicles. Most of the large antral follicles exhibited signs of atresia as indicated by the presence of pynotic granulosa cells. The intraovarian distribution of radioiodinated HFSH and HCG 90 min following an intravenous (tail vein) injection of 1 μCi of labeled hormone was examined by autoradiographic analysis. As illustrated, binding sites for FSH (panel A-1) were present in granulosa cells of many follicles, while binding sites for HCG (panel A-2) were preferentially localized in interstitial and theca cells. Six days following hypophysectomy (surgery was performed on day 24 of age by Hormone Assay Laboratories) ovaries contained many small and

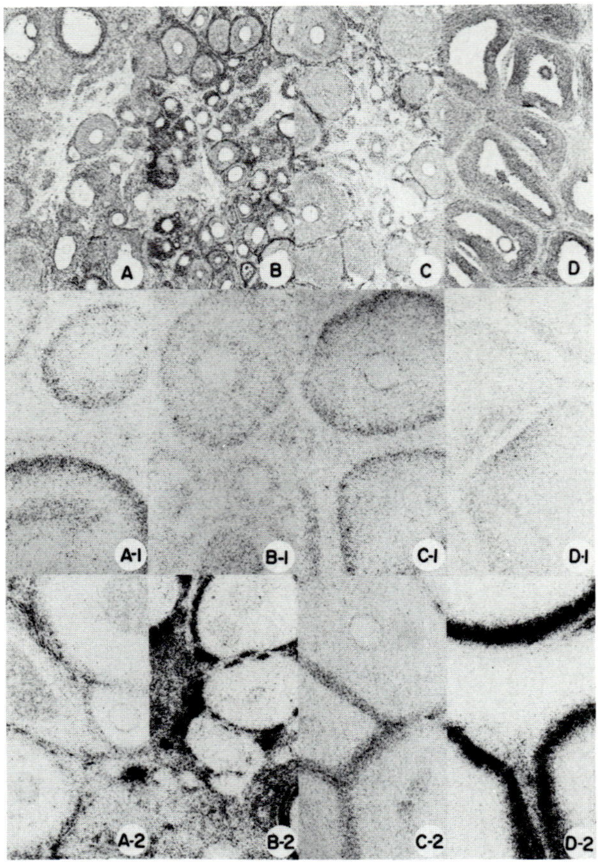

FIG. 4. Ovarian morphology and intraovarian localization of receptor sites for FSH and LH (HCG) during hormonally induced follicular development in the hypophysectomized rat. Panels A to D illustrate the morphological appearance of ovaries of A, intact rat; B, hypophysectomized rat; C, hypophysectomized rat treated with estradiol (2 mg/day × 4); D, hypophysectomized, estradiol-primed rat treated with HFSH (2 μg/day × 2). Panels (A-1 to D-1) represent intraovarian uptake of ^{125}I-HFSH and panels (A-2 to D-2) represent intraovarian uptake of ^{125}I-HCG 90 min after an intravenous injection of labeled hormone into rats treated as described in panels A to D. Note that FSH is localized in the granulosa cell layer of follicles, while HCG (LH) is distributed in the theca and interstitial layers. Only when rats are primed with estra and FSH does receptor for LH appear in the granulosa layer.

a few large preantral follicles with little apparent change in the distribution of binding sites for HFSH and HCG (panels B, B-1, and B-2). Treatment of hypophysectomized rats with estradiol (E; 2 mg/day) for

4 days (HE_4; the subscript indicates the number of days each hormone was given) resulted in an increase in the number of large preantral follicles (panel C). Estradiol treatment appeared to cause no major changes in the distribution of binding sites for FSH or LH (panels C-1 and C-2). However, when hypophysectomized, estradiol-primed rats were given HFSH at 2 μg per day for 2 days (HE_4F_2), large antral follicles developed (panel D), and the granulosa cells contained within these large preovulatory follicles exhibited a marked and newly acquired ability to bind HCG (panel D-2). If estradiol-FSH-treated hypophysectomized rats (HE_4F_2) were subsequently exposed to an ovulating dose of OLH (LER-1374A; 60 μg), the large antral follicles luteinized. The same dose of OLH appeared to increase follicular atresia in ovaries in which LH receptor was restricted to theca and interstitial cells (Richards, 1975). Thus, the increase in LH receptor in granulosa cells of large antral follicles appeared to be related to the ability of these follicles to luteinize in response to LH. As demonstrated by autoradiographic analyses (Stumpf, 1969) and *in vitro* binding studies (Richards, 1975), receptors specific for estradiol have been demonstrated to be preferentially localized in nuclei of granulosa cells.

These descriptive analyses of the intraovarian distribution of binding sites for FSH, LH, and estradiol have been more fully characterized by a series of quantitative studies. As shown in Fig. 5, hypophysectomy causes the nuclear content of estradiol receptor [as measured by the nuclear exchange assay of Anderson et al. (1972) and Richards (1975)] to decrease in granulosa cells. Estradiol administered alone for 1 to 4 days or estradiol treatment followed by HFSH for 2 days (2 μg/day) caused nuclear content of receptor for estradiol to increase in isolated granulosa cells. In contrast, LH given to estradiol-primed or estradiol-FSH-primed hypophysectomized rats caused the nuclear content of estradiol receptor to decrease. Thus, increases in nuclear estradiol receptor in isolated granulosa cells appeared to be associated with follicular growth, while loss of nuclear estradiol receptor was associated with the termination of follicular growth either by atresia (HE_4L) or luteinization (HE_4F_2L).

To determine if estradiol acted to increase follicular responsiveness to gonadotropins by increasing the number of FSH and/or LH receptors per cell, receptor content for FSH and LH was measured in isolated granulosa cells and residual ovarian tissue (containing interstitial cells, theca cells, and granulosa cells of small follicles not capable of being expressed from

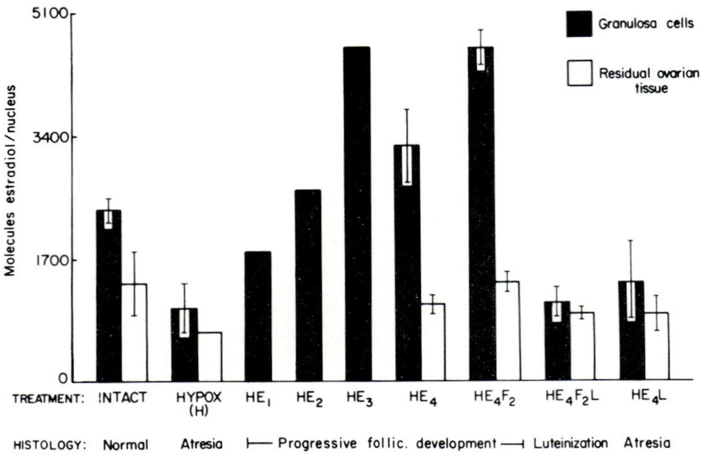

FIG. 5. Hormonal effects on nuclear estradiol receptor. In this and subsequent figures, H refers to hypophysectomy (surgery on day 24 of age; treatment begun 4 to 6 days later); E refers to estradiol; F to HFSH; and L to OLH. The subscripts represent the number of days each hormone was given: E (2 mg/day); HFSH (2 μg/day); OLH (60 μg, single injection).

the ovary by gentle pressure). As shown in Fig. 6 FSH receptor is preferentially localized in granulosa cells. Treatment of hypophysectomized rats for 1 to 4 days with estradiol caused only a slight increase in the content of FSH receptor per granulosa cell (cpm ^{125}I-HFSH specifically bound per microgram DNA of granulosa cells (or residual ovarian tissue). Such a modest increase in FSH receptor would not seem sufficient to account for the marked increase in responsiveness of estradiol-primed granulosa cells to FSH. Thus, our first hypothesis, that estrogen might act to increase the number of FSH receptors per granulosa cells, seems unlikely. Furthermore, the data represented in the lower panel of Fig. 6, indicate that estradiol alone acted to decrease the number of receptors for LH in isolated granulosa cells and in the residual ovarian tissue (data expressed as cpm ^{125}I-HCG specifically bound per microgram DNA of granulosa cells or residual ovarian tissue).

The effects of estradiol given alone were then compared to the effects of HFSH given alone. As shown in Fig. 7, HFSH administered for 1 to 4 days (2 μg/day) caused a progressive and substantial increase in the number of FSH receptors per granulosa cell. These results suggested that FSH, more than estradiol, acted to increase receptor for FSH. In contrast, FSH alone appeared to have little effect on receptor content for

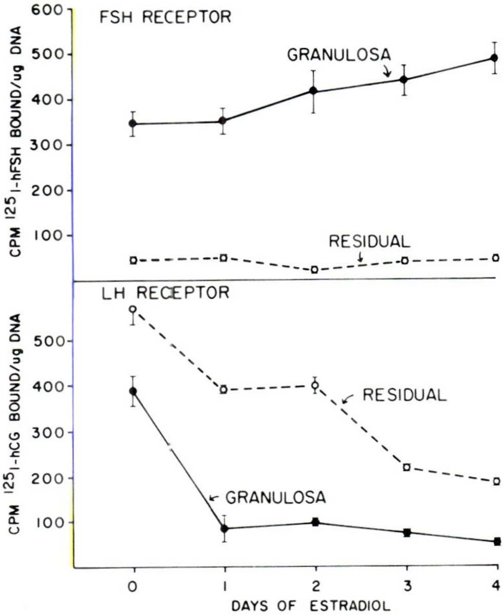

FIG. 6. Effects of estradiol on receptors for FSH and LH in granulosa cells and residual ovarian tissue. Granulosa cells were expressed from ovaries by applying gentle pressure on follicles with a blunt spatula. Cells were collected at 800 × *g*, resuspended in phosphate-buffered isotonic saline (PBS) containing 0.1% gelatin (PBS-gel), and incubated 4 hr at 25° in the presence of ^{125}I-HFSH (LER-1801-3; 90,000 cpm) or ^{125}I-HCG alone with or without an excess of appropriate unlabeled hormone (OFSH (LER-1698) or HCG). Remaining ovarian tissue (theca, interstitial and nonexpressible granulosa cells) was homogenized in PBS, centrifuged at 30,000 × *g* to obtain a crude membrane fraction, resuspended in PBS-gel, and incubated as above. Hormone bound to cells or membranes was measured following removal of free labeled hormone by washing and centrifugation. (From Richards et al., 1976.)

LH in granulosa cells and residual ovarian tissue (bottom panel, Fig. 7). These results using hypophysectomized rats differed markedly from the observation in the *intact*, immature rat in which OFSH alone greatly increased LH receptor in granulosa cells of developing follicles (Zeleznik et al., 1974). However, as was indicated by the autoradiographic studies illustrated in Fig. 4, estradiol pretreatment appeared to facilitate the ability of FSH to increase receptor for LH in granulosa cells of the hypophysectomized rat. Therefore, hypophysectomized rats were treated for 1 to 4 days with estradiol followed by 2 days with HFSH

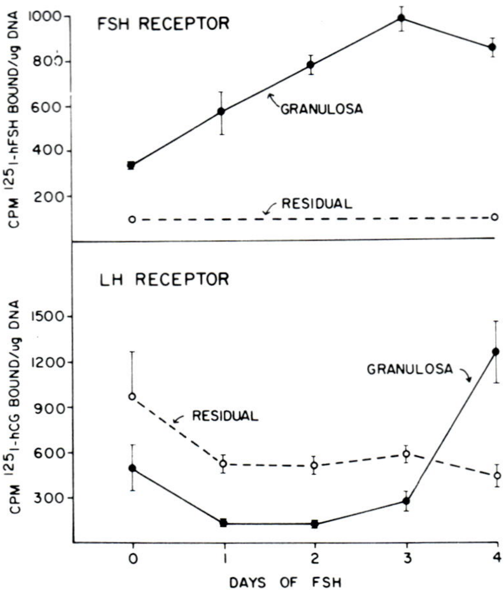

FIG. 7. Effects of HFSH on receptors for FSH and LH in granulosa cells and residual ovarian tissue. HFSH (LER-1577) was given daily at 2 μg. Measurement of receptor content was similar to that described in the legend to Fig. 6. (From Richards et al., 1976.)

(Fig. 8). The ability of HFSH to increase receptor for FSH did not appear to be affected by the duration of estradiol priming. In all groups, with or without estradiol, 2 days of FSH appeared to cause a maximal increase in FSH receptor. In contrast, the ability of FSH to increase receptor for LH was dependent on and directly related to the duration of estradiol priming, a period of at least 3 days being required to achieve a maximal response (Fig. 8). To determine if the rate of increase in receptors for gonadotropins in response to FSH was enhanced by estradiol, hypophysectomized rats were primed for 0 (E_0), 1 (E_1), or 4 (E_4) days with estradiol followed by 1 to 4 days of HFSH (2 μg/day). As shown in Fig. 9, estradiol enhanced the rate at which HFSH stimulated increases in receptor for both FSH and LH, the effect of estradiol being noticeably greater on the ability of FSH to increase receptor for LH. Thus, although 1 day of HFSH stimulated more FSH receptor in granulosa cells primed with estradiol for 4 days than for 1 or 0 days, this facilitory effect of exogenous estradiol was not apparent when rats in all treatment groups had received HFSH for 2 days.

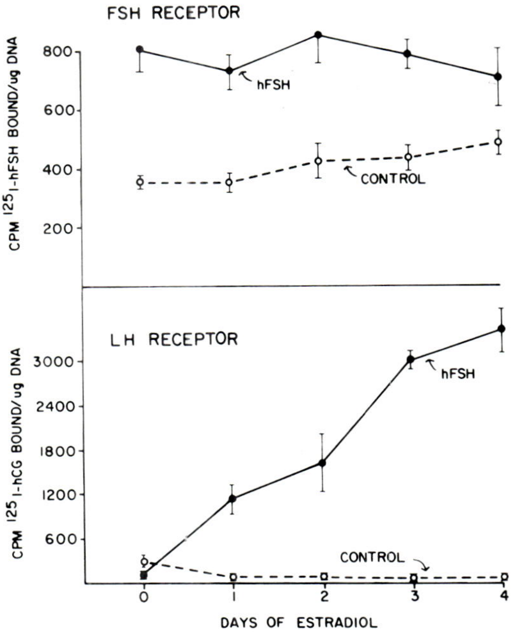

FIG. 8. Effect of estradiol on the ability of FSH to increase receptor for FSH and LH in granulosa cells. See legend to Fig. 6 for details of assay 5.

Furthermore, when the administration of HFSH was continued for 3 or 4 days, the ability of HFSH to stimulate FSH receptor became inversely related to the period of exogenous estradiol exposure: FSH receptor decreased in response to HFSH in rats primed for 4 days with estradiol. More strikingly, estradiol pretreatment markedly enhanced the rate by which FSH stimulated the appearance of LH receptor in granulosa cells from these same rats (Fig. 9). When compared to estradiol pretreatment for 1 day, estradiol for 4 days advanced by 24 hr the time required for HFSH to stimulate the maximal content of LH receptor in granulosa cells. Thus, with estradiol priming for only 1 day, HFSH treatment was required for 3 days, whereas with estradiol priming for 4 days, HFSH was required for only 2 days to achieve maximal LH binding. In the absence of estradiol (E_0), HFSH appeared to be ineffective in stimulating receptor for LH, the increase on day 4 being small and variable.

These results suggested that estradiol might be acting to increase ovarian responsiveness to gonadotropins by facilitating the ability of FSH to increase receptors for both FSH and LH. The mechanisms by which

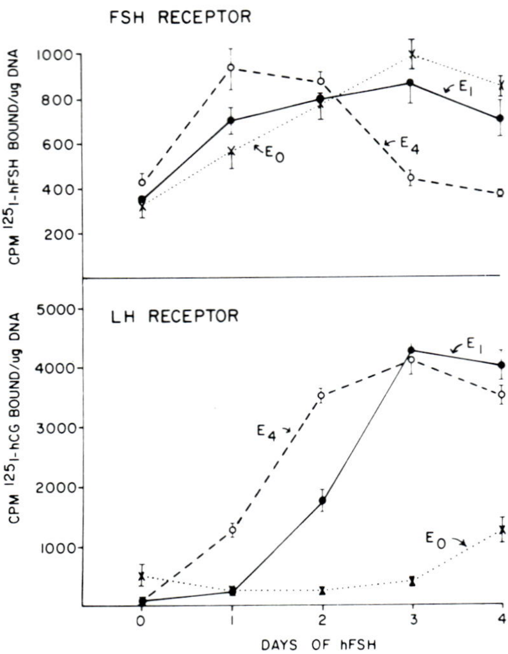

FIG. 9. Effects of estradiol pretreatment on the rate at which FSH increases receptors for FSH and LH in granulosa cells. Estradiol was given for 0, 1, or 4 days followed by 1 to 4 days of HFSH. (From Richards et al., 1976.)

estradiol and FSH interact, however, remain unclear. Since some of the effects of FSH may be mediated by cyclic-3',5'-AMP (Lindner et al., 1974), estradiol might act to modify this FSH response system by increasing the amount of adenylate cyclase capable of being stimulated by FSH or perhaps by increasing the amount of a specific cyclic-AMP-dependent protein kinase (Fig. 10). Alternatively, estradiol may act to effect the synthesis of specific mRNA, possibly for LH and FSH receptor, or for specific proteins that, in the presence of cAMP and protein kinase, may regulate key transcriptional or translational events (Fig. 10).

Once receptor for LH and other events associated with the differentiation of large antral follicles have been stimulated, these follicles can, in the presence of LH, be induced to undergo luteinization. In association with luteinization, functional as well as morphological changes occur. In particular, as shown in Fig. 11, administration of OLH (60 μg) to estradiol, FSH-primed hypophysectomized rats (HE_4F_2) causes receptor content for both FSH and LH to decrease markedly within 24 hr and to remain low. If saline

I. MODIFICATION OF FSH RESPONSE SYSTEM

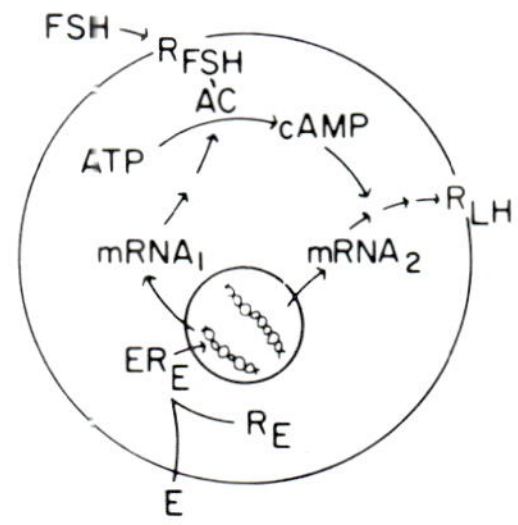

II. REGULATION OF mRNA FOR R_{LH}

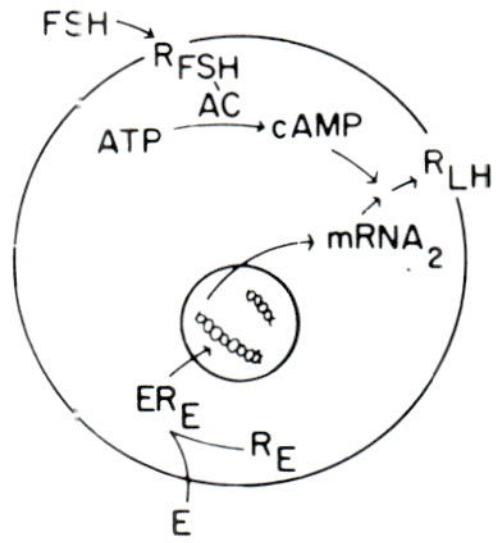

FIG. 10. Two possible mechanisms of action of estradiol in granulosa cells. The first depicts the way in which estradiol may modify the FSH response system; the second, the way estradiol may regulate mRNA for LH receptor. In both, estradiol (E) combines with its receptor (R_E) and is translocated to the nucleus, where it acts to modify this function of the genome. In the first scheme, estradiol is shown to be regulating mRNA ($mRNA_1$) for adenylate cyclase (AC). An increased responsiveness of granulosa cells to FSH might be mediated by increased coupling of adenylate cyclase to FSH binding sites. FSH binding to its receptor (R_{FSH}) would then be expected to stimulate a greater increase in cAMP which, in turn, would affect the appearance (increased synthesis?) of receptor for LH (R_{LH}). The cAMP produced could act either at the transcriptional or translational level. In the second scheme, estradiol is shown to be acting at genomic level to regulate the synthesis of mRNA ($mRNA_2$) for LH receptor. Thus, FSH acting on its receptor would increase cAMP, which, in turn, would increase the translation of preformed $mRNA_2$ for LH receptor. (From Richards and Midgley, 1976.)

rather than OLH is administered, receptors for FSH and LH begin to fall by 48 hr. However, follicular atresia rather than luteinization is observed. When daily injections of HFSH are continued, receptor for LH is maintained. Although FSH receptor decreases, most of these

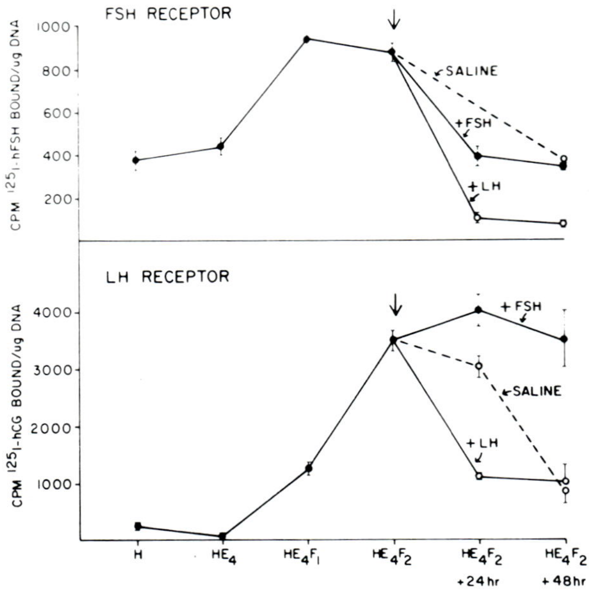

FIG. 11. Effects of FSH and LH on receptors for FSH and LH in granulosa cells of large antral follicles. Hypophysectomized rats were primed with estradiol for 4 days and HFSH for 2 days (HE_4F_2) and divided into three groups: those receiving a single injection of OLH (60 μg), indicated by the *arrow*; those receiving continued daily injections of HFSH (2 μg/day); and those receiving saline. Granulosa cells were examined for FSH and LH receptor content 24 and 48 hr later.

follicles appeared healthy, with no obvious signs of luteinization. Further studies by Rao and Richards (1976) have shown that the LH-mediated decreases in receptor for FSH and LH occur by mechanisms other than the occupation of available hormone-specific binding sites.

In summary, ovarian follicular development, and, more specifically, ovarian granulosa cell differentiation, appears to be associated with hormone-specific regulation of hormone receptors. Results have suggested that estradiol can act on granulosa cells to increase receptor for estradiol; FSH can act on estradiol-primed granulosa cells to increase receptors for FSH and LH; LH can act on estradiol-FSH-primed granulosa cells to effect a decrease in receptors for estradiol, FSH, and LH and concomitantly to effect an increase in receptor for PRL (Richards et al., 1976; Richards and Williams, 1976). This sequence of hormonally induced changes in hormone receptors provides

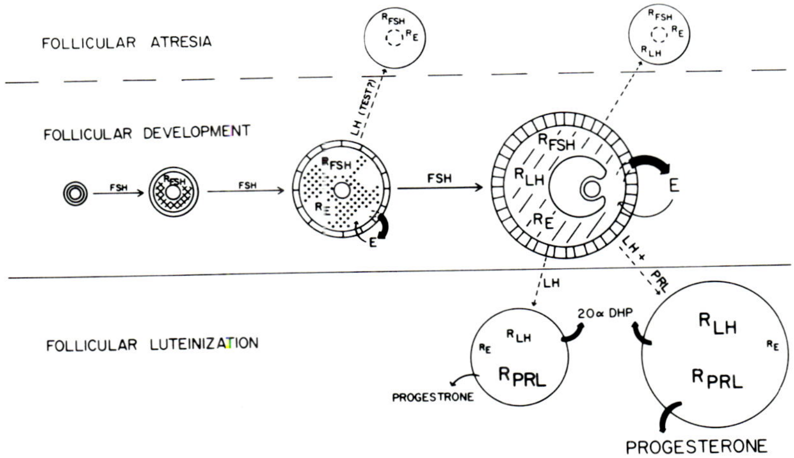

FIG. 12. Hormonal control of follicular development: A 1976 speculation. Once follicles begin to grow they undergo a series of stages of differentiation either in response to the same hormone (such as FSH during very early follicular development) or in response to different hormones acting synergistically (such as estradiol and FSH) or in response to different hormones acting sequentially (as with LH and PRL). At any point, follicles may undergo atresia either as the result of the action of a hormone or hormones (LH and/cr testosterone have been implicated in increasing follicular atresia; Harman et al., 1975) or by the absence of a critical amount of hormone (antral follicles deprived of FSH will rapidly undergo atresia; Richards, 1975). In all cases, atresia appears to be associated with loss of receptor for FSH and estradiol, and, if previously present, LH receptor declines also. The final stage of follicular differentiation, i.e., granulosa cell differentiation, requires both estradiol and FSH and is associated with an increase in FSH and LH receptor (Richards et al., 1976). If luteinization is stimulated, receptor content for LH, FSH, and estradiol declines, while receptor for PRL increases (Richards and Williams, 1976). In the absence of further hormone treatment, corpora lutea produce 20α-dihydroprogesterone (20αDHP) but very little progesterone. However, if PRL is present, corpora lutea produce progesterone and receptor for LH increases markedly. (From Richards and Midgley, 1976.)

evidence for at least three types of receptor regulation: (a) autoregulation, a process by which a hormone effects the content of its own receptor; (b) coordinated regulation, a process by which a steroid and a protein hormone interact to effect the content of receptor for the same or a different protein hormone; and (c) heteroregulation, a process by which one hormone effects the content of an entirely different hormone. Furthermore, the effect of any given hormone appears to be related to cell type and the stage of differentiation of that particular cell type. Thus, although the intraovarian distribution of FSH binding sites suggests that FSH can bind to granulosa cells of follicles at many (if not all) stages of growth (except those undergoing atresia), biochemical data presented here have revealed that the effects of FSH can be dramatically modified by the presence of estradiol. Therefore, as illustrated in Fig. 12, it would appear that the initiation of follicular production of estrogen, and perhaps equally important, the response of granulosa cells to estrogen determines whether or not a given follicle gains the mechanisms necessary to respond to the stimulatory surges of FSH. As a follicle enters the final growth phase, it then gains the mechanisms including the appearance of LH receptor, that ultimately permit it to respond to the surge of LH, to ovulate and to luteinize. Thus, hormone regulation of hormone receptors appears to be an important aspect of follicular maturation.

REFERENCES

1. Anderson, J., Clark, J. H., and Peck, E. J., Jr. (1972): Estrogen and nuclear binding sites. *Biochem. J.*, 126:561-567.
2. Eshkol, A., and Lunenfeld, B. (1972): Gonadotropic regulation of ovarian development in mice during infancy. In: *Gonadotropins*, edited by B. B. Savena, C. G. Beling, and H. M. Gandy, pp. 335-346. John Wiley & Sons, New York.
3. Harman, S. M., Louvet, J. P., and Ross, G. T. (1975): Interaction of estrogen and gonadotropins of follicular atresia. *Endocrinology*, 96:1145-1152.
4. Hirshfield, A. N. (1976): Autoradiographic analysis of the growth of ovarian follicles in the cycling rat. In: Proceedings of the Ninth Annual Meeting of the Society for the Study of Reproduction, abstract 79, p. 55.
5. Katzenellenbogen, B. S., and Greger, N. G. (1974): Ontogeny of uterine responsiveness to estrogen during early development in the rat. *Mol. Cell. Endocrinol.*, 2:31-42.
6. Lindner, H. R., Tsafriri, A., Lieberman, M. E., Zor, U., Koch, Y., Bauminger, S., and Barnes, A. (1974): Gonadotropic action on cultured graafian follicles: Induction of maturation division of the mammalian oocyte and differentiation of the luteal cell. *Recent Prog. Horm. Res.*, 30:79-126.
7. O'Malley, B. W., Rosenfeld, G. C., Comstock, J. P., and Means, A. R. (1972): Induction of specific translatable messenger RNA's by oestrogen and progesterone. In:

5th Karolinska Symposia on Research Methods in Reproductive Endocrinology: Gene Transcription in Reproductive Tissues, edited by E. Diczfalusy, pp. 381-395. Karolinska Institutet, Stockholm.

8. Pederson, T. (1970): Follicle kinetics in the ovary of the cyclic mouse. *Acta Endocrinol. (Kbh.)*, 64:304-323.
9. Pederson, R., and Peters, H. (1968): Proposal for a classification of oocytes and follicles in the mouse ovary. *J. Reprod. Fertil.*, 17:555-557.

9a. Pencharz, R. I. (1940): Effects of estrogens and androgens alone and in combination with chorionic gonadotropin in the ovary of the hypophysectomized rat. *Science*, 91:554-555.

10. Rao, M. C., and Richards, J. L. (1976): Hormone receptor regulation by LH: An active inducible process. In: Proceedings of the 58th Annual Meeting of the Endocrine Society, abstract 285, p. 199.
11. Richards, J. S. (1975): Content of nuclear estradiol receptor complex in rat granulosa cells during follicular development: Modification by estradiol and gonadotropins. *Endocrinology*, 97:1174-1184.
12. Richards, J. S., and Midgley, A. R., Jr. (1976): Protein hormone action: A key to understanding follicular and luteal cell development. *Biol. Reprod.*, 14:82-94.
13. Richards, J. S., Ireland, J. J., Rao, M. C., Bernath, G. A., Midgley, A. R., Jr., and Reichert, L. E., Jr. (1976): Ovarian follicular development in the rat: Hormone receptor regulation by estradiol, follicle stimulating hormone and luteinizing hormone. *Endocrinology*, 99:1562-1570.
14. Richards, J. S., and Williams, J. J. (1976): Luteal cell receptor content for prolactin (PRL) and luteinizing hormone (LH): Regulation by LH and PRL. *Endocrinology*, 99:1571-1581.
15. Schimke, R. T., Rhoads, R. E., Palacios, R., and Sullivan, D. (1973): Ovalbumin mRNA, complimentary DNA and hormone regulation in chick oviduct. In: *6th Karolinska Symposia on Research Methods in Reproductive Endocrinology: Protein Synthesis in Reproductive Tissue*, edited by E. Diczfalusy, pp. 357-379. Karolinska Institutet, Stockholm.
16. Sómjen, D., Sómjen, G., King, R. J. B., Kaye, A. M., and Lindner, H. R. (1973): Nuclear binding of estradiol-17 β and induction of protein synthesis in the rat during postnatal development. *Biochem. J.*, 136:25-33.
17. Stumpf, W. E. (1969): Nuclear concentration of ^{3}H-estradiol in target tissues. Dry-mount autoradiography of vagina, oviduct, ovary, testis, mammary tumor, liver and adrenal. *Endocrinology*, 85:31-37.
18. Williams, P. C. (1945): Studies of the biological action of serum gonadotropin. 1. Decline in ovarian response after hypophysectomy. 2. Ovarian response after hypophysectomy and oestrogen treatment. *J. Endocrinol.*, 4:127-136.

19. Zeleznik, A. J., Midgley, A. R., Jr., and Reichert, L. E., Jr. (1974): Granulosa cell maturation in the rat: Increased binding of human chorionic gonadotropin following treatment with follicle-stimulating hormone *in vivo*. *Endocrinology,* 95:818-825.

Ovarian Follicular Development and Function,
edited by A. R. Midgley and W. A. Sadler.
Raven Press, New York © 1979.

Androgen Receptor and Ovarian Follicular Development

J. R. Schreiber

Department of Reproductive Medicine, School of Medicine,
University of California at San Diego, La Jolla, California

In studies of the role of sex steroid hormones in ovarian follicular development, we have presented data that implicate intraovarian androgen in the control of preantral follicular maturation and possibly atresia (1). Steroid effects on target cells are felt to involve a cytosol receptor and the nuclear translocation of a steroid receptor complex. To further define androgen function, we have reported the presence of a testosterone (T) receptor in the cytosol of ovaries from estrogen-primed hypophysectomized immature female rats (HIFR) (2). This cytosol binding protein is thermolabile, binds T specifically and with high affinity, and sediments at 7 to 8 S on sucrose gradients. Further characterization of the receptor showed it to be present in the cytosol of isolated granulosa cells, to be an acidic protein, and to have cysteine residues that are necessary for binding (3). A comparison of the physicochemical properties of the ovary cytosol binding protein with the prostatic DHT receptor and epididymal extracellular binding protein, ABP, is shown in Table 1. The receptors are quite similar and are clearly different from ABP.

We have demonstrated the selective nuclear accumulation of T by granulosa cells after the *in vivo* administration of ^{3}H-T to estrogen-primed HIFR (3). After the *in vitro* incubation of the ovaries with ^{3}H-T, a thermolabile protein specific for T was extracted from nuclei under high salt, 0.4 M KCl, conditions. The sucrose gradient analysis of the high salt nuclear extract is shown in Fig. 1. The decrease in the binding peak in the presence of 10-fold excess unlabeled T shows the binding to be saturable.

The cytosol and nuclear T receptors may play a role in the regulation of granulosa cell proliferation during preantral follicular growth.

TABLE 1. *Comparison of physicochemical properties of ovary cytosol binding protein with prostate DHT receptor and epididymal extracellular binding protein, ABP* (3)

	Receptor		
	Ovary	Prostate (4)	ABP (5,6)
Sedimentation coefficient	7.8	8.0	4.6
Stokes radius (A)	73	84	48
Molecular weight (daltons)	240,000	270,000	91,000
Frictional ratio	1.8	1.96	1.61
Heat stable	No	No	Yes
Cysteine residue important	Yes	Yes	Yes
Protamine sulfate precipitable	Yes	Yes	Yes

We also have data that are consistent with a direct role for androgens in granulosa cell differentiation (7). We have developed a short-term culture of granulosa cells from preantral follicles of ovaries from estrogen-stimulated HIFR. Cells were grown for 3 days in Ham's F-10 medium supplemented with 10% serum from hypophysectomized male rats. Progesterone production was measured during the last 24 hr of culture. The effect on progesterone (P) production of hormones added

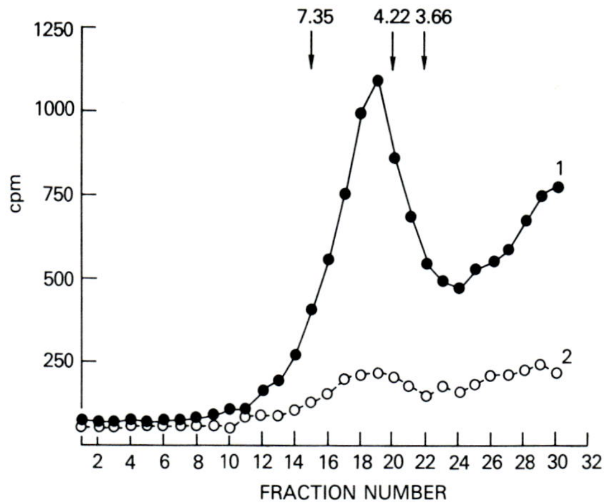

FIG. 1. Analysis by sucrose gradient centrifugation of ^{3}H-T binding by nuclear extracts of ovaries incubated *in vitro*. One group of ovaries was incubated at 32°C for 45 min with 20 nM ^{3}H-T (curve 1), whereas the other group was incubated with 20 nM ^{3}H-T plus 200 nM unlabeled T (curve 2). Top of the gradient is on the right. (3)

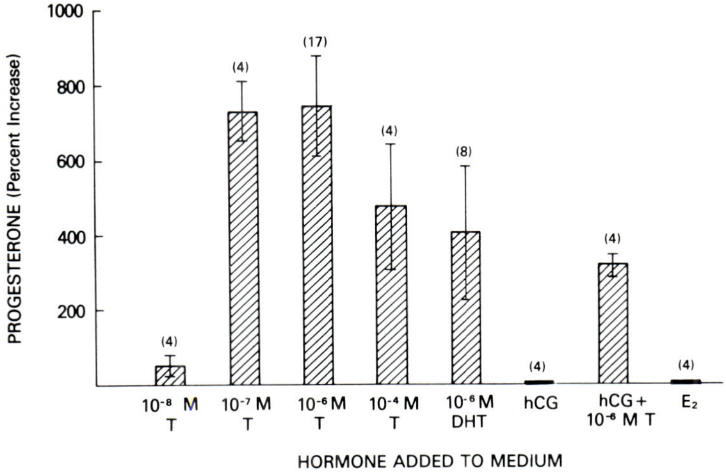

FIG. 2. Percent increase in progesterone (P) production by rat granulosa cells cultured for 3 days in the presence of T, DHT, 100 mIU/ml HCG and 7.8×10^{-10} M E_2. P was measured during the last 24 hr of culture. Parentheses indicate the number of pairs of cultures(7).

to the medium during the 3-day culture is shown in Fig. 2. T at 10^{-7}, 10^{-6}, and 10^{-4} M as well as DHT at 10^{-6} M increased P production 400 to 700% over paired control cultures. Human chorionic gonadotropin (100 mIU/ml) had no effect. Estradiol-17β (7.8×10^{-10} M, the concentration of estradiol produced during the incubation with 10^{-6} M T) also had no effect. P was identified by both a specific radioimmunoassay and Sephadex LH-20 column chromatography. The stimulatory influence of T and DHT on these preantral follicle cells is consistent with a direct role for androgens in granulosa cell differentiation.

REFERENCES

1. Louvet, J.-P., Harman, S., Schreiber, J., and Ross, G. T. (1975): *Endocrinology*, 97:366.
2. Schreiber, J., Reid, R., and Ross, G. T. (1976): *Endocrinology*, 98:1206.
3. Schreiber, J., and Ross, G. T. (1976): *Endocrinology*, 99:590.
4. Mainwaring, W. (1969): *J. Endocrinol.*, 45:531.
5. Ritzen, E., Dobbins, M., Tindall, D., French, F., and Nayfeh, S. (1973): *Steroids*, 21:593.

6. Hansson, V. (1972): *Steroids*, 20:575.
7. Lucky, A. W., Schreiber, J., Hillier, S., Schulman, J., and Ross, G. T. (1977): Endocrinology, 100:128.

Ovarian Follicular Development and Function,
edited by A. R. Midgley and W. A. Sadler.
Raven Press, New York © 1979.

Mechanisms Modulating Gonadotropin Action

Judith L. Vaitukaitis and Barry D. Albertson

Thorndike Memorial Laboratory, Section of Endocrinology and Metabolism, Boston City Hospital, Boston University School of Medicine, Boston, Massachusetts 02118

Cyclic AMP-dependent protein kinases (CDPK) are present in a wide variety of tissues including ovary. Figure 1 schematically depicts the holoenzyme nature of cAMP-dependent protein kinase. The regulatory subunit of CDPK is also known as cAMP binding protein. When cyclic AMP binds to the regulatory subunit, this results in a dissociation of the subunits and an unmasking of the activity of the catalytic subunit.

Cyclic AMP-dependent protein kinases appear to play a significant role in mediating acute hormonal effects *in vivo*. When pseudopregnant rats are given rapid

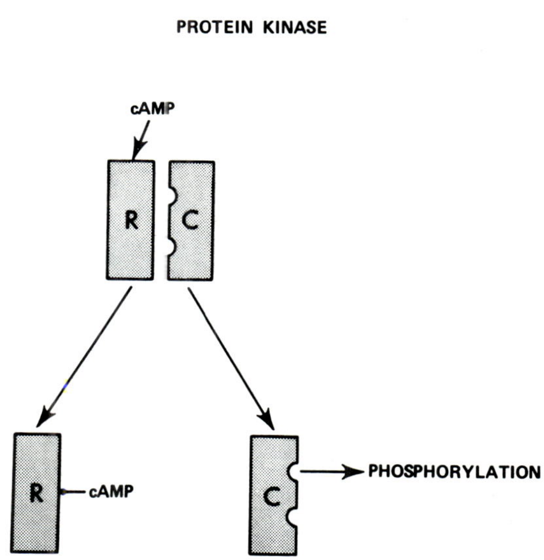

FIG. 1. Schematic representation of cAMP-dependent protein kinase complex. R = regulatory subunit or cAMP binding protein. C = catalytic subunit.

intravenous injections of highly purified human chorionic gonadotropin, maximal activation of cyclic AMP-dependent protein kinases is observed 5 to 30 min after injection (1). That activation is temporally related to both intracellular increase of cAMP and hormonal binding to specific membrane receptors.

Figure 2 depicts the changes in intraovarian concentration of the holoenzyme of cAMP-dependent protein kinase, as well as the change in ovarian intracellular accumulation of cAMP in groups of pseudopregnant Sprague-Dawley rats sacrificed at timed intervals after injection of 5 IU hCG per tail vein. Interestingly, the intraovarian concentration of the cyclic AMP-dependent protein kinase holoenzyme decreased acutely within 5 min of intravenous hCG injection. However, several more minutes were required before a

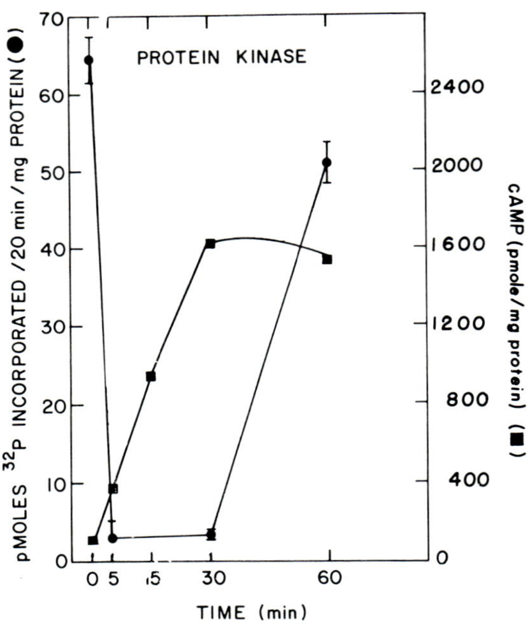

FIG. 2. Effect of time on the *in vivo* activation of the cAMP-dependent protein kinase holoenzyme and intracellular accumulation of cAMP in response to hCG. Pseudopregnant rats were given 5 IU HCG by tail vein injection and groups of 5 animals sacrificed at timed intervals after injection. Partially purified 40,000 × *g* ovarian tissue supernatant was assayed for cAMP by radioimmunoassay and CDPK holoenzyme activity ascertained in an *in vitro* assay in which ^{32}P was incorporated from γ^{32}P-ATP to calf thymus histone IIA in the presence and absence of added cAMP.

significant increase in intracellular accumulation of cyclic AMP occurred. Moreover, the intraovarian concentration of the holoenzyme returned toward basal levels in the face of intraovarian cAMP concentrations greater than those observed with initial acute activation of CDPK. The acute fall in holoenzyme activity reflects *in vivo* dissociation of the holoenzyme complex in response to hCG stimulation. One would have expected a delay in the return of the holoenzyme complex toward basal levels until such time that intraovarian cAMP concentrations approached basal levels. The dissociation of those two events suggests that the protein kinase holoenzyme may be compartmentalized functionally or geographically.

Moreover, when graded doses of hCG were injected intravenously into the tail vein of pseudopregnant rats, the cyclic AMP-dependent protein kinase holoenzyme was totally endogenously dissociated with as little as 500 mIU of hCG intravenously--a dose of hCG that barely elicited a statistically significant increase in intraovarian cyclic AMP. The magnitude of that difference resembled that observed with hCG stimulation of isolated Leydig cells and stimulation of intracellular accumulation of cAMP and testosterone synthesis (2). In both cases, only a fraction of receptors need be occupied to evoke complete dissociation of the CDPK holoenzyme or maximum steroid synthesis.

In separate studies, groups of animals were pretreated with inhibitors of protein synthesis--actinomycin D or cycloheximide--to ascertain whether the acute activation of cyclic AMP-dependent protein kinases or the recovery of holoenzyme activity towards the basal state were dependent on new protein synthesis. Neither actinomycin D nor cycloheximide blocked the acute activation of the holoenzyme; however, cyclic AMP-dependent protein kinase activity of both groups was significantly retarded, suggesting that the return of holoenzyme activity is dependent on new protein substances (1).

When graded doses of highly purified human LH (LER 1417) were given intravenously to groups of pseudopregnant rats (3), essentially the same qualitative patterns were observed with LH, suggesting that hCG and LH are operating at the same ovarian receptors. In experiments designed to further evaluate whether the protein kinase holoenzyme is compartmentalized, groups of pseudopregnant rats were given sequential injections of highly purified hCG or LH (3). Interestingly, the return of holoenzyme activity toward basal levels after the initial hormonal injection was significantly retarded by administration of the second hormonal injection 30 min after the first. On the other hand, intraovarian accumulation of cAMP was unaffected by sequential hormonal injections, suggesting desensitization of the target organ to hormonal stimulation. Figure 3 depicts those effects when 2 μg hCG (CR 119) was injected into pseudopregnant rats. The fact that the CDPK holoenzyme activity was significantly retarded from returning toward basal levels after the second hormonal injection again suggested compartmentalization of the holoenzyme of protein kinase--possibly juxtaposed with the plasma cell membrane where sufficient "new" cAMP was generated to dissociate the holoenzyme complex and unmask the activity of the catalytic subunit. Studies are currently underway in an effort to elucidate the mechanism by which desensitization occurs. The latter may be an important mechanism preventing a target organ from becoming excessively stimulated by persistently elevated circulating levels of hormone.

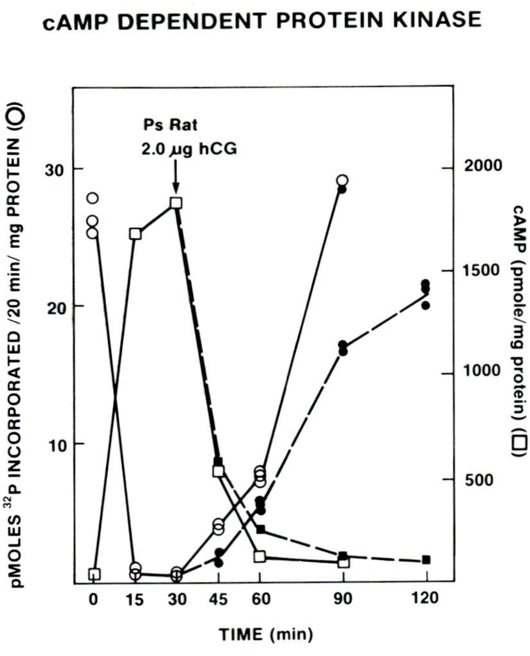

FIG. 3. Effect of sequential injections of highly purified hCG on CDPK holoenzyme and intracellular accumulation of cAMP. Assays were performed as outlined in legend of Fig. 2. Closed circles or squares depict activity in groups of animals given two injections of hormone 30 min apart. Open circles or squares depict activity in groups of animals given a single injection of hCG.

OVARIAN FSH RECEPTORS

In initial studies using tritiated FSH, possessing significant intrinsic biologic activity (4), enhanced specific ovarian FSH binding was found after pretreatment of immature hypophysectomized rats with synthetic or naturally occurring estrogens (5). However, in those studies it was not ascertained whether the increased ovarian binding reflected altered receptor affinity for FSH, an increase in the number of specific FSH binding sites per cell, or simply an increase in the number of granulosa cells with a fixed number of FSH binding sites. The following studies were designed to ascertain which of those mechanisms accounted for enhanced ovarian binding of FSH. Groups of 21-day-old immature hypophysectomized female rats were pretreated with one, two, or four 10 mm silastic capsules containing an aqueous slurry of 17-β-estradiol. Six days after subcutaneous implantation of the silastic capsules, groups of animals were sacrificed and the ovaries removed; granulosa cells were harvested from groups of animals undergoing similar treatment regimen. A direct relationship between the number of silastic capsules implanted and circulating concentrations of estradiol was observed among the groups of rats. Granulosa cells were gently dispersed mechanically. The number of specific FSH binding sites and their equilibrium association constants were ascertained.

Table 1 relates circulating estradiol concentrations, ovarian weight as well as number of specific FSH binding sites and its apparent equilibrium constant in the groups of animals treated with 0, 1, 2, or 4 silastic capsules containing 17-β-estradiol. In short, although estrogen pretreatment enhanced specific ovarian binding of FSH, there was no change in the number of specific FSH binding sites per granulosa cell, nor any significant change in its apparent equilibrium association constant (6). The estrogen mediated enhanced ovarian FSH binding was solely the result of proliferation of granulosa cells with the same number of FSH binding sites with unaltered affinity for FSH. Bradbury had observed a mitogenic effect of estrogen on granulosa cell proliferation (7). However, the mechanism by which estrogen mediates that effect is unknown.

In summary, cyclic AMP-dependent protein kinases appear to play an acute role in mediating the hormonal effects of hCG and possibly FSH. Whether the protein kinase holoenzyme and the adenylate cyclase complex associated with specific FSH and hCG receptors are biochemically identical has not been ascertained. The physiologic substrate for the catalytic subunit has

TABLE 1. *Effect of estradiol on granulosa cell binding of FSH*

Treatment (number of silastic capsules)	Estradiol (pg/ml serum)	Mean ovarian weight	Binding sites per granulosa cell ($\bar{x} \pm$ SE)	Apparent K_a ($\bar{x} \pm$ SE)[a]
0	67	6.8	5453 ± 919	$0.44 \times 10^{10}\ M^{-1}$ ±0.19
1	230	8.7	5762 ± 352	$0.09 \times 10^{10}\ M^{-1}$ ±0.03
2	541	11.2	5102 ± 730	$0.14 \times 10^{10}\ M^{-1}$ ±0.04
4	1093	13.8	5070 ± 600	$0.20 \times 10^{10}\ M^{-1}$ ±0.02

not been identified. Moreover, whether the regulatory subunit of protein kinase mediates any other biochemical effect in addition to its masking the activity of the catalytic subunit in the holoenzyme complex is unknown. Although ovarian LH receptors may be induced through a variety of hormonal manipulations (8-10), induction of ovarian FSH receptors has not been observed.

ACKNOWLEDGMENT

This study was supported in part by Grant No. HD09358-01 from the National Institutes of Health.

REFERENCES

1. Vaitukaitis, J. L., Lee, C. Y., Ebersole, E. R., Lerario, A. C. (1975): New evidence for an acute role of protein kinase in hCG action. *Endocrinology*, 97:215.
2. Catt, K. J., and Dufau, M. L. (1973): Spare gonadotrophin receptors in rat testis. *Nature (New Biol.)*, 244:219.
3. Albertson, B. D., and Vaitukaitis, J. L. (1976): New evidence for compartmentalization of hCG responsive ovarian cAMP-dependent protein kinase activity *in vivo*. *Program of the 58th Annual Meeting of the Endocrine Society*, San Francisco, Calif.
4. Vaitukaitis, J. L., Sherins, R., Ross, G. T., Hickman, J., Ashwell, G. (1971): A new method for the preparation of radioactive FSH with preservation of biologic activity. *Endocrinology*, 89:1356.
5. Goldenberg, R. L., Reiter, E. O., Vaitukaitis, J. L., Ross, G. T. (1972): Interaction of FSH and hCG on follicle development in the ovarian augmentation reaction. *Endocrinology*, 91:533.
6. Louvet J.-P., and Vaitukaitis, J. L. (1976): Induction of Follicle stimulating hormone (FSH) receptors in rat ovaries by estrogen priming. *Endocrinology*, 99:758.
7. Bradbury, J. T. (1961): Direct action of estrogen on the ovary of the immature rat. *Endocrinology*, 68:115.
8. Channing, C. P. (1970): Effect of stage of the estrous cycle and gonadotrophins upon luteinization of porcine granulosa cells in culture. *Endocrinology*, 87:156.
9. Channing, C. P., and Kammerman, S. (1974): Binding of gonadotropins to ovarian cells. *Biol. Reprod.*, 10:179.
10. Kammerman, S., and Ross, J. (1975): Increase in numbers of gonadotropin receptors on granulosa cells during follicle maturation. *J. Clin. Endocrinol. Metab.*, 41:546.

Ovarian Follicular Development and Function,
edited by A. R. Midgley and W. A. Sadler.
Raven Press, New York © 1979.

The Role of Protein Synthesis in the Stimulation by LH of Prostaglandin Accumulation in Rat Preovulatory Follicles *In Vitro*

Martin R. Clark, John M. Marsh, and William J. LeMaire

Departments of Obstetrics and Gynecology and
The Endocrine Laboratory, University of Miami, School
of Medicine, Miami, Florida

One of the questions Dr. Richards' chapter raises is, "How does one hormone alter cellular machinery to produce a change in its own or another hormone's receptors?" Since these changes occur over periods of hours or days, a number of intervening events may take place. Such events could include for LH, for example, acute changes in cyclic nucleotide and steroid metabolism. On the other hand, the long-term effect on receptors may be due to subsequent or parallel alterations in specific RNA or protein synthesis.

Our own studies have shed some light on the mechanism of one delayed aspect of LH action, the stimulation of prostaglandin accumulation in preovulatory follicles. It has been shown that prostaglandins E and F rise in follicles just prior to ovulation, and that this increase is somehow required for follicular rupture (1). The time course of prostaglandin accumulation *in vivo* shows a latency of a few hours and reaches a peak at the time of ovulation (1,2).

In follicles incubated *in vitro* with LH, a similar lag period was observed (Fig. 1). Isolated follicles from PMSG-treated immature rats were incubated with LH (NIH-LH-S18) and PGE was measured by radioimmunoassay (3). The fact that PGE did not increase until 3 hr after LH stimulation caused us to speculate that macromolecular synthesis might be involved in this action of LH. We were able to confirm this suspicion by inhibition of PGE accumulation by puromycin (Fig. 2). It was found that the concentrations of puromycin (100 μM) that are known to inhibit protein synthesis were also able to inhibit PGE accumulation. No inhibition was observed at low puromycin concentrations or by a related compound, puromycin aminonucleoside (PAN), which

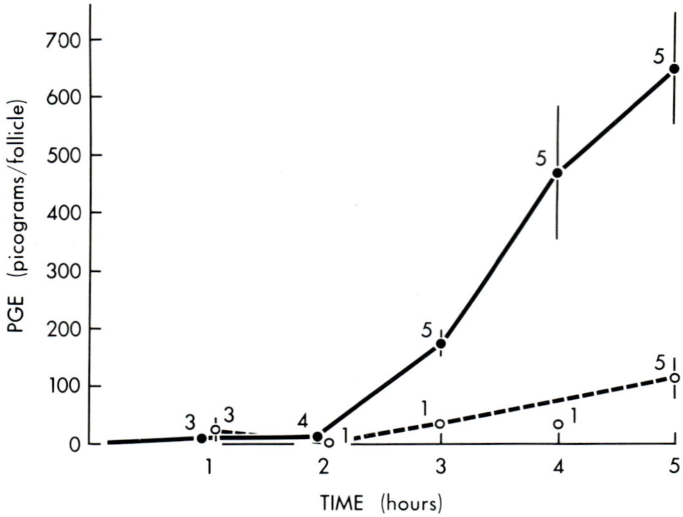

FIG. 1. Time course of prostaglandin accumulation by rat follicles *in vitro*: Effect of LH. Preovulatory follicles were isolated from PMSG-treated immature rats and preincubated for 1 hr in KRBB, pH 7.4, 37°C, under 95% O_2 and 5% CO_2. The follicles were then incubated for the additional times indicated with fresh KRBB, either in the presence (●——●) or absence (o---o) of 5 μg/ml of LH. The total accumulation of PGE was measured by radioimmunoassay. The number of experiments appears adjacent to each point on the graph. Data are presented as mean ± SE. Statistical comparisons: LH at 2 hr vs. LH at 3 hr, $p < 0.001$; LH at 3 hr vs. LH at 4 hr, $p < 0.05$; LH at 4 hr vs. LH at 5 hr, $p < 0.05$; and LH at 5 hr vs. control at 5 hr, $p < 0.001$.

does not inhibit protein synthesis. We have since observed inhibition of PGE accumulation by cycloheximide and actinomycin D. It would appear that PGE accumulation may be dependent on concomitant macromolecular synthesis. This effect does not seem to be secondary to the known inhibition by antibiotics of steroid synthesis, since steroid synthesis is not required for PGE accumulation (4).

It might be worth speculating that this rise in prostaglandins is not only part of the ovulatory process, but also part of LH-induced differentiation of the follicle. In support of this concept is the fact that Grinwich et al. have reported a decrease in LH receptors of 72% in rat corpora lutea 24 hr after administration of $PGF_{2\alpha}$ (5). This correlates well with Dr. Richards' observation of a decrease in LH receptor 18 hr after treatment with LH. It could be hypothesized that the ovulatory increase in prostaglandins is thus responsible for the immediate loss of LH receptors

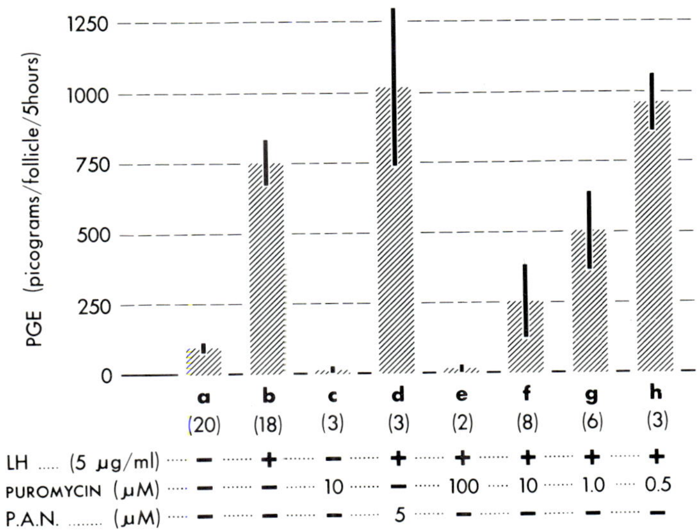

FIG. 2. Effect of puromycin on LH-stimulated accumulation of PGE. Preovulatory follicles were isolated from PMSG-treated immature rats and preincubated for 1 hr in KRBB, pH 7.4, 37°C, under 95% O_2 and 5% CO_2. The follicles were then incubated for an additional 5 hr in fresh KRBB with the appropriate test substances, and total accumulation of PGE was measured by radioimmunoassay. Data are represented as mean ± SE, with a number of experiments given in parentheses below the letter identifying each experiment. Statistical comparisons: a vs. b and h, $p < 0.001$; a vs. d and g, $p < 0.05$; b vs. c, $p < 0.01$; and b vs. f, $p < 0.05$.

following ovulation. It is interesting to note that prolactin can reverse the loss of receptors, observed in both the Richards and Grinwich et al. systems. This hypothesis is obviously incomplete, since it has also been shown that prostaglandins do not appear to be involved in the acute refractoriness of the follicle to LH, as will be discussed by Dr. LeMaire in another section.

Finally, I would like to direct one additional question to Dr. Richards. Is there any evidence that the LH receptor induced by prolactin treatment for 96 hr (Richards, Fig. 6) is functional in terms of stimulation of steroidogenesis? This is an interesting point to consider because progesterone synthesis (Richards, Fig. 7) does not appear to depend on LH at this time.

ACKNOWLEDGMENT

This work was supported in part by NIH grants HD-08747 and HD-03142 and Ford Foundation grant 0338. We thank Dr. J. Pike, from the Upjohn Company, for the gift of prostaglandins, Dr. H. R. Behrman, from Yale University, for the gift of prostaglandin E antibody, and Dr. J. Jewell for the gift of PMSG (Equinex). We appreciate the excellent technical assistance of Adela Gomez and Ana Loynaz, and thank Maria A. Rodriguez for the careful preparation of the manuscript.

REFERENCES

1. LeMaire, W. J., and Marsh, J. M. (1975): Interrelationships between prostaglandins, cyclic AMP, and steroids in ovulation. *J. Reprod. Fertil.*, Suppl. 22:53.
2. Bauminger, S., and Lindner, H. R. (1975): Periovulatory changes in ovarian prostaglandin formation and their hormonal control in the rat. *Prostaglandins*, 9:737.
3. LeMaire, W. J., Yang, N. S. T., Behrman, H. R., and Marsh, J. M. (1973): Preovulatory changes in the concentration of prostaglandins in rabbit graafian follicles. *Prostaglandins*, 3:367.
4. Bauminger, S., Lieberman, M. E., and Lindner, H. R. (1975): Steroid-independent effect of gonadotropins on prostaglandin synthesis in rat graafian follicles *in vitro*. *Prostaglandins*, 9:753.
5. Grinwich, D. L., Hichens, M., and Behrman, H. R. (1976): Control of the LH receptor by prolactin and prostaglandin $F_{2\alpha}$ in rat corpora lutea. *Biol. Reprod.*, 14:212.

Ovarian Follicular Development and Function,
edited by A. R. Midgley and W. A. Sadler.
Raven Press, New York © 1979.

Characterization of Follicles During the Bovine Estrous Cycle

*Eileen A. Merz, †B. G. England, and *E. R. Hauser

*Department of Meat and Animal Science, University of Wisconsin, Madison, Wisconsin 53706 and †Department of Pathology, University of Michigan, Ann Arbor, Michigan 48104

Fertility control is an economically important factor in beef production. We are studying ovarian folliculogenesis, hoping eventually to control the number of ovulations per cow. The resultant increase in birth rate could greatly enhance beef production efficiency.

The experiment reported here is a 3 × 2 × 2 factorial designed to examine the effects of stage of the estrous cycle, side of ovulation, and unilateral ovariectomy on follicular function in beef cows. A total of 36 cows was used. Each unilaterally castrated animal had at least two calves after hemicastration. The control and unilateral groups were each assigned an equal number of left- and right-side ovulations. Ovaries were collected at slaughter on days 6 (early luteal), 12 (midluteal), and 18 (early follicular phase) of the estrous cycle (estrus = day 0; average cycle, 21 days). The estradiol-17β (E_2) concentration in follicular fluid was estimated by radioimmunoassay for each large follicle (≥ 8 mm). E_2 content per follicle was calculated as (0.5236)(diameter)^3(concentration). E_2 values, percent atresia, and number of large follicles were analyzed after $\log_{10}(X + 1)$, arcsin $\sqrt{\text{percentage}}$, and square root transformations, respectively. All data presented are means of transformed data reconverted to original units. Significant day effects were partitioned into the two orthogonal comparisons of day 6 versus days 12 and 18, and day 12 versus day 18. In all cases, only the former comparison was significant (see Table 1). The average number of large follicles per cow and corpus luteum (CL) weight on days 12 and 18 exceeded that on day 6. The average E_2 content of the largest follicle, or of all large follicles per animal, was greater on day 6 than on days 12 and 18.

TABLE 1. *Significant stage of cycle differences*

Orthogonal comparison	Day 6 vs.	day 12 +	Day 18	$p <$
CL weight (g)	3.69	6.22	5.07	0.025
No. follicles ≥ 8 mm per cow	1.28	1.84	2.09	0.025
Content E_2 (ng), largest follicle per cow	88.47	10.91	14.05	0.025
Content E_2 (ng), follicles ≥ 8 mm per ovary	73.13	15.40	25.10	0.05

The main effect of day was not significant for percent atresia of large follicles or diameter of the largest follicle per cow.

The number of large follicles per cow was not significantly different between unilateral and control animals. This equality indicates compensatory hypertrophy of large follicle numbers after unilateral castration. Although the E_2 content of large follicles on the average unilateral ovary (872.19 ng) did not differ statistically from that of the average of the two control ovaries (413.74 neg, $p < 0.1$), the difference could still be of physiological significance. The E_2 content of the largest follicle per cow tended ($p < 0.1$) to be greater in unilaterals (22.06 ng) than in controls (5.62 ng). There was no difference in the number of large follicles between CL-bearing and non-CL-bearing ovaries within control animals. When only the CL-bearing ovaries of the control animals were considered, they had fewer large follicles and a smaller E_2 content in large follicles than the CL-bearing (and only remaining) ovaries of the unilateral castrates.

The weight of the CL-bearing ovary minus the CL was greater in unilaterals than in controls. This could indicate a greater development of small follicles or a build-up of scar tissue as the result of the many previous ovulations. The number of small follicles on the surface of the ovary did not differ by treatment. The presence of more large follicles on the unilateral ovaries could have masked the presence of small follicles at the surface.

There was a greater percent atresia of large follicles on the right (73.8%) than on the left (37.5%) ovaries ($p < 0.025$). However, the differences in the number of large follicles and in their E_2 concentrations and contents were not statistically significant.

That the right ovary had the same number of large follicles as the left, but a greater percent atresia, was interesting, since 60% of the control animals would be expected to ovulate on the right at the next estrus.

There were a number of day-by-treatment interactions. There were no significant differences on day 6 between unilateral and control animals in the number of large follicles per cow, or total E_2 content of large follicles per ovary. However, both endpoints were greater in unilaterals than in controls on day 12, and vice versa on day 18. There was a greater percent atresia of large follicles in unilaterals on day 6, but more atresia of large follicles in controls on days 12 and 18 (see Table 2).

TABLE 2. *Significant stage of cycle by treatment interactions for follicles ≧ 8 mm*

	Day 6			Day 12			Day 18			
	U[a]		C	U		C	U		C	$p <$
% Atretic	62.9	>	16.7	47.2	<	94.7	50.0	<	61.3	0.025
Number per cow	1.29	=	1.26	2.32	>	1.42	1.63	<	2.61	0.01
Content E_2 (ng) per ovary	105.3	=	50.7	36.6	>	6.2	9.6	<	63.6	0.025

[a]U, unilateral castrate; C, normal control.

Working with large follicles enabled us to rate each one individually for atresia, gonadotropin binding, and estradiol concentration in follicular fluid.

The within-large-follicle correlation between atresia rating (1-5, highly mitotic to highly pyknotic granulosa nuclei) and E_2 content was $r = -0.60$ ($p < 0.05$). Fluid from large atretic follicles did not exceed 11.5 ng/ml E_2. Fluid from mitotic follicles ranged from 0.41 to 863.39 ng/ml E_2.

FSH bound to the granulosa of almost all the large follicles. Only two follicles classified as atretic showed slight binding of HCG in the granulosa. All of the other follicles bound HCG only in the theca. The follicles with many mitotic figures in the granulosa nuclei varied from no to extensive HCG binding in those cells (see Fig. 1).

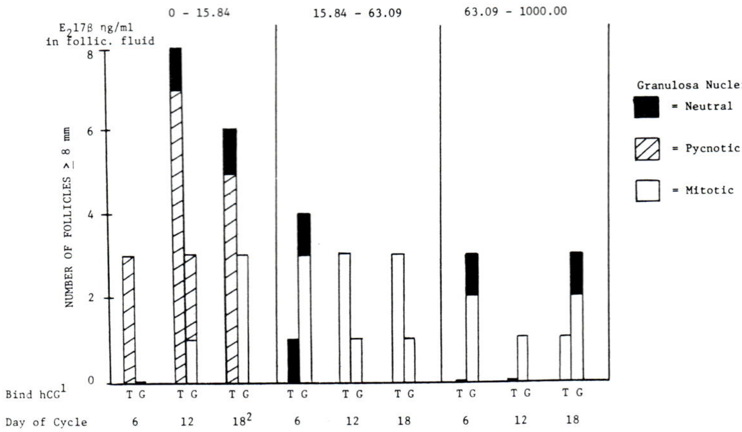

[1]T=theca; G=granulosa

[2]Four follicles in this class (2 mitotic, 2 pycnotic) showed no hCG binding.

FIG. 1. Follicular data classified by day of cycle (estrus = day 0).

Because the binding site data were essentially qualitative, and several data points were lost due to technical difficulties, it was not possible to draw conclusions concerning gonadotropin binding. There were, however, trends that we feel indicate the desirability of continued work in the area with cattle.

Cows have several qualities that make them desirable as experimental animals. The presence of several large ovarian follicles concurrently provides adequate samples for comparison. They usually have only one preovulatory follicle per estrus, which facilitates its differentiation from other large follicles destined for atresia. Also, rectal palpation of the cow's reproductive tract is a simple technique for monitoring ovarian activity.

Ovarian Follicular Development and Function,
edited by A. R. Midgley and W. A. Sadler.
Raven Press, New York © 1979.

Discussion Summary: Session on Follicular Hormone Receptors

J. S. Richards

The specific effects of steroid and protein hormone regulation of granulosa cell differentiation during follicular development and atresia are beginning to be more clearly defined. In particular, it is becoming convincingly clear that hormones selectively act to regulate the appearance and/or disappearance of hormone receptors in granulosa cells at specific stages of follicular growth. Briefly, in the hormonally induced, hypophysectomized immature rat, the steroid estradiol and the glycoprotein FSH act synergistically to increase granulosa cell content of receptors for FSH and LH. LH, which acts to initiate luteinization of granulosa cells, in turn effects a decrease in granulosa cell receptor content for FSH, LH, and estradiol. In many situations, but not necessarily all, an increase or decrease in receptor appears to be related to a corresponding change in target cell responsiveness. Thus, the appearance of LH receptor in granulosa cells is associated with LH-stimuable luteinization, while the decrease in LH receptor appears correlated with a loss of LH-induced activation of adenylate cyclase or "desensitization." The precise time course relationship, however, between loss of receptor and a decrease in hormone stimuable cyclase awaits further study. A notable exception is the marked increase in the responsiveness of granulosa cells to FSH after estradiol pretreatment; a change *not* associated with an increase in FSH receptor.

The estradiol-FSH induced increases in receptor content for FSH and LH might occur as a result of increased synthesis, decreased degradation, or activation, and conversely, the LH-induced decreases by decreased synthesis, increased degradation, inactivation, or internalization. Little evidence is available to support any of these possibilities. However, since the time required for both receptor increases and decreases to occur after hormone stimulation is relatively long (at least 6 hr), mechanisms other than allosteric alteration of membrane components and occupation of available binding sites by hormone *in vivo* seems likely. If receptor loss were the result of the occupation of binding sites, uptake of hormone *in vivo* should continue as binding to available sites *in vitro* is decreasing. Evidence does not support this hypothesis. Rather, uptake of a trace of labeled hormone (HCG) increases during simultaneous uptake of a dose of unlabeled HCG given to initiate luteinization, and only as the uptake of labeled hormone decreases *in vivo* does binding to available sites *in vitro* also decrease. Furthermore, evidence has shown that LH can effect a decrease in FSH (as well as LH) receptor, while very highly purified FSH (containing 0.8% of luteinizing dose of LH) at high concentrations can effect a decrease in LH (as well as FSH) receptor. Since binding sites for these two hormones are hormone specific, it would seem unlikely that either FSH or LH could alter receptor for the other hormone by local allosteric mechanisms. More probably, since granulosa cells undergo similar marked morphological and functional changes in response to these hormones, both hormones share a common intracellular regulatory mechanism (possibly cAMP). Whether FSH and LH receptor interactions activate the same or different molecules of adenylate cyclase and whether or not the cAMP produced activates the same or different protein kinases is unclear. The response of granulosa cells appears to be mediated, in part, by cAMP. However, even here the situation remains unclear. If local, membrane-associated production of cAMP is compartmentalized, a rapid intramembranal elevation of cAMP may be sufficient to cause a physiological event (change in receptor activity?); high concentrations in the cytosol may be generated as a pharmacological event with or without additional effects on cell function. In corpora lutea very small changes in cellular concentrations of cAMP can rapidly (5 min) decrease cAMP-dependent protein kinase activity, which, when returned to basal levels (60 min), appears unaffected by the increasingly high intracellular concentrations of cAMP. Clearly, a dose-response study relating cAMP production and receptor content in luteinizing granulosa cells is needed. A similar study is also needed to examine the mechanism(s) by which FSH acts to increase receptors for FSH and LH in granulosa cells of developing follicles. The dose of HFSH that stimulates the appearance of LH receptor and antral formation has not been shown to be capable of effecting a marked increase in intracellular levels of cAMP *in vivo*. Interestingly, this dose of HFSH also stimulates a marked increase in cAMP binding capacity of granulosa cell cytosol. Presumably this increase reflects, in part at least, an increase in cAMP-dependent protein kinase(s). Thus, the increased responsiveness of large antral follicles to LH may involve not only changes in receptor content but also changes in the intracellular capacity to respond to cAMP.

Additional effects of LH or FSH may be mediated by products generated by the initial hormone-receptor response. For example, some of the effects of LH on granulosa cell function (atresia and luteinization) may be mediated by follicular production of androgen and/or prostaglandins. In the corpus luteum, prostaglandins have been shown to cause a decrease in LH receptor. Thus, a similar prostaglandin-mediated process may occur within the follicle during luteinization. The effect of androgens on granulosa cell receptor content has not been thoroughly studied. However, androgens have been implicated in increasing the process of atresia. The presence of a high affinity androgen receptor in granulosa cell cytosol has been clearly documented, the binding characteristics of this receptor have been described and evidence for the translocation of testosterone to granulosa cell nuclei has been provided. Furthermore, testosterone administered to granulosa cells in culture has been shown to stimulate a very specific response, increased production of progesterone by 400 to 700%. Whether this response occurs at any time during the normal development of follicles (does androgen increase progesterone in follicles undergoing atresia or does it mediate LH-induced luteinization?) is not clear, but the observation is an exciting one, since two steroids have now been shown to modify target cell steroidogenesis: testosterone increasing progesterone production in cultured granulosa cells and estradiol increasing progesterone production in rabbit and rat corpora lutea. The mechanisms by which these steroids act is unclear and probably not via cAMP. If not, perhaps they can act to regulate the activity (synthesis?) of a specific enzyme.

Since the observed changes in granulosa cell receptor content have been induced in an immature, hypophysectomized model system, the doses of hormones used and the relationship of these induced changes to what is occurring during a normal follicular cycle need to be examined. Although the dose of estradiol administered may seem pharmacological, the effects of the steroid at the ovarian level appear physiological: estradiol reduces atresia and supports granulosa cell mitotic activity and differentiation. Since the membrana granulosa of each follicle is avascular, a high systemic level may be required to maintain (for an extended period of time) physiological concentrations within the follicle. Histological preparations fail to reveal abnormal appearances of granulosa cells during 4 days of estradiol treatment. The HFSH was given at a dose that stimulated follicular antral formation and the appearance of LH receptor without causing morphological signs of luteinization. The OLH was given at a dose necessary to cause complete luteinization of large antral follicles. The sequence of hormone treatments promotes growth of large antral follicles within a period of 4 to 5 days. This time span is similar to the length of a normal estrous cycle in the rat. However, it is not clear when estradiol begins to act in the course of normal follicular growth. However, it does appear that in the cow those follicles possessing LH receptor are also the ones containing the most estradiol. Thus, although the time at which any given follicle is selected to be the one to ovulate may vary between species (events may be speeded up or slowed down; LH receptor may be retained for variable periods of time), estradiol appears to play an important role as an intrafollicular hormone.

HORMONE ACTION IN FOLLICLES

Ovarian Follicular Development and Function,
edited by A. R. Midgley and W. A. Sadler.
Raven Press, New York © 1979.

Hormone Action in Ovarian Follicles: Adenylyl Cyclase and Protein Kinase Enzyme Systems

*Mary Hunzicker-Dunn, *Richard A. Jungmann, and †Lutz Birnbaumer

*Department of Biochemistry, Northwestern University Medical School, Chicago, Illinois 60611 and the †Department of Cell Biology, Baylor College of Medicine, Houston, Texas 77030

The mature ovarian follicle is a structure capable of a variety of complex hormonal responses. Tonic levels of gonadotropins induce follicular maturation, consisting of granulosa and theca cell proliferation and steroidogenesis; while ovulatory levels of gonadotropins (about 10 times greater than tonic levels) induce acute steroidogenesis, oocyte maturation, luteinization, ovulation, and ultimately corpus luteum (CL) formation. Although it is becoming increasingly evident that at least some of these physiological responses require, in addition to LH and FSH, adequate levels of ovarian steroids, especially estrogens, this review will be concerned with the actions of the gonadotropins LH and FSH on ovarian follicles.

The initial step in hormonal action requires that the target cell, in this case either granulosa or theca cells, recognize only those hormones for which it contains the cellular "machinery" to respond to. Follicular cells can discriminate among the circulating protein hormones by the presence of specific hormone receptors (as discussed in a separate chapter in this book) located on the outer surface of the plasma membrane. Interaction of a specific hormone with its membrane receptor generally leads to activation of a membrane adenylyl cyclase, promoting cAMP formation and the subsequent biological effect (Robison et al., 1970).

In this chapter we shall first review the current evidence that implicates cAMP in gonadotropin hormone action in follicular cells and then describe the relationship that seems to exist between the responsiveness of the adenylyl cyclase system and the functional capacity of the follicle during various stages of final follicular maturation. Of prime relevance is the concept that hormone action is determined not only by the amount of hormone exposed to the cellular plasma membrane but also by the responsiveness and the sensitivity of the adenylyl cyclase system.

REVIEW OF CURRENT EVIDENCE FOR A ROLE OF cAMP IN GONADOTROPIN ACTION IN FOLLICLES

There is considerable experimental evidence indicating that many of the effects of LH and FSH in follicles are mediated by cAMP, as recently reviewed by Marsh (1975). Initially this postulate was based on observations of gonadotropin-induced increases in follicular cAMP concentrations. LH promotes an increase in intracellular cAMP concentrations in graafian follicles (Ahren et al., 1974; Lamprecht et al., 1973; Lindner et al., 1974; Tsafriri et al., 1972) as soon as 20 sec after exposure of the follicle to LH (Ahren et al., 1974). Furthermore, these elevations in cAMP are only transient (Ahren et al., 1974) as is common to other systems demonstrating hormone-induced increases in cAMP concentrations (Kakiuchi and Rall, 1968; Parker et al., 1974; Powell et al., 1971; Robison et al., 1971; Sutherland et al., 1965). FSH, free of LH contamination, also promotes an increase in intracellular cAMP concentrations in mature follicles (Ahren et al., 1974; Lindner et al., 1974). These gonadotropin-induced elevations in follicular cAMP concentrations seem to be due to activation of the adenylyl cyclase system, measured utilizing an ATP-prelabeling technique that determines turnover rates of cAMP (Kolena and Channing, 1971; Kolena and Channing, 1972; Marsh et al., 1972; Tsafriri et al., 1972; Lamprecht et al., 1973; Marsh et al., 1973), rather than to changes in the activity of other enzyme systems, like phosphodiesterases

Although many of the gonadotropin-induced effects in follicles could be mediated by cAMP, current evidence implicates this nucleotide only in steroidogenesis, luteinization, ovum maturation, and, more speculatively, ovulation.

The possible mediatory role of cAMP in LH-induced steroidogenesis in follicular cells has been studied the longest. Follicles synthetize and secrete steroids throughout the final stages of follicular maturation (Sasamoto and Kennan, 1973; Schwartz, 1974). Prior to the ovulatory surge of gonadotropins, they synthetize primarily estrogens in response to tonic levels of LH and FSH (Butcher et al., 1974; Kalra and Kalra, 1974; Naftolin et al., 1972; Shaikh, 1971; Smith et al., 1975;

Yoshinaga et al., 1969). For a limited time following the ovulatory surge of gonadotropins, they literally pour a variety of steroids--including estrogens, progestins, and androgens--into the ovarian vein and follicular fluid compartments (Hilliard and Eaton, 1971; Hilliard et al., 1974; Mills, 1975; Younglai, 1972; Younglai, 1974). Thereafter, steroidogenesis is terminated for a period of some 36 hr, or until the CL develops to a functional capacity to accommodate steroidogenesis (Butcher et al., 1974; Hilliard and Eaton, 1971; Hilliard et al., 1974; Lieberman et al., 1975; Smith et al., 1975; Younglai, 1972). A number of studies have demonstrated cAMP- or dibutryl cAMP-induced steroidogenesis in isolated granulosa cells or in follicles (Channing, 1970*a*; Channing and Seymour, 1970; Cirillo et al., 1969; Ellsworth and Armstrong, 1973; Ellsworth and Armstrong, 1974). Since few of these studies have demonstrated steroid production in less than 8 hr following exposure to the nucleotide, they may reflect differentiated granulosa cell or luteal cell steroidogenesis rather than follicular steroidogenesis, as suggested by Marsh (1975). If a role for cAMP in the control of follicular steroidogenesis is valid, then steroidogenesis should be elicited within minutes. Two recent studies have demonstrated more rapid steroid formation in *in vitro* experiments utilizing rabbit follicles: Mills (1975) obtained estrogen, progesterone, and testosterone production within 3 hr of cyclic nucleotide presentation, and Younglai (1975*b*) obtained steroidogenesis within 30 min of cyclic nucleotide presentation. Mills also found that the stimulatory effects of LH and cAMP on steroidogenesis were not additive, indicating that the probable action of LH on steroidogenesis was via a cAMP intermediate.

The action of LH on ovum maturation may also be mediated by cAMP, since Tsafriri has shown that both dibutryl cAMP (injected into the follicle) and LH (introduced into the culture medium) can induce meiotic division in follicle-enclosed oocytes *in vitro* (Tsafriri et al., 1972; Tsafriri et al., 1973).

A number of investigators have demonstrated that cAMP or its dibutryl derivative can mimic the effects of LH in inducing morphological as well as functional luteinization (Channing, 1970*a*; Channing, 1970*b*; Channing and Seymour, 1970; Channing, 1973; Channing, 1974; Cirillo et al., 1969; Ellsworth and Armstrong, 1973; Keyes et al., 1972; Miller and Keyes, 1974). Contrary to these findings, LeMaire et al. (1972) not only inhibited LH-induced luteinization of rabbit follicles with the intrafollicular injection of cAMP and dibutryl cAMP but also stimulated luteinization with the intrafollicular injection of cyclic nucleotide phosphodiesterase. The data of LeMaire clearly indicate an inhibitory role for cAMP in follicular luteinization and are difficult to reconcile with the above evidence for a stimulatory role of cAMP in follicular luteinization.

cAMP has also been postulated to mediate LH-induced ovulation; however, its role seems, at best, to be indirect. While cAMP and/or its dibutryl derivative can mimic the effect of LH by increasing the distensibility of the follicular wall (Rondell, 1974) and can overcome an antiadrenergic blockade of spontaneous ovulation in chicken follicles (Kao and Nalbandov, 1972), their effects seem to be via cAMP-mediated increases in steroid or prostaglandin (PG) synthesis. Thus, Rondell reported that a blockade of steroidogenesis prevented the cAMP-induced increase in follicular wall distensibility (Rondell, 1974) and Baranczuk et al. (1974) reported that progesterone, the synthesis of which is more than likely a cAMP-mediated event, can induce ovulation. A possible route to ovulation via PG, elevations of which have been unequivocally linked with the actual rupture of the follicle (Armstrong et al., 1974; Grinwich et al., 1972; LeMaire et al., 1973), was postulated when Marsh et al. (1974) showed that exogenous cAMP could induce elevations in PG levels in granulosa cell cultures. Intrafollicular injections of cAMP or dibutryl cAMP, however, are unable to mimic the effect of intrafollicular LH in causing ovulation (Hunzicker-Dunn, *personal observation*; LeMaire et al., 1972).

In summary, there is convincing evidence for a mediatory role of cAMP in LH-induced steroidogenesis and ovum maturation, and probably in luteinization; however, its role in ovulation is not yet tenable.

THE ADENYLYL CYCLASE SYSTEM IN DEVELOPING FOLLICLES

In view of the evidence for a mediatory role of cAMP in gonadotropic hormone action in follicles, we investigated the presence and responsiveness of the adenylyl cyclase system in follicles during various stages of final follicular maturation (Bockaert et al., 1976; Hunzicker-Dunn and Birnbaumer, 1976*a*, 1976*b*, and 1976*c*). If cAMP levels are increased by enhanced synthesis, then the ovarian follicle should contain an adenylyl cyclase system that can be stimulated by LH and/or FSH. Immature antral follicles do contain an adenylyl cyclase system, but it is barely responsive to LH stimulation (less than twofold). It is during the final stages of follicular maturation (prior to the ovulatory surge of gonadotropins) that follicular cells acquire an adenylyl cyclase system that is highly responsive to stimulation by LH. The acquisition of a highly LH-sensitive adenylyl cyclase system during follicular maturation will be discussed in three different follicular systems.

Rats

The 4-day cycling rat is a useful model in which to follow final follicular maturation. Following ovulation on the morning of estrus, a new cohort of follicles begins the final stages of follicular maturation in preparation for ovulation and CL formation. Follicles obtained at 1000 hr on estrus, metestrus, and diestrus contain an adenylyl cyclase system, measured as basal cyclase activity, but it is only poorly responsive to LH stimulation (Fig. 1). Between 1000 hr on diestrus and 1000 hr on protestrus, LH- and FSH-stimulated adenylyl cyclase activities rise about threefold and basal activities decrease, resulting in a relative stimulation of about 12-fold for LH. Thus, just prior to receiving the ovulatory surge of gonadotropins, follicles in the cycling rat acquire an adenylyl cyclase system that is highly responsive to stimulation by LH.

The same pattern of follicular development and concomitant "maturation" of the adenylyl cyclase system can be seen in follicles dissected from immature rats that are primed with an injection of pregnant mare serum gonadotropin (PMSG) in order to induce both follicular development and subsequent ovulation. The injection of 3 IU PMSG i.v. promotes a rapid increase in LH-stimulated adenylyl cyclase activity and a decrease in basal activity (Fig. 2). By the second day following PMSG injection, LH-stimulated adenylyl cyclase activity has increased to levels equivalent to those measured in proestrus rat follicles on the morning before the surge of gonadotropins. Thus, on the day before ovulation, the adenylyl cyclase system has become highly responsive to LH, exhibiting more than a 12-fold stimulation.

Pigs

The pig follicle also offers a useful model system for studying follicular maturation, since antral

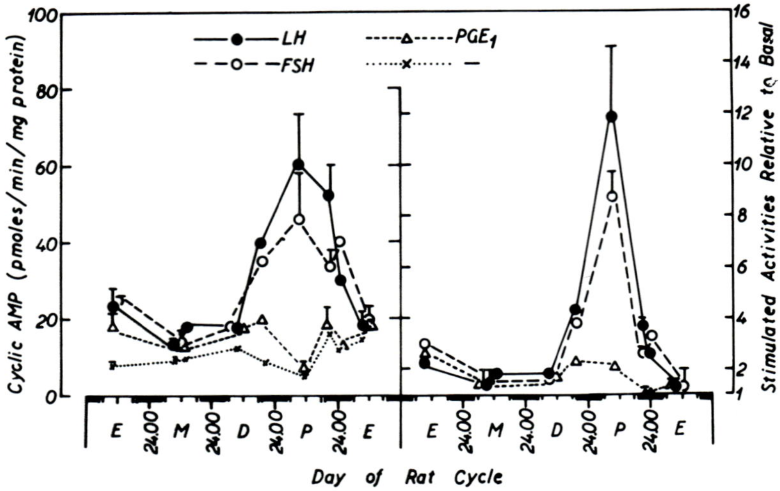

FIG. 1. Adenylyl cyclase activities measured in homogenates of ovarian follicles obtained from rats with a 4-day estrous cycle. Rats (Charles River, outbred) exhibiting at least two consecutive 4-day estrous cycles as determined by vaginal smears were used (lights on from 0500 to 1900). Rats were sacrificed at the indicated times on each day of the estrus cycle. Ovaries were placed in iced Krebs-Ringer bicarbonate buffer (KRB) and follicles were dissected with the aid of a dissecting microscope and homogenized (Dounce) in 10 volumes of a medium containing 27% (w/w) sucrose, 1.0 mM EDTA, and 10 mM Tris-HCl, pH 7.5. The homogenate was filtered through No. 12 silk screen, and 20-μl aliquots were assayed for basal and hormonally responsive adenylyl cyclase activities. Adenylyl cyclase incubations were for 10 min at 37°C in 50 μl of a medium containing 3.0 mM $[\alpha\text{-}^{32}P]$ATP, 5.0 mM $MgCl_2$, 1.0 mM EDTA, 1.0 mM$[^3H]$cAMP, 25 mM Tris-HCl, pH 7.0, an ATP-regenerating system consisting of 20 mM creatine phosphate and 0.2 mg/ml creatine kinase at pH 7.0, homogenate protein, and when present 10 μg/ml LH (NIH-LH-B8), FSH (NIH-FSH-P1), PGE_1 (Upjohn), or NaF. The reaction was terminated with the addition of 0.1 ml of 10 mM cAMP, 10 mM ATP, and 1.0% sodium dodecyl sulfate, followed by boiling. $[^{32}P]$cAMP was isolated according to the method of Salomon et al. (1974) using Dowex (Krishna et al., 1968) and alumina (Ramachandran, 1971; White and Zenser, 1971) chromatography. Single points represent one assay in which follicles from five rats were used. Mean ± SEM is shown where two or three such assays were performed. (From Hunzicker-Dunn and Birnbaumer, 1976*b*.)

follicles of different sizes, and presumably in different stages of development, can easily be dissected from porcine ovaries. The adenylyl cyclase system in small, 1- to 2-mm follicles is only minimally stimulated by LH (less than twofold by LH) (Fig. 3).

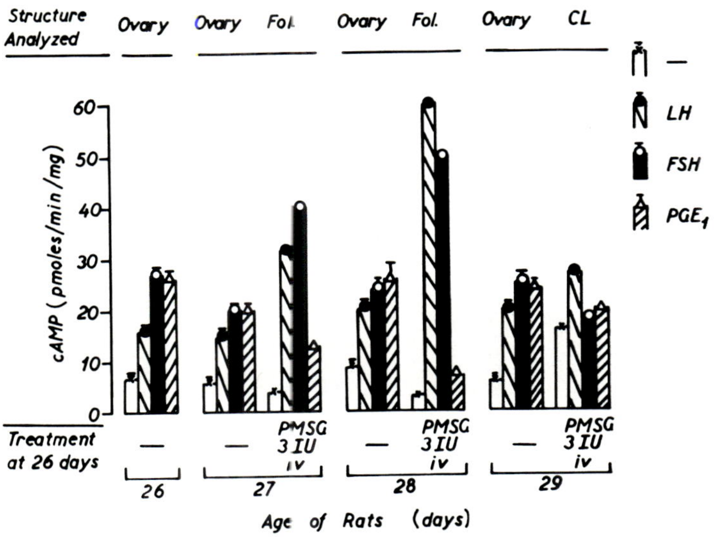

FIG. 2. Effect of pregnant mares serum gonadotropin (PMSG) on adenylyl cyclase activities in ovarian follicles and resulting CL from prepubertal rats. Prepubertal rats (Charles River, outbred, received at 20 days of age) were injected at 26 days with 3 IU PMS i.v. (0800 hr). Animals were sacrificed at 1000 hr on days 27, 28, and 29. Follicles and/or CL were dissected and handled as previously described in the legend to Fig. 1. The adenylyl cyclase activities in homogenates of follicles and CL are compared to those of total ovaries from rats which received no injection (-). Single points represent one assay in which follicles and/or CL from 10 rats were used. Mean ± SEM is shown where triplicate assays in which ovaries from three rats were used in each assay. For rest of conditions, see legend to Fig. 1. (From Hunzicker-Dunn and Birnbaumer, 1976*b*.)

LH-stimulated cyclase activity has risen slightly in medium-sized follicles, and in large, 6- to 8-mm preovulatory follicles, LH-stimulated adenylyl cyclase activity has increased about sixfold over that in small follicles, with no significant change in basal cyclase activity during follicular growth.

Rabbits

In this third follicular system, the different stages of follicular maturation cannot easily be discerned, yet two different sizes of follicles are easily obtained from mature ovaries--small, unovulable ones (< 1.5 mm in diameter) and ovulable ones (2 mm in

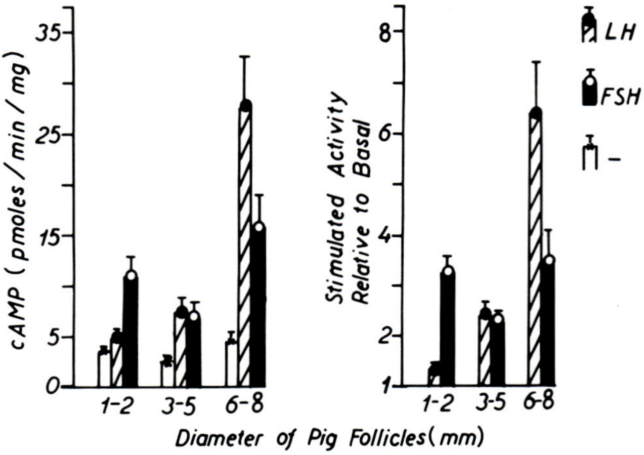

FIG. 3. Adenylyl cyclase activities in homogenates of antral pig follicles of various sizes. Ovaries were obtained on the assembly line from nonpregnant, slaughterhouse pigs and were immediately placed in iced KRB. Follicles were dissected from the ovaries less than 3 hr after animal slaughter and were popped, homogenized, and assayed as described in the legend to Fig. 1 for rat follicles. (From Bockaert et al., 1976, and *unpublished*.)

diameter). The adenylyl cyclase system in follicles measuring less than 1.5 mm in diameter is stimulated less than twofold by LH (Table 1). In 2-mm, preovulatory follicles, LH-stimulated adenylyl cyclase activity has risen twofold and basal activity had decreased, resulting in a relative stimulation of about eightfold for LH.

Possible Physiological Implications

Thus, in four different system in which follicular growth and maturation can be monitored, the responsiveness of the cyclase system to LH increases *only* during the latest stages of follicular maturation, i.e., just prior to the ovulatory surge of gonadotropins. Similarly, HCG-specific binding sites in granulosa cells from porcine and rat graafian follicles, and presumably in follicles from other species, increase as the follicle approaches ovulability (Channing and Kammerman, 1973; Lee and Ryan, 1973; Lee, 1976; Rajaniemi et al., 1974; Zeleznik et al., 1974). It is not known whether the development of LH receptors precedes the development of a potentially LH-responsive adenylyl cyclase

TABLE 1. *Hormonal responsiveness of small and large graafian follicles of the rabbit*

Diameter of follicles[a] (mm)	Addition to incubation	Adenylyl cyclase activity: Absolute[b]	Adenylyl cyclase activity: Relative to basal
Less than 1.5	None	20 ± 0.8	--
	LH	35 ± 1.2	1.75
	FSH	31 ± 1.0	1.55
	PGE_1	33 ± 0.8	1.65
More than 1.5	None	6.3 ± 3.8	--
	LH	70.2 ± 11.9	8.46
	FSH	58.3 ± 7.0	7.02
	PGE_1	20.0 ± 8.0	2.41

[a]When present LH, FSH, and PGE_1 were 10 μg/ml each.
[b]Values reported are pmoles per min per mg homogenate protein. *Follicles of less than 1.5 mm*: mean ± SD of triplicate determinations of one assay in which follicles from four rabbits were used. *Follicles of more than 1.5 mm:* mean ± SEM of five assays in each of which follicles from three rabbits were used.

system or whether the two events occur concurrently. It is conceivable that a potentially LH-responsive adenylyl cyclase system is already present in small antral follicles and that it is the acquisition of LH receptors followed by coupling of these receptors to the adenylyl cyclase that enables the follicle to respond to the surge of gonadotropins by ovulation. Experiments to decide one way or the other are currently in progress.

Investigators have demonstrated that, although small follicles contain only few LH receptors, they contain many more FSH receptors (Channing and Kammerman, 1973; Richards and Midgley, 1976; Stouffer et al., 1976; Zeleznik et al., 1974). Furthermore, although the cyclase system in these small follicles is unresponsive to LH stimulation, it appears to be coupled functionally to FSH receptors, as evidenced by the following observation. FSH[1]-stimulated adenylyl cyclase activity is considerably higher than LH-stimulated cyclase activity in 1- to 2-mm porcine follicles (Fig. 3; Lee, 1976), in follicles obtained from cycling rats on the morning of estrus (Fig. 1), and in follicles obtained

[1]FSH (NIH-FSH-P1) is contaminated with 0.0075 U/mg LH (Reichert and Wilhelmi, 1973).

from PMSG-primed rats 24 hr after PMSG injection (Fig. 2). Similarly, Koch et al. (1973) reported that FSH (free of LH contamination), but not LH, promotes an 11-fold increase in the concentration of cAMP in ovaries obtained from 27- to 29-day old rats (these ovaries presumably contain only small antral follicles). Thus, there is substantial evidence indicating that the adenylyl cyclase system is in fact mature in small follicles, and that it can be coupled in a functional manner (to FSH receptors).

We have also observed that the acquisition of LH responsiveness is accompanied by decreases in basal cyclase activity. This is seen in follicles dissected from PMSG-primed prepubertal rats (Fig. 2), in cycling rats (Fig. 1), and in rabbits (Table 1) as ovulability approaches. A similar finding with respect to fluoride-stimulated cyclase activity was observed by Robison et al. (1970) in developing rat brain. One may interpret these data to mean that it is the coupling of the LH receptors to the active cyclase system that produces the observed restriction of basal cyclase activity.

From these lines of evidence, we conclude: (a) that small antral follicles contain an active adenylyl cyclase system; (b) that the cyclase system in small follicles is coupled to FSH receptors and is capable of forming cAMP upon FSH stimulation; and (c) that the appearance of an LH responsive adenylyl cyclase system is due first to the "formation" of LH receptors followed by coupling of these receptors to the already active cyclase system. We are assuming that, as in rat adipose tissue, a single adenylyl cyclase system is coupled to more than one group of hormone-specific receptors (Birnbaumer and Rodbell, 1969; Rodbell et al., 1970).

We do not know which hormone or groups of hormones are responsible for the induction of the gonadotropin-responsive adenylyl cyclase system. It will be interesting to determine if hormonal treatments promoting hormonal receptor development, as discussed in another chapter in this book, also promote "maturation" of a responsive adenylyl cyclase system.

These results on the consistent induction of a highly LH-responsive adenylyl cyclase system *only* during the latest stages of follicular maturation would indicate that in order to "properly respond" to the preovulatory surge of gonadotropins, the follicle must contain an adenylyl cyclase system that is maximally responsive to LH. When follicles are presented with an ovulatory surge of gonadotropins *prior* to final follicular maturation--and presumably prior to the acquisition of a highly gonadotropin-responsive adenylyl cyclase system--they do not ovulate, but rather become atretic, luteinizing with entrapped ova (Reed and Hounslow, 1971; Richards, 1975). It would therefore appear that some segment of the ovulatory response itself requires a sufficiently responsive adenylyl cyclase system. Although these authors did not report on the presence or absence of steroidogenesis in these luteinized follicles, follicular steroidogenesis should not require a highly LH-responsive cyclase system, since steroids are being synthesized in response to tonic levels of gonadotropins long before the LH-stimulated cyclase activities greatly increase. Of course, proof that a highly LH-responsive adenylyl cyclase system is a prerequisite for the follicle to initiate the ovulatory response awaits studies correlating the effects of ovulatory doses of gonadotropins on follicles with adenylyl cyclase systems of differing LH responsiveness.

THE EFFECT OF AN OVULATORY DOSE OF GONADOTROPINS ON THE ADENYLYL CYCLASE SYSTEM IN OVULABLE FOLLICLES

We have shown that mature, ovulable follicles contain a fully responsive, LH-sensitive adenylyl cyclase system. Exposure of these follicles to ovulatory doses of LH, such as occurs in mated rabbits and in cycling rats, as well as to ovulatory doses of exogenous gonadotropins results in a rapid loss in the capacity of the adenylyl cyclase system to be stimulated by LH, i.e., in gonadotropin-induced desensitization of the adenylyl cyclase system to further gonadotropin stimulation. This phenomenon was first observed in ovarian tissues by Marsh et al. (1973) and by Lamprecht et al. (1973). These authors, both of whom used an *in vitro* ATP-prelabeling technique to demonstrate hormone-stimulated cAMP accumulation, showed that graafian follicles exhibit a decreased ability to accumulate cAMP with time following exposure to LH. We confirmed that the gonadotropin-induced decline in cAMP accumulation was indeed due to desensitization of the adenylyl cyclase system itself, and studied some of the requirements and characteristics of this phenomenon (Hunzicker-Dunn and Birnbaumer, 1976*a*, 1976*b*).

Rabbits

To this end we determined adenylyl cyclase activities in homogenates of rabbit ovarian follicles obtained before and at varying times after an ovulatory surge of gonadotropins. Estrus follicles, as previously described, contain a highly LH-responsive adenylyl cyclase system (Table 1). The intravenous injection of an ovulatory dose of HCG results within minutes in a decline in the ability of LH and FSH to stimulate the cyclase system (Fig. 4). By 5 min, both LH-stimulated adenylyl cyclase activity and relative responsiveness have declined more than 50%. As ovulation approaches, the adenylyl cyclase becomes even less responsive to LH and to FSH so that at the time of ovulation--10 to 11 hr after HCG injection--the

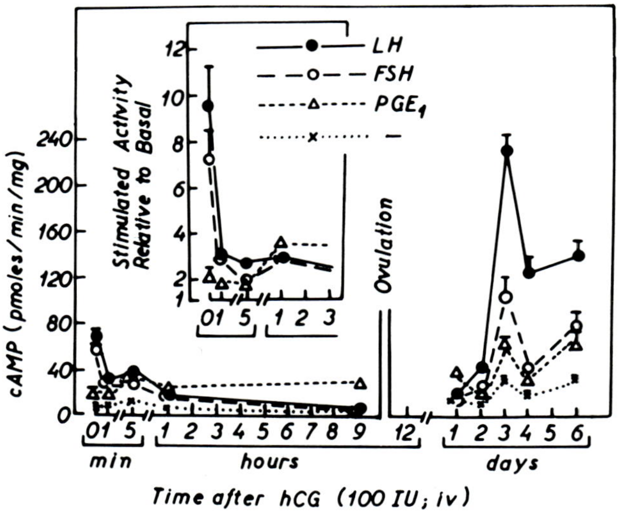

FIG. 4. Effect of the administration of an ovulatory dose of HCG on the adenylyl cyclase activities in rabbit ovarian follicles and resulting CL formed after ovulation. Estrous rabbits (3.5 to 4.5 kg) were given 100 IU HCG (Ayerst, dissolved in 0.9% saline) i.v., and sacrificed by cervical dislocation at the times indicated on the figure. Their ovaries were removed and immediately cooled in iced KRB as described in the legend to Fig. 1 for rat follicles. The "0 min" point represents graafian follicles dissected from ovaries of estrous rabbits, which received no injection. Single points represent one assay (duplicate incubations) in which a minimum of two rabbits was used. Mean ± SEM is shown where two to six such assays were performed. (From Hunzicker-Dunn and Birnbaumer, 1976*a*.)

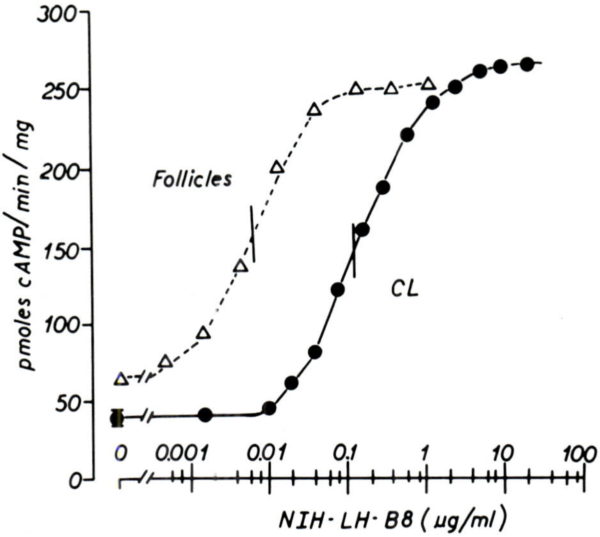

FIG. 5. Sensitivity of rabbit follicles and CL to LH (NIH-LH-B8). Follicles and CL were dissected and homogenized as described in the legend to Fig. 1. Membranes were obtained by filtering the homogenate first through gauze and then through No. 12 silk screen, and then by centrifuging the filtrate at 12,000 × *g* for 15 min. The membranes were washed twice by repelleting and were finally resuspended in the original homogenization buffer, quick frozen in dry-ice acetone bath, and stored at -70°C for not more than 2 months. Adenylyl cyclase assays were carried out at 30°C in 50 μl of a mixture containing 1.5 mM [α-^{32}P]ATP, 5.0 mM $MgCl_2$, 1.0 mM EDTA, 1.0 mM EGTA, 20 mM creatine phosphate (pH 7.0), 0.2 mg/ml creatine kinase, 25 mM Bis-Tris-propane-HCl (BTP) buffer, pH 7.0, 15 μg membrane protein, and the indicated concentrations of LH. (From Birnbaumer et al., 1976.)

cyclase is no longer stimulated by LH or FSH. The cyclase remains desensitized to LH and FSH until 48 hr after HCG injection (equivalent to the time needed for CL formation). By 72 hr the cyclase in the newly formed CL is again sensitive to gonadotropins and, in fact, exhibits an LH-stimulated adenylyl cyclase activity which is four times greater than that of estrus follicles, while the FSH-, PGE_1-, and NaF-stimulated activities of this luteal cyclase equal those in follicles. Although the cyclase system in CL can be stimulated to a greater extent, it is less sensitive to LH stimulation than that in follicles, i.e., higher concentrations of LH are required to stimulate half-maximally

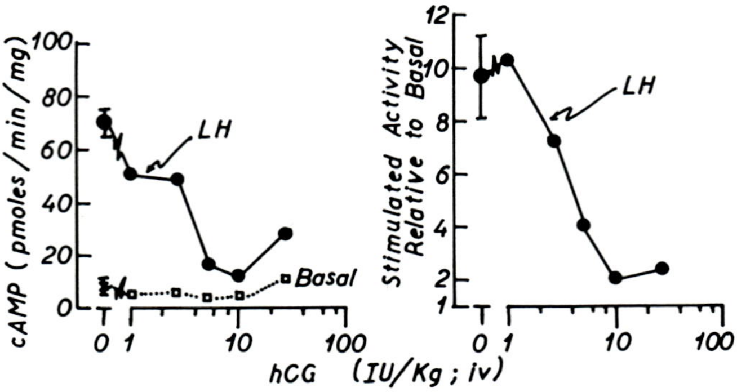

FIG. 6. Effect of the dose of HCG on the responsiveness of adenylyl cyclase to LH. Rabbit follicles were dissected and homogenized 2 hr after injection of the indicated doses of HCG. "0" HCG represents follicles from estrous rabbits which received no HCG injection. For rest of conditions, see legends to Figs. 1 and 4. (From Hunzicker-Dunn and Birnbaumer, 1976*a*.)

the adenylyl cyclase system in CL compared to that in follicles (Fig. 5).

Desensitization of the adenylyl cyclase system to LH is also elicited with endogenous LH, as obtained in rabbits after mating, and follows a time course that is equivalent to that seen following HCG injection (Hunzicker-Dunn and Birnbaumer, 1976*a*).

Both LH- and HCG-induced desensitization of the adenylyl cyclase system in follicles exhibits specific characteristics. Desensitization is an actual loss in the ability of the enzyme to be stimulated by LH and FSH (Fig. 4). It is selective for gonadotropins, since the responsiveness to PGE_1 and to NaF (not shown but giving a ninefold stimulation) is not changed (Fig. 4). It requires ovulatory doses of HCG (> 5 IU HCG/kg body weight, *personal observation*), with subovulatory doses stimulating rather than desensitizing the cyclase system (Fig. 6). Desensitization is obtained after an indomethacin treatment that blocks PG synthesis to the extent of blocking ovulation, suggesting that desensitization is not mediated by PG (Fig. 7). And finally, desensitization can be induced *in vitro* and well as *in vivo* if estrus rabbit follicles are incubated 2 hr in Krebs-Ringer bicarbonate buffer in the presence of 0.8 or 8 μg/ml LH (Fig. 8). At these concentrations, LH promotes a loss of LH- and FSH-stimulated adenylyl cyclase activity that is equal to that in the counterpart *in vivo* experiment, while BSA or a lower concentration of LH does not reduce LH- or FSH-stimulated cyclase activities.

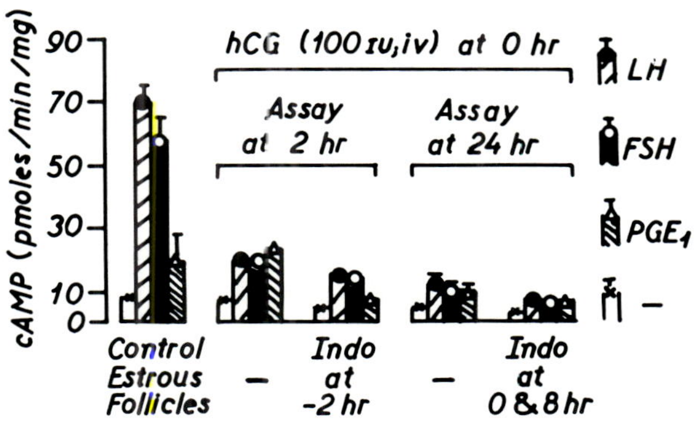

FIG. 7. Lack of effect of indomethacin on the loss of responsiveness of the adenylyl cyclase system to LH in rabbit follicles obtained 2 or 24 hr after the injection of HCG. Indomethacin (70 mg/ml per 3.5 to 4.5 kg rabbit s.c., Sigma Chemical Company) was injected either 2 hr before HCG injection (100 IU i.v.) or with and 8 hr after HCG injection. Basal and LH-stimulated adenylyl cyclase activities measured in follicles obtained from rabbits that received saline and indomethacin injections were equivalent to those measured in control estrous follicles. For rest of conditions, see legends to Figs. 1 and 4. (From Hunzicker-Dunn and Birnbaumer, 1976*a*.)

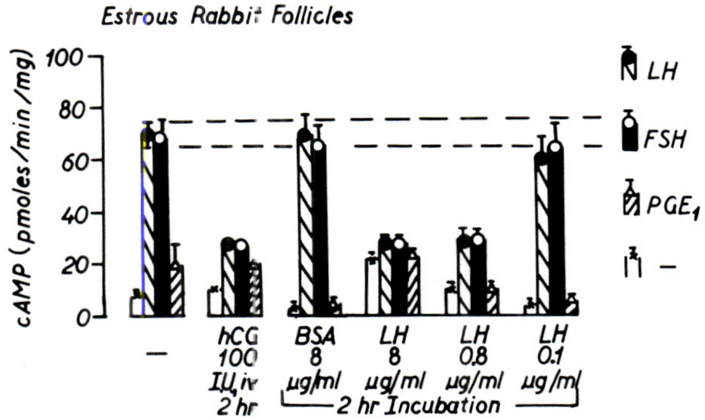

FIG. 8. Effect of *in vitro* exposure of rabbit follicles to BSA or LH on the responsiveness of adenylyl cyclase. Ovaries from estrous rabbits were placed in 0.9% saline at room temperature. Unpopped follicles (about 2 mm in diameter) were dissected and incubated (four to five follicles per flask) at 37°C for 2 hr with continuous gassing (95% O_2/5% CO_2) in 2.5 ml KRB and 1 mg/ml glucose containing either BSA or LH at the concentrations shown on the figure. Follicles were then homogenized and adenylyl cyclase activity was determined in the presence of 10 µg/ml LH, FSH, and PGE_1. "--" represents follicles from estrus rabbits that were not incubated. For rest of conditions see legend to Fig. 1. (From Hunzicker-Dunn and Birnbaumer, 1976*a*.)

Since an ovulatory dose of LH or HCG also induces ovulation of follicles contained in ovaries of pregnant and pseudopregnant rabbits, we investigated whether the adenylyl cyclase system in these follicles is desensitized by a surge of gonadotropins (Hunzicker-Dunn and Birnbaumer, 1976*c*). Prior to receiving an ovulatory dose of gonadotropin, these follicles contain an adenylyl cyclase system that is as responsive to LH as that in follicles from estrus rabbits (> sevenfold stimulation) (Table 2). Within 2 hr after an HCG injection, LH-stimulated adenylyl cyclase activity had declined about 50%. Thus, follicles contained in

TABLE 2. *Effect of HCG injection on adenylyl cyclase activities in homogenates of follicles larger than 2 mm in diameter from ovaries of pregnant and pseudopregnant rabbits*

Treatment[a]	Adenylyl cyclase activities[b]	
	Basal	LH-stimulated
Estrous follicles		
None (n = 5)	8.3 ± 3.8	70.2 ± 11.0
HCG at -2 hr (n = 3)	12.2 ± 5.1	15.6 ± 2.1
Follicles of pregnancy[c]		
None (n = 16)	3.6 ± 0.4	26.7 ± 2.5
HCG at -2 hr (n = 10)	8.8 ± 0.9	17.7 ± 1.6
Follicles of pseudopregnancy[d]		
None (n = 11)	3.7 ± 0.4	29.9 ± 3.6
HCG at -2 hr (n = 10)	6.5 ± 0.8	15.9 ± 1.0

[a]When indicated, 100 IU HCG was injected i.v. per 3.5 to 4.5 kg rabbit 2 hr prior to sacrifice.

[b]Values represent pmoles cAMP formed per min per mg homogenate protein ± SD of the mean. When added to the assay, LH (NIH-LH-B9) was 10 μg per ml.

[c]Follicles were dissected from ovaries of 1- to 28-day pregnant rabbits. There was no significant variation in the activities obtained at various times of pregnancy.

[d]Follicles were dissected from ovaries of rabbits between 1 and 18 days after the induction of pseudopregnancy with HCG (100 IU HCG per 3.5 to 4.5 kg rabbit, i.v.). These activities did not vary significantly throughout pseudopregnancy.

ovaries of pseudopregnant and pregnant rabbits are at least partially desensitized to ovulatory doses of HCG, but appear to do so at a slower rate than follicles from estrous rabbits, which lost more than 70% of their LH-sensitive adenylyl cyclase activity by 2 hr after HCG injection (Fig. 4). It is interesting to note that, while there is no apparent difference in the functional capacity of these luteal follicles as compared to those in estrus rabbits, at least in terms of their ovulability, the adenylyl cyclase in follicles obtained during the luteal phase is considerably less active in terms of LH-stimulated adenylyl cyclase activity even though it has maintained a high responsiveness to LH stimulation.

Rats

We further determined whether the highly responsive LH-sensitive adenylyl cyclase measured in proestrous rat follicles is desensitized following the endogenous surge of gonadotropins that occurs in the afternoon of proestrus in cycling rats. By 2100 hr on proestrus, the responsiveness of the cyclase to LH and FSH has declined about 70%; however, this is due more to an initial increase in basal activity rather than to a decrease in LH-stimulated cyclase activity (Fig. 1). By 2345 hr, LH-stimulated cyclase activity has lost about 50% of its activity. On the morning of estrus, the adenylyl cyclase in freshly ovulated follicles has become completely desensitized (unresponsive) to gonadotropin stimulation. This loss of LH-stimulated adenylyl cyclase activity is not accompanied by a loss of NaF-stimulated cyclase activity, a finding recently confirmed in rat ovaries by Nugent et al. (1975).

We investigated whether ovulation in gonadotropin-treated prepubertal rats is associated with desensitization of the follicular adenylyl cyclase system. As previously discussed, treatment of 26-day old prepubertal rats with PMSG promotes both follicular growth and a highly LH-responsive adenylyl cyclase system within 2 days (Fig. 2). By the third day, the follicles have synchronously ovulated. Concomitant with ovulation is a loss of responsiveness of the adenylyl cyclase system to LH and FSH, comparable to that measured in freshly ovulated follicles on the morning of estrus in cycling rats. Treatment of rats on day 26 with HCG and on day 27 with estradiol benzoate also results in desensitization of a highly responsive LH-sensitive adenylyl cyclase system in ovulated follicles by the morning following the ovulation (Fig. 9).

When the ovulatory surge of gonadotropins is prevented either by Nembutal treatment at 1200 hr on proestrus in cycling rats (Table 3) or by omission of estradiol benzoate injection in HCG-treated prepubertal rats (Fig. 9), follicles do not ovulate and the follicular adenylyl cyclase system retains its responsiveness

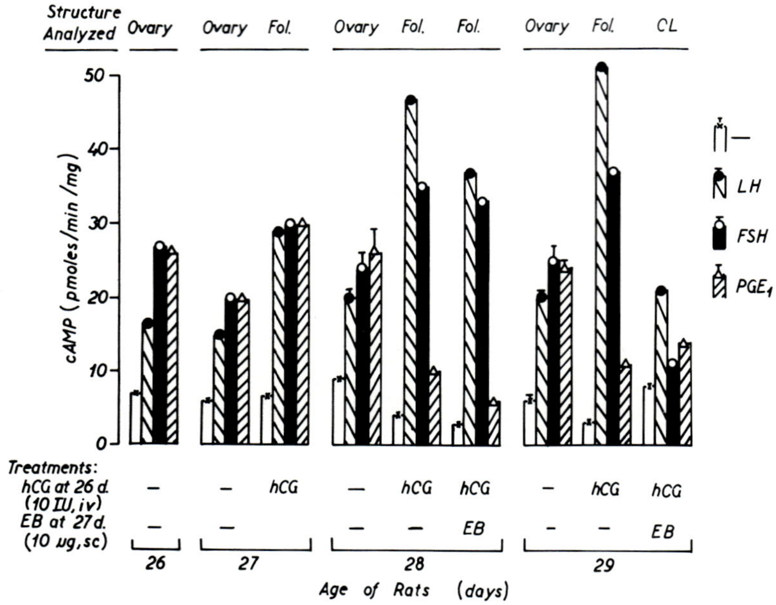

FIG. 9. Effect of HCG or HCG plus estradiol benzoate (EB) on adenylyl cyclase activities in ovarian follicles and resulting CL dissected from prepubertal rat ovaries. Rats received either 10 IU HCG i.v. on day 26 (0800 hr) or 10 IU HCG i.v. on day 26 (0,800 hr) plus 10 µg EB s.c. on day 27 (0800 hr). For rest of conditions, see legends to Figs. 1 and 2. A group of control rats received saline i.v. on day 26 (0800 hr) or saline on day 26 and EB, 10 µg s.c. on day 27. The LH- and FSH-stimulated adenylyl cyclase activities in ovaries from rats that received saline or saline plus EB did not differ from the adenylyl cyclase activity in ovaries from uninjected rats on days 27, 28, or 29 (not shown). Follicles could not be dissected from unstimulated ovaries. (From Hunzicker-Dunn and Birnbaumer, 1976*b*.)

to LH and FSH on the expected day of ovulation. Thus, the adenylyl cyclase system in follicles *only* becomes desensitized to gonadotropin stimulation following a surge of gonadotropins.

Possible Physiological Implications

Desensitization of the follicular adenylyl cyclase system to LH and FSH stimulation would seem to be an obligatory response of a mature follicle to an ovulatory surge of gonadotropins. However, the physiological role of desensitization in the control of follicular

TABLE 3. *Effect of Nembutal on follicular and luteal adenylyl cyclase activity in rat ovaries at proestrus and estrus*

Treatment on proestrus[a]	Adenylyl cyclase activities (pmoles/min/mg protein)											
	Proestrous follicles (2100 hr)				Estrous CL (1000 hr)				Estrous follicles (1000 hr)			
	--	LH	FSH	PGE_1	--	LH	FSH	PGE_1	--	LH	FSH	PGE_1
None[b]	16.3 ±4.2	51.9 ±8.2	38.7 ±5.5	17.7 ±5.0	15.0 ±2.0	18.5 ±4.5	18.6 ±3.2	17.3 ±1.5	--	--	--	--
Nembutal[c]	7.7	45.0	37.6	13.3	--	--	--	--	5.4	53.6	37.5	12.0

[a]Activities were determined in duplicate and agreed within 10%.

[b]Mean ± SEM; n (number of assays) was 3 and 2 for follicles and CL, respectively. Follicles and CL of five rats were used for each assay.

[c]The animals were kept under Nembutal anesthesia from 1200 hr (initiated with 30 mg/kg i.p.) until 1700 hr.

function is, as we shall discuss, not clear, especially since the process is significantly slower in rat follicles than in rabbit follicles.

In Follicular Steroidogenesis

There is considerable evidence indicating that follicular steroidogenesis is mediated by cAMP, as we have previously reviewed. Thus, an ovulatory dose of LH or HCG would be expected to stimulate the adenylyl cyclase system, producing large increases in cAMP levels and in steroid output. This sequence of events has been shown to occur in mature follicles from rabbits where an ovulatory dose of LH or HCG initially stimulates the adenylyl cyclase system (Fig. 4), promoting large increases in cAMP accumulation (Marsh et al., 1973). Concomitantly, steroid levels in follicles and in the ovarian vein rise dramatically, reaching a maximum about 4 hr after the ovulatory surge of gonadotropins (Hilliard and Eaton, 1971; Hilliard et al., 1974; Younglai, 1972; Younglai, 1974). Steroid levels then decline to unmeasurable levels prior to ovulation (Hilliard and Eaton, 1971; Hilliard et al., 1974; Mills and Savard, 1973; Younglai, 1972). Once steroidogenesis is terminated in the follicle, it cannot be reinstituted by exogenous gonadotropins (Mills et al., 1972; Mills and Savard, 1973).

We have shown that within minutes after activation of the adenylyl cyclase system in rabbit follicles by an ovulatory surge of LH or HCG, the enzyme is becoming unresponsive to its own stimulant. It seems reasonable, therefore, that the result of this desensitization of adenylyl cyclase is both a cessation in cAMP formation, as measured by Marsh et al. (1973), and a consequent cessation in steroid output, as described below. The time span during which the follicle is unable to produce steroids--when ovulation and differentiation of follicular cells into luteal cells is occurring--correlates with the time when the adenylyl cyclase system is completely unresponsive to gonadotropin stimulation. Possibly, the desensitized adenylyl cyclase system is the cause of the impaired steroid synthetic capacity of the rabbit follicle, since gonadotropins could not stimulate cAMP formation from an unresponsive adenylyl cyclase system. The inability of gonadotropins to stimulate the cyclase system and ultimately steroid synthesis could also be due to a loss of HCG binding sites. However, the data of Mills and McPherson (1974) suggest that there is no loss of HCG binding sites between 0 and 12 hr after coitus.

This basic sequence of events following the ovulatory surge of gonadotropins has been shown to occur in rat follicles, with some modifications. The initial steps, that of cyclase activation, cAMP formation, and steroidogenesis, appear to be the same; however, later stages seem to be delayed. Desensitization of the cyclase system is slower, with a 50% loss occurring 8 to 10 hr after the gonadotropin surge (Fig. 1), and steroidogenesis persists at peak rates until 6 to 8 hr after the surge of gonadotropins (Butcher et al., 1974; Kalra and Kalra, 1974; Schwartz, 1974; Smith et al., 1975). However, around the time of ovulation in rat follicles, as in rabbit follicles, steroidogenesis is nil and remains low until the CL forms. Thus, an unresponsive cyclase system is accompanied by steroidogenic inactivity also in rat follicles.

Ovulation

While desensitization of the cyclase system appears to be associated with the period of steroidogenic quiescence in the preovulatory and postovulatory follicle, it remains to be seen whether desensitization is a necessary step in the complex sequence of reactions that precludes ovulation. Although the adenylyl cyclase system in rabbit follicles was found to be completely unresponsive to gonadotropin stimulation long before ovulation, the cyclase system in rat follicles was found to decline only about 50% when tested between 8 and 10 hr after the endogenous gonadotropin surge. It seems reasonable to assume, therefore, that *complete* desensitization may not be a prerequisite for ovulation to occur. It remains to be seen whether incomplete desensitization would interfere with the ovulatory process.

However, lowered cAMP levels, the consequence of a desensitized adenylyl cyclase system, may be required for ovulation to proceed, since high intrafollicular levels of cAMP can prevent ovulation (Hunzicker-Dunn, *personal observation*; LeMaire et al., 1972; Ryan and Coronel, 1969). The finding by Challis et al. (1974) that dibutryl cAMP inhibits PGF synthesis by rabbit granulosa cells in culture provides an explanation for the observed inhibition of ovulation by cAMP. Since PG synthesis is a necessary prerequisite for ovulation, then these data imply that the decline in cAMP--via desensitization of adenylyl cyclase--may be a necessary prerequisite for PG accumulation and therefore for ovulation. Contrary to these data of Challis, however, Marsh and co-workers have shown that both cAMP and LH promote increases in PG levels in rabbit follicles *in vitro* (Marsh et al., 1974). PG synthesis was not an immediate response to LH, and did not increase significantly until 5 hr after exposure. Similarly, cAMP promoted what seems to be an equally delayed response in PG synthesis; however, PG levels were measured only at 5 hr after exposure to cAMP. It may be that careful dissection of the time-dependent effects of PG will be necessary in order to clarify these apparent contradictions introduced by the data of Challis and Marsh, since PG affects cell function in more than one way. For example, only a delayed synthesis of PG could associate them with follicular rupture, since blockade of PG synthesis at the time of an ovulatory surge of gonadotropins does not block ovulation (O'Grady et al., 1972).

While desensitization of the follicular adenylyl cyclase system may play an important role in the processes leading up to ovulation, it is not always followed by ovulation, as seen in the following examples: (a) large, 2- to 3-mm unovulated (atretic?) follicles dissected from rabbit ovaries 16 hr after an HCG injection contained an adenylyl cyclase system that was unresponsive to gonadotropin stimulation (Table 4); (b) unovulated follicles dissected 24 hr after an HCG

TABLE 4. *Adenylyl cyclase activity in homogenates of ovulated and unovulated follicles dissected before and 16 hr after an HCG injection*

		Adenylyl cyclase activities (pmoles cAMP/min/mg protein)	
Rabbit tissue	Treatment[a]	Basal	LH-stimulated
Follicles	None	8.3 ± 1.7[b]	70.2 ± 5.3[b]
Ovulated follicles	HCG at -16 hr	32.3 ± 10.2[c]	26.5 ± 2.1[c]
Intact follicles	HCG at -16 hr	29.7 ± 3.1[c]	38.0 ± 4.2[c]

[a]When indicated, 100 IU HCG was injected i.v. per 3.5 to 4.5 kg rabbit 16 hr prior to sacrifice.

[b]Values are mean ± SEM (n = 5) of assays in each of which tissues from two rabbits were analyzed in duplicate.

[c]Values are mean ± SD of an assay in which tissues from two rabbits were assayed in triplicate.

injection in which ovulation was blocked by indomethacin treatment contained an adenylyl cyclase that was desensitized (Fig. 7). Both of these instances suggest that the presence of an unresponsive adenylyl cyclase in follicles means only that the follicle has received a surge of gonadotropins. Any interruption in the sequence of biochemical events preceding ovulation could prevent ovulation, as is the case in the second example, in which an inhibition of PG synthesis prevented ovulation.

Luteal Steroidogenesis

The adenylyl cyclase system becomes sensitive to gonadotropin stimulation between 38 and 62 hr after ovulation in the rabbit (Fig. 4) and about 34 hr after ovulation in the rat (Hunzicker-Dunn and Birnbaumer, 1976*b*). This acquisition of hormonal responsiveness by the cyclase system corresponds in both species to the time when the newly formed CL begins to synthetize steroids (Butcher et al., 1974; Challis et al., 1973; Hilliard and Eaton, 1971; Smith et al., 1975). The striking temporal correlation between these two events clearly suggests that a close functional relationship exists between the LH-responsive cyclase system and steroidogenesis.

PROTEIN KINASES IN OVULABLE FOLLICLES

It is generally accepted that cAMP mediates many of the effects of peptide hormones by regulating the activity of cAMP-dependent protein kinases. cAMP has been shown to activate protein kinase by binding to the regulatory subunit (R) of the inactive holoenzyme (RC), thereby releasing the active (cAMP-insensitive) catalytic subunit (C) which catalyzes the transfer of phosphate from ATP to cellular proteins (Krebs, 1972; Langan, 1973). Presumably, the phosphorylation of specific cellular protein(s) alters cellular function in such a way that the known hormonal response occurs. Two studies have demonstrated that preovulatory rat follicles contain protein kinase activity whose phosphorylative capacity is enhanced by cAMP (Lamprecht et al., 1973; Tsafriri et al., 1972). Moreover, the studies of Channing and Seymour (1970) *imply* that a cAMP-dependent protein kinase(s) may be present and functional even in small (1 to 2 mm) porcine follicles, since they were able to elicit morphological and functional (steroidogenic) luteinization with dibutryl cAMP (but not with LH and/or FSH as they could in larger porcine follicles). We have expanded the studies of Tsafriri et al. (1972) and Lamprecht et al. (1973) by investigating the activity and multiplicity of follicular protein kinases during the final stages of follicular maturation.

HCG-Induced Activation of Protein Kinases

Estrous rabbit follicles contain cAMP-dependent protein kinases that are rapidly activated (dissociated into regulatory and catalytic subunits) following exposure of the follicles to an ovulatory dose of HCG (100 IU HCG/3.5 to 4.5 kg rabbit, i.v.). By determining the protein kinase activity ratio (protein kinase activity -cAMP/+cAMP), one obtains an index of the relative degree of dissociation or activation of cAMP-dependent protein kinase (Soderling et al., 1974). Thus, within 1 min following the injection of HCG, protein kinase activity ratios, measured in 10,000 × *g* supernatant fractions obtained from homogenates of dissected follicles have increased significantly ($p < 0.005$), reaching peak levels of activity within 10 min ($p < 0.001$) (Fig. 10). Thereafter, protein kinase activity ratios slowly decline. Follicular protein kinases are also activated following the injection of an ovulatory dose of LH ($p < 0.001$) and, to a lesser extent, following the injection of FSH ($p < 0.05$); ACTH, PRL, and epinephrine are without effect (Table 5).

TABLE 5. *In vivo effect of various hormones on the activation of follicular protein kinases*

Treatment[a]	Protein kinase activity ratio (-cAMP/+cAMP)[b]
None	0.57 ± 0.04
Saline	0.58 ± 0.03
HCG (100 IU)	0.92 ± 0.02[c]
LH (50 μg)	1.02 ± 0.05[c]
FSH (125 μg)	0.80 ± 0.08[d]
ACTH (200 μg)	0.60 ± 0.04
PRL (200 μg)	0.59 ± 0.01
Epinephrine (200 μg)	0.52 ± 0.01

[a]Injections were given i.v. in 0.5 ml saline and rabbits were sacrificed after 10 min.

[b]Values are mean ± SEM of assays in each of which tissues from two rabbits were analyzed in duplicate.

[c]Differs from control follicles at $p < 0.001$.

[d]Differs from control follicles at $p < 0.05$.

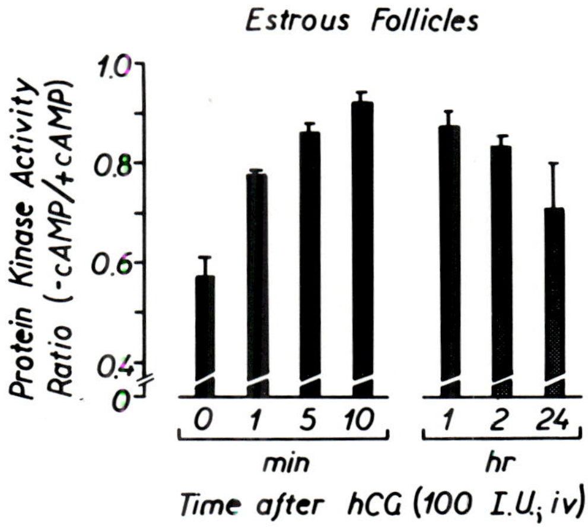

FIG. 10. Effect of the administration of an ovulatory dose of HCG on the protein kinase activity ratios in rabbit ovarian follicles. Estrous rabbits were given 100 IU HCG i.v. and were sacrificed at the times indicated on the figure as described in the legend to Fig. 4. Ovaries were removed and immediately cooled in an iced solution containing 0.5 M NaCl, 10 mM EDTA, 0.5 M 3-isobutyl-l-methyl xantine, and 10 mM potassium phosphate buffer, pH 6.5. Graafian follicles or CL were dissected and homogenized (Tenbroeck-type, Kontes) in 10 volumes of the above buffered 0.5 M NaCl solution. The homogenate was centrifuged at 10,000 × *g* for 10 min and 10 µl aliquots of the supernatant were assayed for protein kinase activity in incubations for 10 min at 37°C in 225 µl of a medium containing 44 µM [γ-^{32}P]ATP, 22 mM α-glycerol phosphate, 4.4 mM NaF, 4.4 mM magnesium acetate, 0.89 mM theophylline, 0.4 mM dithiothreitol, and 100 µg protamine sulfate, pH 7.0, in the absence and presence of 4.5×10^{-6} M cAMP. The reaction was terminated with the addition of 1 ml of 15% trichloroacetic acid containing 1% sodium dodecyl sulfate, and the [^{32}P]protamine sulfate was precipitated on Millipore filters (0.3 µm). Filters were washed and counted in PCS liquid scintillation fluid (Amersham Searle). Values are mean ± SEM of assays in each of which follicles from two or three rabbits were analyzed in duplicate.

This HCG-induced activation of protein kinases can also be seen in profiles obtained after Sephadex G-100 chromatography, a technique that separates the catalytic subunit C from the inactive RC holoenzyme. Chromatography of a 10,000 × *g* supernatant fraction from homogenates of estrous rabbit follicles demonstrates a major peak of protein kinase activity that elutes with the void volume and that is sensitive to cAMP stimulation (Fig. 11). When follicles are exposed to an

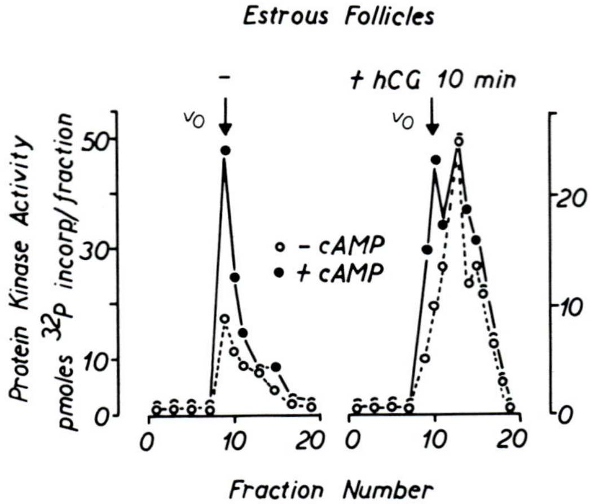

FIG. 11. Effect of the administration of an ovulatory dose of HCG on the elution profiles of protein kinase activity by Sephadex G-100 chromatography. Rabbits were sacrificed 10 min after the HCG injection (100 IU HCG, i.v.), their ovaries were removed, and their follicles were dissected, homogenized, and centrifuged as described in the legend to Fig. 10. A volume of 0.47 ml (about 1.5 mg protein) of the 10,000 × *g* supernatant was filtered on a Sephadex G-100 (1 × 30 cm) column, eluting 1 ml per fraction with 0.5 M NaCl. The void volume (V_O) of each column was determined by the elution of blue dextran. Each fraction (25 μl) was assayed for protein kinase activity in the presence (solid circles) and absence (open circles) of cAMP, as described in the legend to Fig. 10. Points on the graph represent results from single experiments, each of which required two to three rabbits. Each experiment was repeated at least three times to assure reproducibility. "--" represents the filtration of the supernatant of follicles obtained from estrous rabbits that did not receive injections. Control saline injections did not alter the elution profiles seen in follicles obtained from uninjected estrous rabbits.

ovulatory dose of HCG (100 IU HCG, i.v.) for 10 min, and the 10,000 × *g* supernatant fraction from the homogenates of these follicles is filtered on Sephadex G-100, a second peak of protein kinase activity appears. This second peak of activity elutes after the void volume and represents the catalytic subunit, since it is not stimulated by cAMP but is inhibited (> 70%) by saturating concentrations of the heat-stable protein kinase inhibitor protein isolated from rabbit skeletal muscle. This heat-stable protein kinase inhibitor has been shown to interact specifically with the free catalytic subunit of cAMP-dependent protein kinase

(Ashby and Walsh, 1972). Accordingly, the studies shown in Figs. 10 and 11 both indicate that exposure of follicles to an ovulatory dose of HCG results in the dissociation of follicular cAMP-dependent protein kinases into regulatory and catalytic subunits. Homogenization of follicles was conducted in buffer containing 0.5 M NaCl, which has been previously shown to prevent the reassociation of the Type II protein kinase subunits (Soderling et al., 1974; *personal observation*). In the absence of 0.5 M NaCl we were unable to demonstrate a significant degree of activation of follicular protein kinases following HCG injection, probably due to the rapid reassociation of R and C in the absence of 0.5 M NaCl.

Multiplicity of Protein Kinases in Preovulatory and Postovulatory Follicles

Subcellular fractionation of estrous rabbit follicles reveals that 88.0 ± 1.9% (n = 6) of the protein kinase activity is located in the cytosol (105,000 × *g* supernatant fraction), and that it is primarily cAMP-dependent [cAMP-sensitive activity is inhibited 79.8 ± 0.7% (n = 5) by saturating concentrations of the heat-stable inhibitor]. Because most of the protein kinase activity is found in the cytosol of follicles, we investigated the multiplicity of cytosol protein kinases before and at varying times after the injection of an ovulatory dose of HCG.

DEAE cellulose chromatography of the cytosol reveals that whole estrous rabbit follicles possess primarily one major type of protein kinase (PK) activity which elutes as a split peak (PK 3 on Fig. 12, upper left panel) between 0.2 M and 0.3 M potassium phosphate. As judged by the sensitivity both to inhibition by saturating concentrations of the heat-stable inhibitor and to stimulation by cAMP, both forms of PK 3 appear to be cAMP-dependent protein kinases (cAMP-sensitive activity is inhibited > 80% by the heat stable inhibitor).

The injection of an ovulatory dose of HCG promotes the appearance of a transitory peak of protein kinase activity, PK 1 (Fig. 12, lower left panel), which elutes between 0.04 and 0.05 M potassium phosphate (around fractions 6 to 9). PK 1 is inhibited about 45% by saturating concentrations of the heat-stable protein kinase inhibitor, indicating that the activity in PK 1 consists both of cAMP-dependent and cAMP-independent protein kinases. PK 1 appears within 10 min following HCG injection, and is still present at 60 min

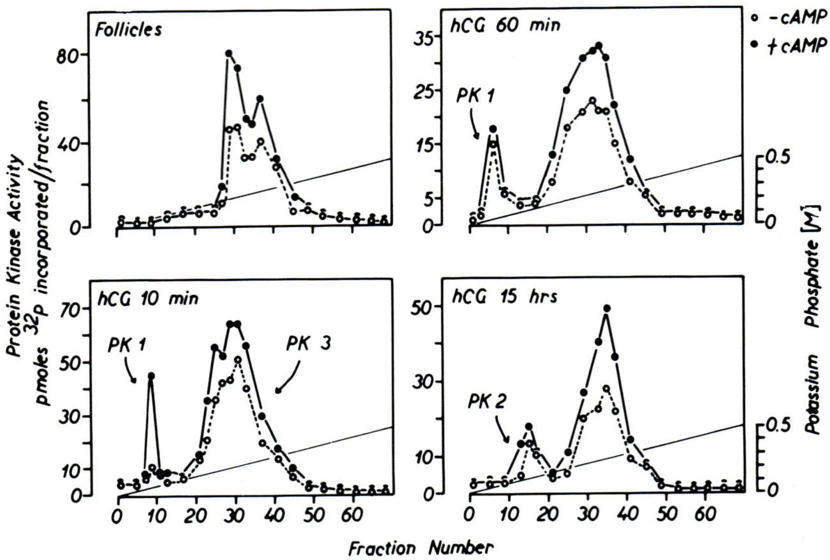

FIG. 12. Early effect of the administration of an ovulatory dose of HCG (100 IU HCG i.v.) on the elution profiles of protein kinase by DEAE cellulose chromatography. Rabbits were sacrificed at the times indicated on the figure. Their ovaries were removed and immediately cooled in an iced solution containing 10 mM Tris-HCl, pH 7.5, 3 mM $MgCl_2$, and 0.32 M sucrose. Follicles were dissected and homogenized (Tenbroeck-type, Kontes) in 10 volumes of the Tris-Mg-sucrose solution. The homogenate was centrifuged at 1,000 × *g* for 10 min to remove nuclei and cell debris. The resulting supernatant was centrifuged at 10,000 × *g* for 30 min to remove mitochondria and lysosomes. A cytosol was obtained by centrifuging the 10,000 × *g* supernatant at 105,000 × *g* for 60 min. The cytosol (about 5 to 10 mg protein) was chromatographed on a 1 × 3 cm DEAE cellulose column, eluting about 0.75 ml fraction with a 50-ml salt gradient of potassium phosphate extending from 10 mM to 500 mM. Fifty-microliter aliquots of every fraction were assayed for protein kinase activity in the presence (solid circles) and absence (open circles) of 4.5×10^{-8} M cAMP as described in the legend to Fig. 10. Follicles: PK 3 elutes in fractions 29 to 40. HCG 10 min: PK 1 elutes in fractions 7 to 10 and PK 3 elutes in fractions 25 to 35. HCG 60 min: PK 1 elutes in fractions 5 to 8 and PK 3 elutes in fractions 25 to 40. HCG 15 hr: PK 2 elutes in fractions 11 to 17 and PK 3 elutes in fractions 29 to 40. Points on the graph represent results from single experiments each of which required 8 to 10 rabbits. Each experiment was repeated at least three times to assure reproducibility.

(Fig. 12, upper right panel), but has disappeared following ovulation, i.e., 15 hr post-HCG injection (Fig. 12, lower right panel; Fig. 13, lower left panel).

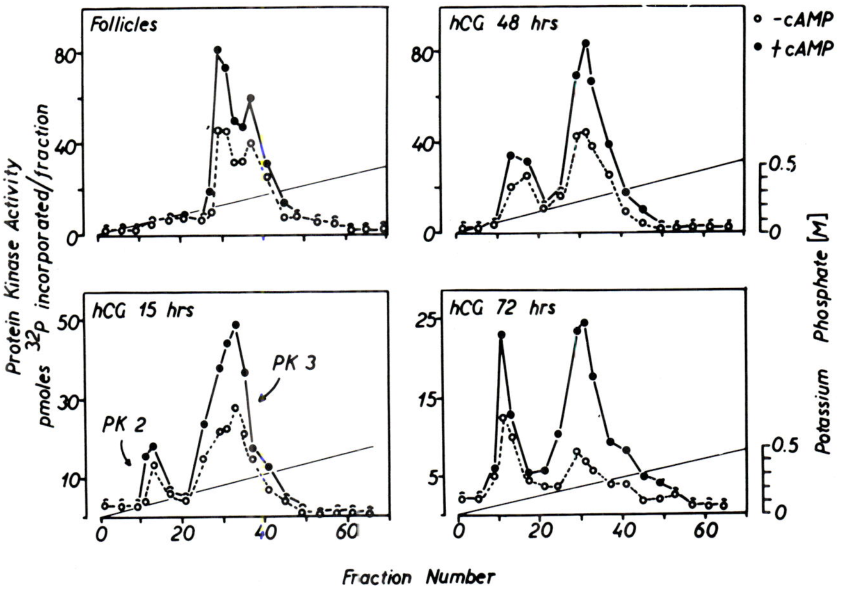

FIG. 13. Delayed effect of the administration of an ovulatory dose of HCG on the elution profiles of protein kinase by DEAE cellulose chromatography. Follicles: PK 3 elutes in fractions 29 to 40. HCG 15 hr: PK 2 elutes in fractions 11 to 17 and PK 3 elutes in fractions 29 to 40. HCG 48 hr: PK 2 elutes in fractions 10 to 17 and PK 3 elutes in fractions 27 to 35. HCG 72 hr: PK 2 elutes in fractions 11 to 15 and PK 3 elutes in fractions 25 to 35. For rest of conditions, see legend to FIG. 12.

A second, more delayed peak of protein kinase activity is apparent by 15 hr after HCG injection, PK 2 (Fig. 12, lower right panel; Fig. 13, lower left panel). As follicular differentiation into CL procedes, the relative activity of PK 2 increases, reaching levels nearly equivalent to those of the major peak--PK 3--by 72 hr post-HCG injection. Throughout luteinization, PK 2 has also become increasingly sensitive to stimulation by cAMP and to inhibition by saturating concentrations of the heat-stable protein kinase inhibitor protein. By 72 hr, PK 2 is almost completely activated by cAMP and is inhibited 60 to 70% by the heat-stable inhibitor. PK 2 is occasionally detected as a very minor peak of protein kinase activity (< 10% of the area of PK 3) in estrus follicles and in follicles obtained at 10 and 60 min following HCG injection. However, PK 2 is neither stimulated by cAMP nor inhibited by saturating concentrations of the heat-stable protein kinase inhibitor, indicating that PK 2 in preovulatory follicles is a cAMP-independent protein kinase.

Throughout follicular differentiation and luteinization, the relative activity and cAMP-dependency of PK 3 has remained unaltered (Figs. 12 and 13), suggesting that its function relates to a step or steps in cellular regulation that are not restricted to those occurring specifically within follicles or CL. However, the observation that both PK 1 and PK 2 appear only in specific structures and at specific times suggests that PK 1, on the one hand, may be involved in an early step of the preovulatory process while PK 2, on the other hand, may be involved in CL function. We are currently investigating the roles of PK 1 and PK 2 in follicular and luteal functions.

CONCLUDING REMARKS

We have described two separate effects that LH imposes on the adenylyl cyclase system: stimulation and desensitization. The distinction seems to be in the concentration of LH which pertubates the receptor-coupled adenylyl cyclase system. Thus, subovulatory levels of LH stimulate the adenylyl cyclase system, promoting increases in cAMP levels and initiation of the physiological response--steroidogenesis. Ovulatory levels of LH also stimulate the adenylyl cyclase system, at least initially, again producing increases in cAMP levels and initiation of the physiological response--steroidogenesis, ovum maturation, and luteinization. Shortly after exposure to ovulatory hormone levels, however, the cyclase system slowly becomes unresponsive to its stimulant, resulting in a decline in cAMP levels that may terminate a physiological response--steroidogenesis--and/or may facilitate a response--PG synthesis and ultimately follicular rupture.

The biochemical processes that occur between exposure of the follicle to an ovulatory surge of gonadotropins and CL formation are complex and are not understood. Many require RNA and protein synthesis (Gorski and Pandos, 1966; Keyes et al., 1972; Lieberman et al., 1975; Pool and Lipner, 1966; Pool and Lipner, 1969; Rondell, 1974; Tsafriri et al., 1972; Younglai, 1975). It is not unreasonable to suspect that cAMP-dependent protein kinases serve as intermediates in some of these processes. We have demonstrated that a cAMP-dependent protein kinase(s) is rapidly activated by the ovulatory surge of gonadotropins, presumably via adenylyl cyclase-stimulated increases in cAMP synthesis. We have also shown that the ovulatory surge of HCG promotes, within minutes, the appearance of a transitory peak of follicular protein kinase activity (PK 1). We do not know if the appearance of this protein kinase is dependent upon cyclase-mediated increases in cAMP. In addition, we have shown that a second, seemingly new species of cAMP-dependent protein kinase activity (PK 2) appears with HCG-induced luteinization of follicular cells. As with PK 1, we do not know if the appearance of PK 2 is related to cyclase-mediated increases in cAMP.

In conclusion, an ovulatory dose of LH promotes the following sequence of events in preovulatory follicles:

1. LH binds to a specific plasma membrane receptor site.
2. A highly LH-responsive adenylyl cyclase system is rapidly activated, resulting in increased concentrations of intracellular cAMP levels.
3. cAMP then activates cAMP-dependent protein kinases, unmasking the active catalytic subunit.
4. The catalytic subunit, through phosphorylation of specific cellular proteins at the nuclear and/or cytoplasmic level, initiates the biochemical processes that stimulate the "normal" hormonal response.
5. A transitory peak of protein kinase activity appears, an enzyme that may well be necessary for early preovulatory follicular functions.
6. The adenylyl cyclase system slowly becomes unresponsive (desensitized) to its own stimulant, cAMP synthesis ceases, and cAMP-stimulated processes are terminated, while cAMP-inhibited processes are activated.
7. A luteal-specific species of protein kinase activity appears and presumably mediates CL function via adenylyl cyclase-stimulated cAMP activation.

We believe that the responsiveness of the adenylyl cyclase system to gonadotropic stimulation is a prime determinant of the ultimate response evoked by the stimulatory hormone. Although a number of hormone-specific alterations in protein kinase activities and multiplicities have been demonstrated, the mechanism by which these kinases regulate the final hormonal response(s) is not known and clearly requires further investigation.

ACKNOWLEDGMENT

This work was supported in part by USPHS grants HD-06513, HD-09541, and HD-07495, and by grants from the Clayton Foundation and the Research and Educational Fund, Northwestern Memorial Hospital.

REFERENCES

1. Ahrén, K., Herlitz, H., Nilsson, L., Perklev, T., Rosberg, S., and Selstam, G. (1974): Gonadotropins and cyclic AMP in various compartments of the rat ovary. In: *Gonadotropins and Gonadal Function*, edited by N. R. Moudgal, pp. 364-375. Academic Press, New York.
2. Armstrong, D. T., Moon, Y. S., and Zamecnik, J. (1974): Evidence for a role of ovarian prostaglandins in ovulation. In: *Gonadotropins and Gonadal Function,* edited by N. R. Moudgal, pp. 345-356. Academic Press, New York.
3. Ashby, C. D., and Walsh, D. A. (1972): Characterization of the interaction of a protein inhibitor with adenosine 3',5'-monophosphate-dependent protein kinases. *Journal of Biological Chemistry,* 247:6637-6642.
4. Baranczuk, R., Cheatum, S. G., and Fainstat, T. (1974): Cultivated corpus luteum and ovarian follicles. Abstracts of the Seventh Society for the Study of Reproduction Meeting, p. 66 (Abstract 52).
5. Birnbaumer, L., and Rodbell., M. (1969): Adenyl cyclase in fat cells. II. Hormone receptors. *Journal of Biological Chemistry,* 244:3477-3482.
6. Birnbaumer, L., Yang, P.-C., Hunzicker-Dunn, M., Bockaert, J., and Duran, J. M. (1976): Adenylyl cylcase activities in ovarian tissues. I. Homogenization and conditions of assay in Graafian follicles and corpora lutea of rabbits, rats and pigs; regulation by ATP, and some comparative properties. *Endocrinology,* 99:163-184.
7. Bockaert, J., Hunzicker-Dunn, M., and Birnbaumer, L. (1976): Hormone-stimulated desensitization of hormone-dependent adenylyl cyclase. *Journal of Biological Chemistry,* 251:2653-2663.
8. Butcher, R. L., Collins, W. E., and Fugo, N. W. (1974): Plasma concentration of LH, FSH, prolactin, progesterone and estradiol-17β throughout the 4-day estrous cycle of the rat. *Endocrinology,* 94:1704-1708.
9. Challis, J. R. G., Davies, I. J., and Ryan, K. F. (1973): The concentrations of progesterone, estrone and estradiol-17β in the plasma of pregnant rabbits. *Endocrinology*, 93:971-976.
10. Channing, C. P. (1970*a*): Influences of the *in vivo* and *in vitro* hormonal environment upon luteinization of granulosa cells in tissue culture. *Recent Progress in Hormone Research,* 26:589-622.
11. Channing, C. P. (1970*b*): Effects of stage of estrous cycle and gonadotropins upon luteinization of porcine granulosa cells in culture. *Endocrinology*, 87:156-164.
12. Channing, C. P. (1973): Regulation of luteinization in granulosa cell cultures. In: *The Regulation of Mammalian Reproduction*, edited by S. J. Segal, P. A. Corfman, and P. G. Condliffe, pp. 508-518. Charles C. Thomas, Springfield, Illinois.

13. Channing, C. P. (1974): Temporal effects of LH, hCG, FSH and dibutryl cyclic 3',5'-AMP upon luteinization of Rhesus monkey granulosa cells in culture. *Endocrinology*, 94:1215-1223.
14. Channing, C. P., and Kammerman, S. (1973): Characteristics of gonadotrophin receptors of procine granulosa cells during follicle maturation. *Endocrinology*, 92:531-540.
15. Channing, C. P., and Seymour, J. F. (1970): Effects of dibutryl cyclic 3',5'-AMP and other agents upon luteinization of porcine granulosa cells in culture. *Endocrinology*, 87:165-169.
16. Cirillo, V. J., Anderson, O. F., Ham, E. A., and Geatkin, R. B. L. (1969): The effect of luteinizing hormone and adenosine 3',5'-monophosphate on granulosa cells in culture. *Experimental Cell Research*, 57:139-142.
17. Ellsworth, L. R., and Armstrong, D. T. (1973): Luteinization of transplanted ovarian follicles in the rat induced by dibutryl cyclic AMP. *Endocrinology*, 92:840-846.
18. Ellsworth, L. R., and Armstrong, D. T. (1974): Inhibition of luteinization of transplanted rat ovarian follicles by polyphoretic phosphate. *Endocrinology*, 94:892-896.
19. Gorski, J., and Padnos, D. (1966): Translational control of protein synthesis and the control of steroidogenesis in the rabbit ovary. *Archives in Biochemistry and Biophysics*, 113:100-106.
20. Grinwich, D. L., Kennedy, T. G., and Armstrong, D. T. (1972): Dissociation of ovulation and steroidogenic actions of luteinizing hormone in rabbits with indomethacin, an inhibitor of prostaglandin biosynthesis. *Prostaglandins*, 1:89-96.
21. Hilliard, J., and Eaton, L. W., Jr. (1971): Estradiol-17β, progesterone and 20α-hydroxypregn-4-en-3-one in rabbit ovarian venous plasma. II. From mating through implantation. *Endocrinology*, 89:522-527.
22. Hilliard, J., Scaramuzzi, R. J., Pang, C.-N., Penardi, R., and Sawyer, C. H. (1974): Testosterone secretion by rabbit ovary *in vivo*. *Endocrinology*, 94:267-271.
23. Hunzicker-Dunn, M., and Birnbaumer, L. (1976*a*) Adenylyl cyclase activities in ovarian tissues. II. Regulation of responsiveness to LH, FSH and PGE_1 in the rabbit. *Endocrinology*, 99:185-197.
24. Hunzicker-Dunn, M., and Birnbaumer, L. (1976*b*): Adenylyl cyclase activities in ovarian tissues. III. Regulation of responsiveness to LH, FSH and PGE_1 in the prepubertal, cycling, pregnant and pseudopregnant rat. *Endocrinology*, 99:198-210.
25. Hunzicker-Dunn, M., and Birnbaumer, L. (1976): Adenylyl cyclase activities in ovarian tissues. IV. Gonadotrophin-induced desensitization of the luteal adenylyl cyclase throughout pregnancy and pseudopregnancy in the rabbit and the rat. *Endocrinology*, 99:211-222.

26. Kakiuchi, S., and Rall, T. W. (1968): The influence of chemical agents in the accumulation of adenosine 3',5'-phosphate in slices of rabbit cerebellum. *Molecular Pharmacology,* 4:367-378.
27. Kalra, S. P., and Kalra, P. S. (1974): Temporal interrelationship among circulating levels of estradiol, progesterone and LH during the rat estrous cycle: Effects of exogenous progesterone. *Endocrinology,* 95:1711-1718.
28. Kao, L. W. L., and Nalbandov, A. V. (1972): The effect of antiadrenergic drugs on ovulation in hens. *Endocrinology,* 90:1343-1349.
29. Keyes, P. L., Canastar, G. D., and Miller, J. B. (1972): Studies on luteinization of transplanted rabbit follicles exposed to LH, actinomycin D, or adenosine 3',5'-cyclic monophosphate *in vitro*. *Endocrinology,* 91:197-205.
30. Koch, Y., Zor, U., Pomerantz, S., Chobsieng, P., and Lindner, H. R. (1973): Intrinsic stimulatory action of follicle stimulating hormone on ovarian adenylate cyclase. *Journal of Endocrinology,* 58:677-678.
31. Kolena, J., and Channing, C. P. (1971): Stimulatory effects of gonadotropins on the formation of cyclic adenosine 3',5'-monophosphate by porcine granulosa cells. *Biochimica et Biophysica Acta,* 252:601-606.
32. Kolena, J., and Channing, C. P. (1972): Stimulatory effects of LH, FSH and prostaglandins upon cyclic 3',5'-AMP levels in porcine granulosa cells. *Endocrinology,* 90:1543-1550.
33. Krebs, E. G. (1972): Protein kinases. In: *Current Topics in Cellular Regulation,* edited by B. L. Hornecker and E. R. Stadtman, pp. 99-133. Academic Press, New York.
34. Krishna, G., Weiss, B., and Brodie, B. B. (1968): A simple, sensitive method for assay of adenylyl cyclase. *Journal of Pharmacology and Experimental Therapeutics,* 163:379-385.
35. Lamprecht, S. A., Zor, U., Tsafriri, A., and Lindner, H. R. (1973): Action of prostaglandin E_2 and of luteinizing hormone on ovarian adenylate cyclase, protein kinase and ornithine decarboxylase activity during postnatal development and maturity in the rat. *Journal of Endocrinology,* 57:217-233.
36. Langan, T. A. (1973): Protein kinases and protein kinase substrates. In: *Advances in Cyclic Nucleotide Research,* Vol. 3, edited by P. Greengard and G. A. Robison, pp. 99-153. Raven Press, New York.
37. Lee, C. Y. (1976): The porcine ovarian follicle: III. Development of gonadotropin receptors associated with increase in adenyl cyclase activity during follicle maturation. *Endocrinology,* 99:42-48.
38. Lee, C. Y., and Ryan, R. J. (1973): Interaction of ovarian receptors with human chorionic gonadotrophin. *Biochemistry,* 23:4609-4614.
39. LeMaire, W. J., Mills, T., Ito, Y., and Marsh, J. M. (1972): Inhibition by 3',5'-cyclic AMP of luteinization induced by intrafollicular injection of luteinizing hormone. *Biology*

of Reproduction, 6:109-116.

40. LeMaire, W. J., Yang, N. S. T., Behrman, H. H., and Marsh, J. M. (1973): Preovulatory changes in the concentration of prostaglandins in rabbit Graafian follicles. *Prostaglandins*, 3:367-376.
41. Lieberman, M. E., Barnea, A., Bauminger, S., Tsafriri, A., Collins, W. P., and Lindner, H. R. (1975): LH effect on the pattern of steroidogenesis in cultured Graafian follicles of the rat: Dependence on macromolecular synthesis. *Endocrinology*, 96:1533-1542.
42. Lindner, H. R., Tsafriri, A., Lieberman, M. W., Zor, U., Koch, Y., Bauminger, S., and Barnea, A. (1974): Gonadotropin action in cultured Graafian follicles: Induction of maturational division of the mammalian oocyte and differentiation of the luteal cell. *Recent Progress in Hormone Research*, 30:79-130.
43. Marsh, J. M. (1975): The role of cyclic AMP in gonadal function. In: *Advances in Cyclic Nucleotide Research*, Vol. 6, edited by P. Greengard and G. A. Robison, pp. 137-199. Raven Press, New York.
44. Marsh, J. M., and LeMaire, W. J. (1974): The role of cyclic AMP and prostaglandins in the actions of luteinizing hormone. In: *Gonadotropins and Gonadal Function*, edited by N. R. Moudgal, pp. 376-389. Academic Press, New York.
45. Marsh, J. M., Mills, T. M., and LeMaire, W. J. (1972): Cyclic AMP synthesis in rabbit Graafian follicles and the effect of luteinizing hormone. *Biochimica et Biophysica Acta*, 273:389-394.
46. Marsh, J. M., Mills, T. M., and LeMaire, W. J. (1973): Preovulatory changes in the synthesis of cyclic AMP by rabbit Graafian follicles. *Biochimica et Biophysica Acta*, 304:197-202.
47. Marsh, J. M., Yang, N. S. T., and LeMaire, W. J. (1974): Prostaglandin synthesis in rabbit Graafian follicles *in vitro*. Effect of luteinizing hormone and cyclic AMP. *Prostaglandins*, 7:269-283.
48. Miller, J. B., and Keyes, P. L. (1974): Initiation of luteinization in rabbit Graafian follicles by dibutryl cyclic AMP *in vitro*. *Endocrinology*, 95:253-259.
49. Mills, T. M. (1975): Effect of luteinizing hormone and cyclic adenosine 3',5'-monophosphate on steroidogenesis in the ovarian follicle of the rabbit. *Endocrinology*, 96:440-445.
50. Mills, T. M., and McPherson, J. C., III (1974): The binding of ^{131}I-hCG to mature follicles isolated from the rabbit ovary. *Proceedings of the Society for Experimental Biology and Medicine*, 145:446-449.
51. Mills, T. M., Telegdy, G., and Savard, K. (1972): The synthesis and secretion of progesterone and 20α-hydroxy-4-pregn-3-one by the rabbit ovary at various periods after a single injection of HCG. *Steroids*, 19:621-637.
52. Naftolin, F., Brown-Grant, K., and Corker, C. S. (1972): Plasma and pituitary luteinizing hormone and peripheral

plasma oestradiol concentrations in the normal oestrous cycle of the rat and after experimental manipulation of the cycle. *Journal of Endocrinology,* 53:17-30.

53. Nugent, C. L., Lopata, A., and Gould, M. K. (1975): The effects of exogenous gonadotropins on ovarian adenylate cyclase activity. *Endocrinology,* 97:581-587.
54. O'Grady, J. P., Caldwell, B. V., Auletta, F. J., and Speroff, L. (1972): The effects of an inhibitor of prostaglandin synthesis (indomethacin) on ovulation, pregnancy and pseudopregnancy in the rabbit. *Prostaglandins,* 1:97-106.
55. Parker, C. W., Sullivan, T. J., and Wedner, H. J. (1974): Cyclic AMP and the immune response. In: *Advances in Cyclic Nucleotide Research,* Vol. 4, edited by P. Greengard and G. A. Robison, pp. 1-79. Raven Press, New York.
56. Pool, W. R., and Lipner, H. (1966): Inhibition of ovulation by antibiotics. *Endocrinology,* 79:858-864.
57. Pool, W. R., and Lipner, H. (1969): Radioautography of newly synthetized RNA and protein in preovulatory follicles. *Endocrinology,* 84:711-717.
58. Powell, J. A., Duell, E. A., and Voorhus, J. J. (1971): Beta adrenergic stimulation. *Archives of Dermatology,* 104:359-365.
59. Rajaniemi, H. J., Bajpai, P. K., and Reichert, L. W., Jr. (1974): LH-receptor activity and granulosa-luteal cell differentiation. *Federation Proceedings,* 33:204.
60. Ramachandran, J. (1971): A new simple method for separation of adenosine 3',5'-cyclic monophosphate from other nucleotides and its use in the assay of adenyl cyclase. *Analytical Biochemistry,* 43:227-239.
61. Reed, M., and Hounslow, W. F. (1971): Induction of ovulation in the guinea pig. *Journal of Endocrinology,* 49:203-211.
62. Reichert, L. E., Jr., and Wilhelmi, A. E. (1973): A summary of the biological activities of the various lots of pituitary hormones produced under the programs of the national institute of arthritis, metabolism and indigestive diseases. *Endocrinology,* 92:1301-1304.
63. Richards, J. S. (1975): Estradiol receptor content in rat granulosa cells during follicular development: Modification by estradiol and gonadotropins. *Endocrinology,* 97:1174-1184.
64. Richards, J. S., and Midgley, A. R., Jr. (1976): Protein hormone action: A key to understanding ovarian follicular and luteal cell development. *Biology of Reproduction,* 14:82-94.
65. Robison, G. A., Butcher, R. W., and Sutherland, E. W. (1970): On the relation of hormone receptors to adenyl cyclase. In: *Fundamental Concepts in Drug-Receptor Interactions,* edited by J. F. Danielle, J. F. Moran, and D. J. Triggle, pp. 59-91. Academic Press, New York.
66. Robison, G. A., Butcher, R. W., and Sutherland, E. W., editors (1971): *Cyclic AMP,* Academic Press, New York.

67. Robison, G. A., Schmidt, M. J., and Sutherland, E. W. (1970): On the development and properties of the brain adenyl cyclase system. In: *Advances in Biochemical Psychopharmacology*, Vol. 3, edited by P. Greengard and E. Costa, pp. 11-30. Raven Press, New York.
68. Rodbell, M., Birnbaumer, L., and Pohl, S. L. (1970): Adenyl cyclase in fat cell. III. Stimulation by secretin and the effects of trypsin on the receptor lipolytic hormones. *Journal of Biological Chemistry*, 245:718-722.
69. Rondell, P. (1974): Role of steroid synthesis in the process of ovulation. *Biology of Reproduction*, 10:199-215.
70. Ryan, W. L., and Coronel, D. M. (1969): Adenosine 3',5'-monophosphate as an inhibitor of ovulation and reproduction. *American Journal of Obstetrics and Gynecology*, 105:121-123.
71. Salomon, Y., Londos, C., and Rodbell, M. (1974): A highly sensitive adenylate cyclase assay. *Analytical Biochemistry*, 58:541-548.
72. Sasamoto, S., and Kennan, A. L. (1973): Endogenous gonadotropin requirements for follicular growth and maintenance in immature rats treated with PMS. *Endocrinology*, 93:292-296.
73. Schwartz, N. B. (1974): The role of FSH and LH and of their antibodies on follicle growth and on ovulation. *Biology of Reproduction*, 10:236-272.
74. Shaikh, A. A. (1971): Estrone and estradiol levels in the ovarian venous blood from rats during the estrous cycle and pregnancy. *Biology of Reproduction*, 5:297-307.
75. Smith, M. S., Freemen, M. E., and Neill, J. D. (1975): The control of progesterone secretion during the estrous cycle and early pseudopregnancy in the rat: Prolactin, gonadotrophin and steroid levels associated with rescue of the corpus luteum of pseudopregnancy. *Endocrinology*, 96:219-226.
76. Soderling, T. R., Corbin, J. D., and Park, C. R. (1974): Techniques for the study of protein kinase activation in intact cells. In: *Methods in Enzymology*, Vol. 38, Hormone Action, Part C, Cyclic Nucleotides, edited by J. C. Hardman and B. W. O'Malley, pp. 358-367. Academic Press, New York.
77. Stouffer, R. L., Tyrey, L., and Schomberg, D. W. (1976): Changes in [^{125}I] labeled human chorionic gonadotrophin (hCG) binding to porcine granulosa cells during follicle development and cell culture. *Endocrinology*, 99:516-525.
78. Sutherland, E. W., Øye, I., and Butcher, R. W. (1965): The action of epinephrine and the role of the adenyl cyclase system in hormone action. *Recent Progress in Hormone Research*, 21:623-646.
79. Tsafriri, A., Lieberman, M. E., Barnea, A., Bauminger, S., and Lindner, H. R. (1973): Induction by luteinizing hormone of ovum maturation and steroidogenesis in isolated Graafian follicles of the rat: Role of RNA and protein synthesis. *Endocrinology*, 93:1378-1385.
80. Tsafriri, A., Lindner, H. R., Zor, U., and Lamprecht, S. A. (1972): *In vitro* induction of meiotic division in follicle-enclosed rat oocytes by LH, cyclic AMP, and prostaglandin E_2.

Journal of Reproductive Fertility, 31:39-50.

81. White, A. A., and Zenser, T. V. (1971): Separation of cyclic 3',5'-nucleotide monophosphates from other nucleotides on aluminum oxide columns. *Analytical Biochemistry*, 41:372-396.
82. Yoshinaga, L., Hawkins, R. A., and Stocker, J. F. (1969): Estrogen secretion by the rat ovary *in vivo* during the estrous cycle and pregnancy. *Endocrinology*, 83:103-112.
83. Younglai, E. V. (1972): Effect of mating on follicular fluid steroids in the rabbit. *Journal of Reproductive Fertility*, 30:157-159.
84. Younglai, E. V. (1974): Steroid production by the isolated rabbit ovarian follicle. I. Effects of ovine LH. *Journal of Reproductive Fertility*, 40:95-103.
85. Younglai, E. V. (1975*a*): Steroid production by the isolated rabbit ovarian follicle. III. Actinomycin D-insensitive stimulation of steroidogenesis by LH. *Endocrinology*, 96:468-474.
86. Younglai, E. V. (1975*b*): Steroid production by the isolated rabbit ovarian follicle. IV. Effects of cyclic nucleotides. *Journal of Steroid Biochemistry*, 6:1353-1357.

Ovarian Follicular Development and Function,
edited by A. R. Midgley and W. A. Sadler.
Raven Press, New York © 1979.

Hormone Action in Ovarian Follicles

William J. LeMaire

Department of Obstetrics and Gynecology and The
Endocrine Laboratory, University of Miami,
School of Medicine, Miami, Florida 33152

This chapter will discuss one point of Dr. Hunzicker-Dunn's presentation, namely the possible involvement of prostaglandins in the development of the follicular refractoriness. It has been suggested that the increase of prostaglandin observed prior to ovulation in isolated graafian follicles might be responsible for the development of the refractoriness to LH stimulation *in vitro* (1). Dr. Dunn presented some evidence to suggest that prostaglandins are not involved in this process. In our laboratory with Drs. Davies, Clark, and Marsh, we have carried out a number of studies that would agree with Dr. Dunn's conclusion (2).

First, we carried out the usual *in vitro* incubation of isolated follicles from estrous rabbits in the presence of [^{3}H]adenine (3). After isolation and purfication of the [^{3}H]cyclic AMP formed during a 1 hr incubation, the radioactivity was measured. As can be seen in Fig. 1, 5 μg/ml of LH added to the medium (second column from left) caused the usual marked increased of cyclic AMP accumulation. The results are expressed as a percent increase of the control and the height of the bars shows the mean with the SE indicated by the vertical lines. The addition of PGE_2 and $PGF_{2\alpha}$ at a concentration of 20 μM (third and fourth columns from left) did not abolish the LH stimulation. This concentration of prostaglandins is about 10 times what we would expect in terms of increase of prostaglandins in follicles just before ovulation. The addition of 20 μM lysolecithin, a byproduct of prostaglandin symthesis from phospholipids, did not make the follicles refractory to LH stimulation (right-hand column). At much higher concentrations of lysolecithin (500 to 2,000 μM), nonspecific detergent effects were seen (data not shown).

We then undertook the measurement of LH-stimulated cyclic AMP accumulation in follicles obtained at various

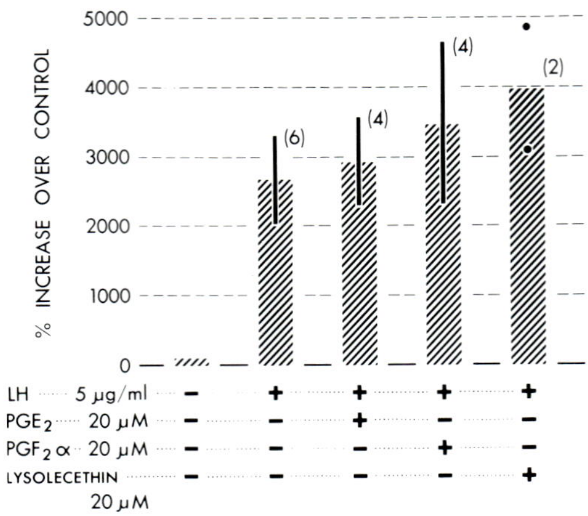

FIG. 1. [^{3}H]cyclic AMP accumulation *in vitro*. Effect of PG on LH stimulation of graafian follicles from estrous rabbits. Isolated follicles from estrous rabbits were incubated *in vitro* and the measurement of [8-^{3}H]cyclic AMP accumulation was performed as described previously (3). Each bar represents the percent stimulation over the control produced by the addition of 5 µg/ml of LH-NIH-S16. The height of the bars represents the mean (number of experiments indicated above each bar) and the vertical lines represent the standard error of the mean. The prostaglandins and the lysolecithin were added together with LH after the 1-hr preincubation and were present throughout the entire 1-hr incubation.

times after the *in vivo* injection of an ovulatory dose of LH and after inhibition of prostaglandin synthesis by indomethacin. This is shown in Fig. 2. The open bars represent the control incubation and the solid bars the LH stimulation. At estrous, the follicles show the usual marked sensitivity to LH. One-half hour prior to the injection of LH, the rabbits received an intravenous injection of either phosphate buffer or indomethacin (20 mg/kg suspended in phosphate buffer). It can be seen that LH preceded by buffer induced the usual refractoriness at 0.5, 5, and 9 hr to *in vitro* LH stimulation. When prostaglandin synthesis was inhibited by pretreatment with indomethacin the refractoriness developed also. Although at 0.5 and 5 hr the decline in LH responsiveness was not as marked as in the control, the refractoriness was nevertheless complete at 9 hr.

As shown in Table 1, completeness of prostaglandin synthesis inhibition in these experiments was verified by measuring PGF levels in representative follicles during these experiments. The *in vivo* injection of LH

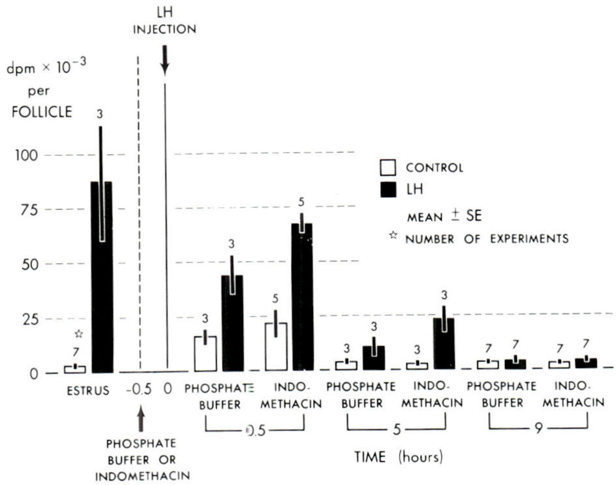

FIG. 2. [8-^{3}H]Cyclic AMP accumulation in isolated rabbit graafian follicles after LH injection *in vivo*. Effect of indomethacin. Rabbit graafian follicles were isolated before (estrous) or at various times (0.5, 5, and 9 hr) after an intravenous injection of 50 μg of NIH-LH-B7. One-half hour preceding the LH injection, the animals received an intravenous injection of either phosphate buffer or indomethacin (20 mg/kg) suspended in phosphate buffer. The measurement *in vitro* [8-^{3}H]cyclic AMP accumulation was performed as described previously (3). The height of the bars represents the mean (number of experiments indicated above each bar) and the vertical lines represent the standard error of the mean. Each experiment consists of a set of control incubations (open bars) and a set of incubations to which LH (5 μg/ml of LH-NIH-S16) has been added (solid bars). At the various times after the *in vivo* LH injection (0.5, 5, and 9 hr) the experiments carried out on animals not pretreated (phosphate buffer) and the animals pretreated with indomethacin (indomethacin) are given side by side.

indeed caused a marked increase in PGF in follicles isolated at 9 hr, and the pretreatment with indomethacin completely abolished this increase.

The experiments shown in Table 2 were carried out to rule out the possibility that indomethacin might have, in itself, an effect on LH responsiveness. It can be seen that at 1 or 9 hr after indomethacin treatment alone, isolated follicles showed the same responsiveness *in vitro* to LH stimulation as the follicles of estrous animals without pretreatment.

Taken together, these data suggest that the increase in follicular prostaglandin, observed in rabbit graafian follicles as ovulation approaches, is not responsible for the development of the refractoriness of these

TABLE 1. *Preovulatory changes in PGF levels in rabbit graafian follicles: effect of indomethacin*

Time of sacrifice	PGF (pg/follicle) mean ± SE
Estrous (n = 5)	321 ± 111
9 hr after LH (n = 4)	1,726 ± 133[a]
9 hr after LH (n = 7) + indomethacin	177 ± 32.5

[a]This value is significantly different from the two others ($p < 0.0001$).

From the follicles isolated at estrous and 9 hr after *in vivo* administration of LH, and submitted to the measurement of cyclic AMP accumulation as described in Fig. 2, several representative follicles were used for the measurement of PGF by radioimmunoassay as described previously (4). One-half hour preceding the *in vivo* injection of 50 μg of NIH-LH-B7, the animals received an intravenous injection of phosphate buffer or 20 mg/kg of indomethacin suspended in phosphate buffer as described in the legend to Fig. 2.

TABLE 2. *Effect of indomethacin on LH-stimulated accumulation of [8-^{3}H]cyclic AMP in rabbit graafian follicles in vitro*

Time of sacrifice	[8-^{3}H]Cyclic AMP accumulation % Increase [a]
Estrous[b]	4,080 ± 1,100 (7)[c]
1 hr after indomethacin (20 mg/kg)	4,470 ± 554 (5)
9 hr after indomethacin (20 mg/kg)	3,260 ± 1,680 (3)
9 hr after LH (50 μg)	174 ± 55.9 (7)

[a]Percent increase is the stimulation of [8-^{3}H] accumulation produced *in vitro* by LH (5μ g/ml) expressed as a percent of the control incubation. [8-^{3}H]Cyclic AMP accumulation was measured as described (3).

[b]Statistical analysis of the data using a nonpaired comparison, student's *t* test: 1 vs. 2 and 1 vs. 3 = ns; 1 vs. 4 = $p < 0.005$; 2 vs. 4 = $p < 0.0001$; and 3 vs. 4 = $p < 0.01$.

[c]Numbers in parentheses indicate the number of experiments in each group.

follicles to *in vitro* stimulation by LH of cyclic AMP accumulation. The nature and mechanism of this refractoriness remains tc be determined.

ACKNOWLEDGMENTS

This work was supported in part by NIH grants HD-08747 and HD-03142 and Ford Foundation Grant 0338. We thank Dr. J. Pike from the Upjohn Company for the gift of prostaglandins, Dr. H. R. Behrman and Dr. C. G. Van Arman from the Merck Institute for Therapeutic Research for the gift of the prostaglandin antibody and indomethacin, and the NIH for the gift of the LH preparations. We appreciate the excellent technical assistance of Adela Gomez, Ana Loynaz, and Adalgisa Rojo, and thank Maria A. Rodriguez for the careful preparation of the manuscript.

REFERENCES

1. LeMaire, W. J., and Marsh, J. M. (1975): Interrelationships Between Prostaglandins, Cyclic AMP and Steroids in Ovulation. *J. Reprod. Fert.*, (Suppl.) 22:53.
2. LeMaire, W. J., Davies, P. J. A., and Marsh, J. M. (1976): The Role of Prostaglandins in the Development of Refractoriness to LH Stimulation by Graafian Follicles. *Prostaglandins*, 12:271.
3. Marsh, J. M., Mills, T. M., and LeMaire, W. J. (1972): Cyclic AMP Synthesis in Rabbit Graafian Follicles and the Effect of Luteinizing Hormone. *Biochim. Biophys. Acta*, 273:389.
4. LeMaire, W. J., Yang, N. S. T., Behrman, H. R., and Marsh, J. M. (1973): Preovulatory Changes in the Concentration of Prostaglandins in Rabbit Graafian Follicles. *Prostaglandins*, 3:367.

Ovarian Follicular Development and Function,
edited by A. R. Midgley and W. A. Sadler.
Raven Press, New York © 1979.

Mechanism of Desensitization of Ovarian Adenylate Cyclase to Continued Hormonal Stimulation: Dependence on *De Novo* Protein Synthesis

S. Lamprecht, U. Zor, Y. Koch, Y. Salomon, and H. R. Lindner

Department of Hormone Research, The Weizmann Institute of Science, Rehovot, Israel

Rat graafian follicles maintained in organ culture respond to three different agonists, luteinizing hormone (LH), follicle-stimulating hormone (FSH), and prostaglandin E_2, with increased cyclic AMP accumulation, oocyte maturation, and a change in the pattern of steroidogenesis (1). In previous studies (2-5) we have shown that rat ovarian slices or graafian follicles, upon prolonged incubation with LH, became refractory to renewed exposure to this hormone, although they remained fully responsive to PGE_2 and FSH in terms of cyclic AMP formation. Likewise, follicles rendered refractory to PGE_2 retained their sensitivity to LH. The refractory state was thus shown to be specifically restricted to the homologous hormone. Desensitization to LH and to PGE_2 was dose dependent, and in the case of LH required hormone concentrations in excess of the minimal dose effective in stimulating cyclic AMP accumulation (Table 1 and ref. 5).

Such desensitization to hormones appears to be a widespread phenomenon (6-12), yet the precise mechanism responsible for its occurrence remains unclear. It has been suggested that the diminished response to the hormone is due (a) to a decrease in the number of hormone binding sites, (b) to alteration of the catalytic component of adenylate cyclase, (c) to an enhanced rate of cyclic AMP degradation by hormone-induced phosphodiesterase, or (d) to the action of an endogenous inhibitor. The present study was intended to test some of these hypotheses.

TABLE 1. *Dose-dependence of desensitization of graafian follicles to LH*[a]

LH concentration (μg/ml)		
Initial culture (24 hr)	Challenge (30 min)	Response Cyclic AMP (pmole/follicle)
None	None	4.2 ± 0.2
	0.4	65.3 ± 6.2
	0.8	129.2 ± 3.1
	2.0	160.4 ± 6.3
0.4	0.4	68.7 ± 11.5
	2.0	154.2 ± 10.5
0.8	0.8	62.5 ± 4.2
	2.0	64.6 ± 3.1
2.0	2.0	62.5 ± 10.4

[a]Follicles were cultured for 24 hr, washed, and incubated for 30 min in KRB containing IBMX (100 μg/ml) and LH at concentration shown (challenge), or in hormone-free medium (none). Means ± SEM for 12 determinations.

MATERIALS AND METHODS

Materials

LH (NIH-LH-S18) and FSH (NIH-FSH-S10) were kindly made available by the National Institute of Arthritis and Metabolic Diseases, NIH. Solutions of PGE_2 (courtesy of Dr. J. Pike of the Upjohn Co., Kalamazoo, Michigan) were prepared as previously described (5). 3-Isobutyl-1-methylxanthine (IBMX) was purchased from the Aldrich Chemical Co., Inc., Milwaukee, Wisconsin. Creatine phosphate, creatine phosphokinase, 3':5'-cyclic AMP and guanosine 5'-triphosphate were the products of Sigma Chemical Co., St. Louis, Missouri. Actinomycin D (Merck, Sharpe and Dohme, Rahway, N.J.) was stored as a stock solution of 1 mg/ml in ethanol and diluted to the desired concentration with culture medium; cycloheximide (Sigma Chemical Co.) was freshly dissolved in the culture medium. Guanylylimidodiphosphate [Gpp(NH)p] was the product of ICN Pharmaceuticals Inc.; (α-^{32}P)-ATP (5 to 10 Ci/mmole) and (8-^{3}H)-cyclic AMP (27 Ci/mmole) were purchased from Amersham Radiochemical Centre, Buckinhamshire, England.

Animals and Organ Culture

Three-month-old Wistar-derived rats with 4-day cycles, as verified by daily vaginal smears, were sacrificed between 8 and 12 A.M. on the day of proestrus. Large graafian follicles were isolated and cultured (13) for variable periods of time with the appropriate hormones. The follicles were then thoroughly washed with saline and incubated in a shaking bath for 30 min at 37°C in Krebs-Ringer bicarbonate buffer (KRB) containing glucose, IBMX (100 μg/ml) and the hormone indicated (5).

Assay of Adenylate Cyclase Activity

Follicles were homogenized at 0 to 4°C (approximately 10 follicles/ml) in Tris acetate buffer, 10 mM, pH, 7.6, containing dithiothreitol (DTT, 1.0 mM). The incubation mixture contained the following in a final volume of 100 μl: Tris acetate, 25 mM, pH, 7.6; magnesium acetate 2 mM; DTT, 1.0 mM; creatine phosphate, 5.0 mM; creatine phosphokinase, 50 U/ml; BSA, 100 μg/ml; unlabeled cyclic AMP, 0.1 mM; and (^{32}P)-ATP, 1.0 mM (3 to 5 × 10^6 cpm). The reaction was started by the addition of homogenate (5 to 25 μg/protein). Termination of the reaction with sodium dodecysulfate and purification of (^{32}P)-cyclic AMP formed was done as previously described (14), using (^{3}H)-cyclic AMP (2 × 10^4 cpm) to monitor cyclic AMP recovery.

Assay of Cyclic AMP

The assay was performed by a modification (3) of the Gilman procedure (15).

Phosphodiesterase activity was assayed in follicles homogenized in 40 mM Tris-HCl buffer, pH 7.8, containing 3.5 mM MgCl (16), using unlabeled cyclic AMP (100 pmole/flask) as substrate.

RESULTS AND COMMENT

Effect of Continued Exposure of Graafian Follicles to LH or PGE_2 on Cyclic AMP accumulation and Phosphodiesterase Activity

Follicles cultured for 24 hr in a hormone-free medium responded to subsequent addition of LH (10 μg/ml) or PGE_2 (10 μg/ml) to the medium within 30 min with a sharp rise in cyclic AMP accumulation (Fig. 1). Follicles cultured for 24 hr in a medium enriched with LH (10 μg/ml) showed a diminished response upon subsequent challenge with fresh LH (30-min incubation in

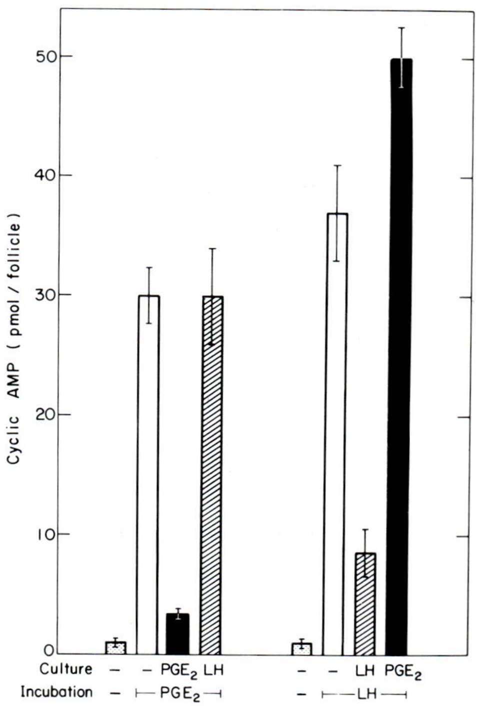

FIG. 1. Effect of continued exposure of graafian follicles to LH and PGE_2 on cAMP production. Follicles were cultured for 24 hr in medium containing the hormone indicated: LH, PGE_2 (10 μg/ml) or without added hormone (-). The follicles were then thoroughly washed with saline and incubated for 30 min in KRB containing IBMX (100 μg/ml) with the hormone specified. Vertical bars represent mean values (± SE) for 18 observations.

KRB-IBMX medium), but remained fully responsive to PGE_2 (Fig. 1) and FSH (not shown; ref. 5). Likewise, follicles cultured with PGE_2 (10 μg/ml) became refractory to PGE_2 within 24 hr, while retaining their sensitivity to LH (Fig. 1) and FSH (ref. 5).

The inclusion of a phosphodiesterase inhibitor (IBMX) in the challenge medium seemed to preclude increased degradation of cyclic AMP in the refractory follicles by this enzyme as a cause of the diminished response. This conclusion was borne out by the observation that follicles desensitized to LH, FSH, or PGE_2 did not differ significantly from control follicles in the rate of degradation of exogenous cyclic AMP (data not shown).

Time Course of Desensitization and Recovery

Follicles were cultured with LH or PGE_2 (10 μg/ml) for variable periods of time (0 to 24 hr), rinsed and exposed in KRB-IBMX medium to fresh hormone. Cyclic AMP accumulation was measured during a 30-min incubation. Refractoriness to LH and PGE_2, respectively, developed soon after exposure of the tissue to either agonist (Figs. 2 and 3), but showed different kinetic characteristics for the two hormones: maximal desensitization toward LH was attained only after 9 to 12 hr (Fig. 2), while desensitization to PGE_2 was maximal after 3 to 6 hr exposure (Fig. 3) to the homologous hormone.

When follicles refractory to PGE_2 stimulation were transferred to hormone-free medium, they progressively regained their responsiveness to the prostaglandin in terms of cyclic AMP formation. Recovery began within 1 hr and was complete 4 hr after the change of medium (Fig. 4). By contrast, LH-desensitized follicles did not recover their responsiveness to the hormone during an 8-hr incubation period in LH-free medium or in medium containing anti-LH serum and subsequent rinsing.

Adenylate Cyclase Activity in Control and LH-Desensitized Follicles

Adenylate cyclase in homogenates of control follicles cultured for 24 hr in hormone-free medium responded with enhanced cyclic AMP formation when challenged with LH (3 μg/ml), GTP (10 μM), Gpp (NH)p (10 μM), or NaF (10 mM) (Table 2). The stimulatory action of LH was significantly potentiated by simultaneous

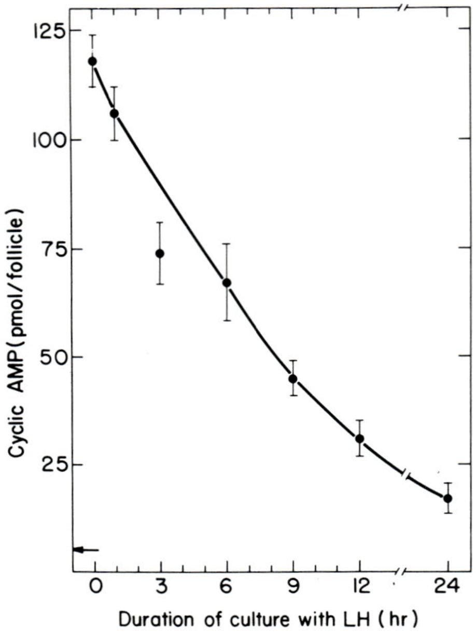

FIG. 2. Time course of desensitization of graafian follicles to LH. Follicles were cultured and processed as described in Fig. 1.

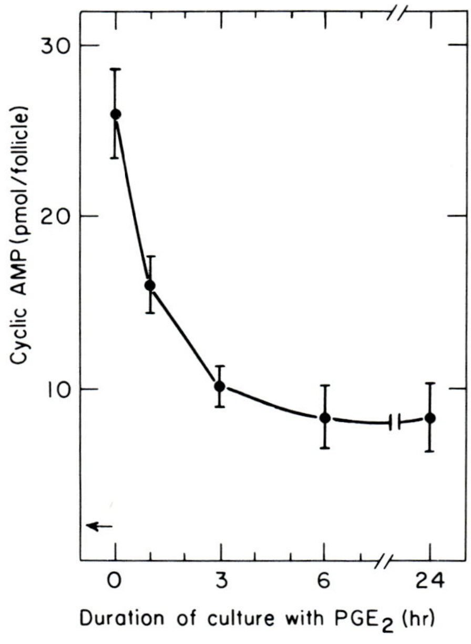

FIG. 3. Time course of desensitization of graafian follicles to PGE_2. Follicles were cultured and processed as described in Fig. 1.

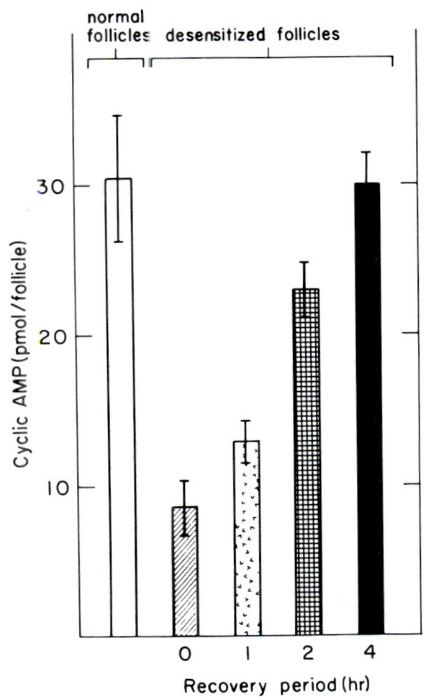

FIG. 4. Recovery of cAMP response of rat graafian follicles desensitized by culturing with PGE_2 (10 μg/ml) for 6 hr. The medium was withdrawn, the follicles rinsed with saline, and incubated with fresh, hormone-free medium for 0-4 hr. The follicles were reincubated at the indicated times for 30 min with PGE_2 (10 μg/ml) and IBMX (100 μg/ml). Normal follicles were cultured for 6 hr in hormone-free medium and processed as described in Fig. 7. Basal cAMP level was 2 ± 0.3 pmole/follicle.

addition of GTP or Gpp(NH)p (Table 2), indicating that the guanine nucleotide regulatory site is operative in follicular adenylate cyclase (17), as it is in enzyme preparations derived from lipocytes and hepatocytes (18). By contrast, homogenates derived from follicles rendered refractory to LH failed to respond to LH alone or to LH in combination with GTP or its imido analog, in comparison to controls treated with the appropriate guanine nucleotide (Table 2). The LH-desensitized preparation responded, however, with increased cyclic AMP formation to NaF, GTP, or Gpp(NH)p when these were added to the incubation medium on their own (Table 2).

It may be concluded that the decline in LH-stimulable adenylate cyclase activity in refractory follicles is not related to a perturbation at the level of the catalytic site of the enzyme, or at the guanine nucleotide regulatory site. Furthermore, since the cyclase assay was conducted under conditions in which ATP concentration was not limiting, yet the enzyme proved

TABLE 2. *Desensitization of LH-sensitive adenylate cyclase in homogenates of rat ovarian graafian follicles*

Assay conditions (15 min)	Preincubation with OLH (24 hr) (Adenylate cyclase, % of basal activity)		
	None	1 μg/ml	10 μg/ml
Basal	100	100	100
LH (3 μg/ml)	383	118	100
GTP (10 μM)	141	157	140
GTP + LH	516	152	130
Gpp(NH)p (10 μM)	308	257	175
Gpp(NH)p + LH	616	237	205
NaF (10 mM)	291	228	220
Basal activity Cyclic AMP formed (pmole/15 min/mg protein)	100	175	166

unresponsive to LH, it is unlikely that the diminished cyclic AMP accumulation in intact refractory follicles can be ascribed to depletion of ATP in a relevant cellular compartment.

Binding of ^{125}I-HCG by LH-Desensitized Follicles

In preliminary experiments, designed to test the hypothesis that refractoriness was due to loss of available binding sites for LH, follicles desensitized by culture for 24 hr with LH were incubated first for 1 hr with anti-LH serum and thereafter for 2 hr with ^{125}I-HCG. The follicles became refractory, but the amount of radioactivity bound was not reduced in comparison to controls cultured for 24 hr in hormone-free medium (Fig. 5A), and was higher than in refractory follicles from which bound LH had not been stripped by antibody (Fig. 5A, normal serum control). It appears that LH-binding capacity is not significantly reduced in desensitized follicles, but it is possible that the binding sites in these follicles are not functionally coupled to the catalytic subunit of adenylate cyclase. Occupation of binding sites in desensitized follicles by partly degraded LH is another possibility that has not been entirely excluded.

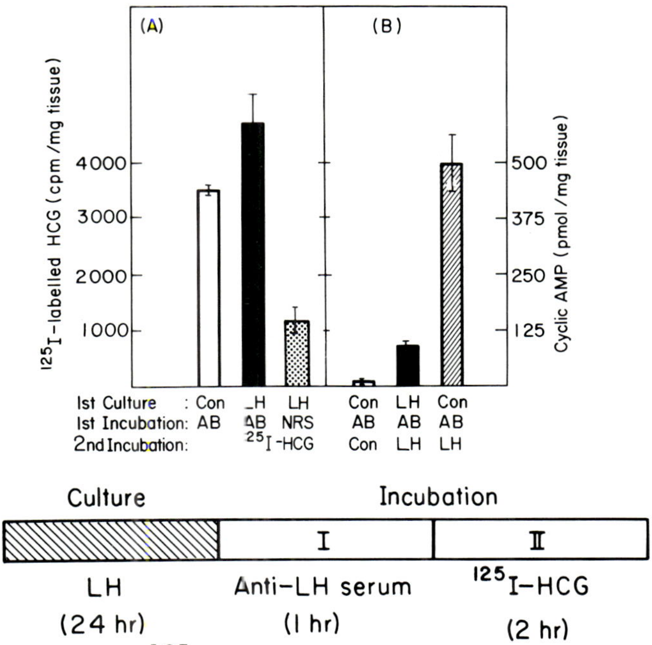

FIG. 5. Binding of ^{125}I-labeled human chorionic gonadotropin by follicles rendered refractory to LH. AB, antiserum to LH: NRS, normal rabbit serum; Con, control follicles cultured in hormone-free medium. Vertical brackets, ± SEM.

Effect of Actinomycin D and of Cycloheximide on the Development of Refractoriness to PGE_2 and LH

Actinomycin D (8 µM) and cycloheximide (30 µM) prevented the development of refractoriness to PGE_2 during a 6-hr incubation period when either of these inhibitors was added to the medium together with the hormone (Fig. 6). At these dose levels the inhibitors themselves had no significant effect on follicular cyclic AMP formation during 6 hr incubations. Actinomycin D also prevented desensitization of follicles to LH during a 9-hr incubation with the hormone (Fig. 7), but cycloheximide was ineffective in this respect: prolonged incubation with cycloheximide (>6 hr) on its own impaired the cyclic AMP response to LH or PGE_2.

The results suggest that synthesis of a protein inhibitor triggered by PGE_2 or LH may be important for the development of the refractory state. The rapid recovery of responsiveness after removal of PGE_2 from the medium implies that the putative inhibitor has a short half-life and that its synthesis depends on continued occupation of the receptor by PGE_2. The finding that desensitization of graafian follicles is

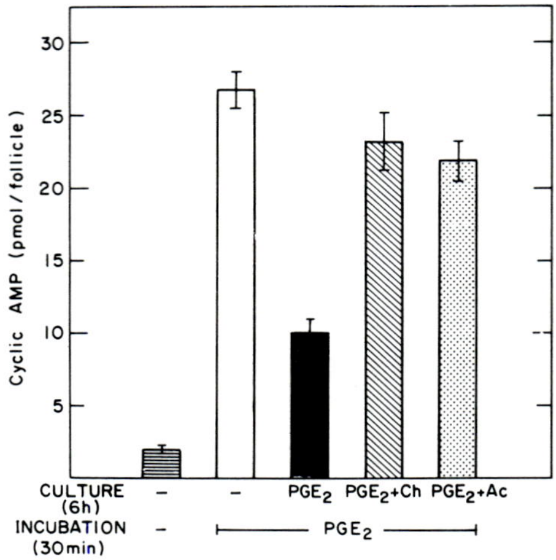

FIG. 6. Reversal by cycloheximide (Ch) and actinomycin D (Ac) of PGE_2-induced refractoriness in graafian follicles cultured for 6 hr. Ch (30 μM) and Ac (8 μM) were present throughout the culture period. Vertical bars represent mean values (± SE). For number of observations see Table 1.

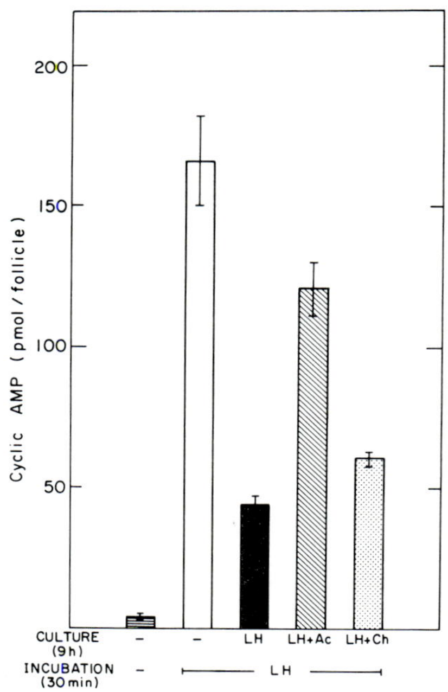

FIG. 7. Effect of cycloheximide (Ch) and actinomycin D (Ac) on LH-induced refractoriness in graafian follicles cultured for 9 hr. CH (30 μM) and Ac (8 μM) were present throughout the culture period. Vertical bars represent mean values (± SE) for 12 determinations.

restricted to the homologous hormone suggests that the inhibitor may be synthesized in distinct PGE_2- and LH-sensitive cells. Evidence referred to in the preceding paragraph argues against the inhibitor being a phosphodiesterase (contrast ref. 11, 19-21). Rather, the inhibitor appears to act by interfering with the coupling of the hormone-receptor complex to the catalytic subunit of adenylate cyclase.

ACKNOWLEDGMENT

This work was supported by the Ford Foundation and the Population Council, New York. One of us (H.R.L.) was the recipient of a travel grant from Serono Laboratories Ltd.

REFERENCES

1. Lindner, H. R., Tsafriri, A., Lieberman, M. E., Zor, U., Koch, Y., Bauminger, S., and Barnea, A. (1974): Gonadotropin action on cultured Graafian follicles: Induction of maturation division of the mammalian oocyte and differentiation of the luteal cell. *Recent Progr. Horm. Res.*, 30:79-138.
2. Zor, U., Lamprecht, S. A., Kaneko, T., Schneider, H. P. G., McCann, S. M., Field, J. B., Tsafriri, A., and Lindner, H. R. (1972): Functional relations between cyclic AMP, prostaglandins and luteinizing hormone in rat pituitary and ovary. In: *Advances in Cyclic Nucleotide Research,* Vol. 1, edited by P. Greengard, R. Paoletti, and G. A. Robison, pp. 503-520. Raven Press, New York.
3. Lamprecht, S. A., Zor, U., Tsafriri, A., and Lindner, H. R. (1973): Action of prostaglandin E_2 and of luteinizing hormone on ovarian adenylate cyclase, protein kinase and ornithine decarboxylase activity during postnatal development and maturity in the rat. *J. Endocrinol.*, 57:217-233.
4. Zor, U., Lamprecht, S. A., Koch, Y., and Lindner, H. R. (1975): Desensitization of Graafian follicles by gonadotropins or prostaglandin E_2 to stimulation of adenylate cyclase by the homologous hormone. *Endocrinology* (Suppl) 96 (Abstr) 628:365.
5. Zor, U., Lamprecht, S. A., Misulovin, Z., Koch, Y., and Lindner, H. R. (1976): Refractoriness of ovarian adenylate cyclase to continued hormonal stimulation. *Biochim. Biophys. Acta,* 428:761-765.
6. Ho, R.-J., and Sutherland, W. (1971): Formation and release of a hormone antagonist by rat adipocytes. *J. Biol. Chem.*, 246:6822-6827.
7. de Vellis, J., and Brooker, G. (1974): Reversal of catecholamine refractoriness by inhibitors of RNA and protein synthesis. *Science*, 186:1221-1222.

8. Remold-O'Donnell, E. (1974): Stimulation and desensitization of macrophage adenylate cyclase by prostaglandins and catecholamines. *J. Biol. Chem.*, 249:3615-3621.
9. Ciosek, C. P., Jr., Fahey, J. V., Ishikawa, Y., and Newcombe, D. S. (1975): PGE_1-mediated cyclic AMP refractoriness: Effects of cycloheximide and indomethacin. *J. Cycl. Nucl. Res.*, 1:229-235.
10. Mukherjee, C., Caron, M. C., and Lefkowitz, J. (1975): Catecholamine-induced subsensitivity of adenylate cyclase associated with loss of β-adrenergic receptor binding sites. *Proc. Nat. Acad. Sci. USA,* 72:1945-1949.
11. Newcombe, D. S., Ciosek, C. P., Jr., Ishikawa, Y., and Fahey, J. V. (1975): Human synoviocytes: Activation and desensitization by prostaglandins and l-epinephrine. *Proc. Nat. Acad. Sci. USA,* 72:3124-3128.
12. Bockaert, J., Hunzicker-Dunn, M., and Birnbaumer, L. (1976): Hormone-stimulated desensitization of hormone-dependent adenylyl cyclase. Dual action of luteinizing hormone on pig Graafian follicle membranes. *J. Biol. Chem.,* 251:2653-2663.
13. Tsafriri, A., Lindner, H. R., Zor, U., and Lamprecht, S. A. (1972): *In-vitro* induction of meiotic division in follicle-enclosed rat oocytes by LH, cyclic AMP and prostaglandin E_2. *J. Reprod. Fert.*, 31:39-50.
14. Salomon, Y., Londos, C., and Rodbell, M. (1974): A highly sensitive adenylate cyclase assay. *Anal. Bioch.*, 58:541-548.
15. Gilman, A. G. (1970): A protein binding assay for adenosine 3',5'-cyclic monophosphate. *Proc. Nat. Acad. Sci. USA,* 67:305-312.
16. Zor, U., Koch, R., and Naor, Z. (1976): Differential effects of prostaglandin synthetase inhibitors on prostaglandin E_2 binding and on prostaglandin- or cholera toxin-induced cyclic AMP accumulation in the rabbit uterus. In: *Advances in Prostaglandin and Thromboxane Research,* Vol. 1, edited by B. Samuelsson and R. Paoletti, pp. 331-335. Raven Press, New York.
17. Salomon, Y., Yanovsky, A., and Lindner, H. R. (1976): Modulation of LH-sensitive ovarian adenylate cyclase by guanosine nucleotides. Proceed. Xth Int'l Congr. of Biochemistry, Hamburg, July 1976, abstr. 08-2-314, p. 444.
18. Rodbell, M., Lin, M. C., Salomon, Y., Londos, C., Harwood, J. P., Martin, B. R., Rendell, M., and Berman, M. (1974): The role of adenine and guanine nucleotides in the activity and response of adenylate cyclase systems to hormones: Evidence for multi-site transition states. *Acta Endocrinol.* (Suppl) 191:11-37.
19. Manganiello, V., and Vaughan, M. (1972): Prostaglandin E_1 effects on cAMP concentration and phosphodiesterase activity in fibroblasts. *Proc. Natl. Acad. Sci. USA,* 69:269-273.
20. Schultz, J., and Daly, J. W. (1973): Cyclic adenosine 3',5'-monophosphate in guinea pig cerebral cortical slices. III. Formation, degradation, and reformation of cyclic adenosine

3',5'-monophosphate during sequential stimulations by biogenic amines and adenosine. *J. Biol. Chem.*, 248:860-866.

21. Browning, E. T., Brostrom, C. O., and Groppi, V. E., Jr. Altered adenosine cyclic 3',5'-monophosphate synthesis and degradation by C-6 astrocytoma cells following prolonged exposure to norepinephrine. *Mol. Pharmacol.*, 12:32-40.

Ovarian Follicular Development and Function,
edited by A. R. Midgley and W. A. Sadler.
Raven Press, New York © 1979.

Hormone Receptor Regulation by Luteinizing Hormone: An Active, Inducible Process

Mrinalini C. Rao

Reproductive Endocrinology Program, Department of Pathology, University of Michigan, Ann Arbor, Michigan 48109

Among the many actions of hormones on their target tissue is their ability to modulate the content of their own receptor as well as that of other hormones. Recent evidence has shown that luteinizing hormone (LH) can act on granulosa cells of preovulatory follicles to induce a decrease in gonadotropin receptors and an increase in prolactin (PRL) receptor in association with luteinization (4,5). LH has also been demonstrated to alter the responsiveness of its target cell (granulosa and luteal) by causing a desensitization of the adenylyl cyclase system to further LH stimulation (1-3). If the receptor content of a target tissue reflects the ability of the tissue to respond to a hormonal signal, then it is conceivable that the sensitivity of the target cell to a hormone may be subject to alteration by hormone-induced changes of receptor content. The following study was undertaken to further elucidate the mechanism by which LH causes a decline in its own receptor content. Our results show that LH causes an inducible decline in its own receptor by mechanisms other than occupancy.

MODEL SYSTEM

To examine the mechanism by which LH decreases LH receptor in granulosa cells of preovulatory follicles, the immature, hypophysectomized rat, pretreated with estradiol and FSH, was used as a model system and is represented in Fig. 1. To study the effects of LH in this model system, 60 μg of ovine LH or saline was superimposed over continuous FSH treatment on Day 34 of age. As shown in Fig. 1, treatment with estradiol and FSH had by this time caused a maximal stimulation

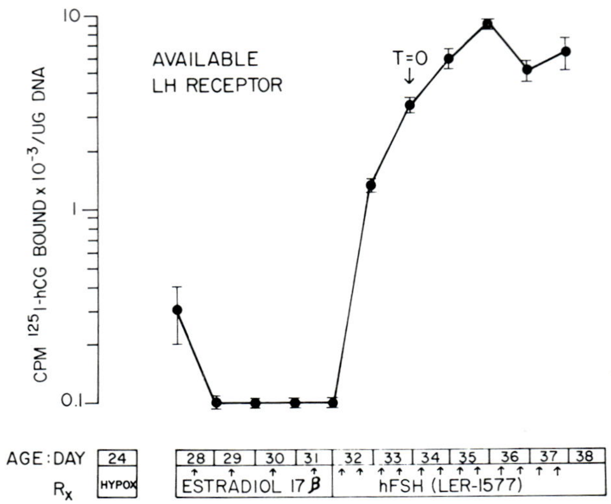

FIG. 1. The experimental model system. Female rats hypophysectomized on Day 24 of age were administered 1.5 mg of estradiol daily from Days 28 to 31 of age. This was followed by administration of 1 μg of HFSH (LER-1577) given twice daily from Days 32 to 37. Available receptor content for LH was measured in granulosa cells on each day of the regimen. Cells were obtained as described in the text. Data are expressed as cpm ^{125}I-HCG bound per microgram DNA on a logarithmic scale. T = 0 depicts the initial time in the model system used for all subsequent figures. In all figures, values represent mean ± SEM with n = 4. Each experiment was repeated at least once.

of LH receptor. For this study the time of administration of LH (Day 34) is depicted as control T = 0. LH and human chorionic gonadotropin (HCG) were used interchangeably after we had determined that in our model system 60 μg of LH and 5 IU of HCG had a similar effect on receptors. Granulosa cells were isolated from these ovaries by application of gentle pressure with a blunt spatula and were collected by centrifugation. Gonadotropin receptors were measured by determining the amount of iodinated hormone specifically bound to these cells *in vitro* and results have been expressed as specific counts bound per microgram of DNA of granulosa cells.

RESULTS AND DISCUSSION

To demonstrate that the effects of LH on decreasing receptors for gonadotropins are not due to adverse effects of the hormone, the available receptor for another

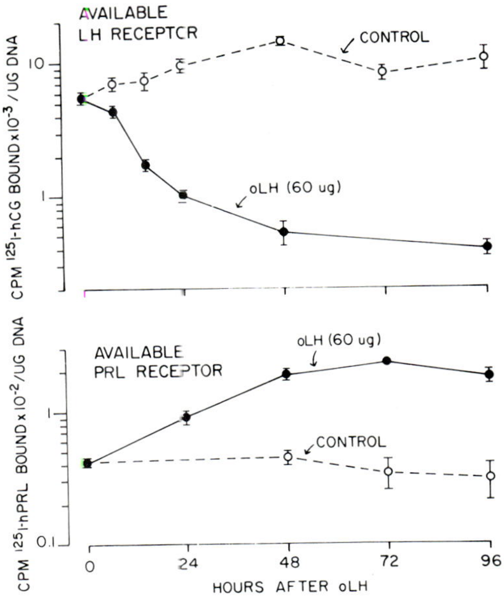

FIG. 2. Effect of ovine LH on available receptors for LH and PRL. The top panel represents available LH receptors and the bottom panel available PRL receptors. Day 0 depicts Day 34 in the model system described as T = 0 in Fig. 1. Solid lines with closed circles represent LH-treated groups and broken lines with open circles represent saline-treated controls. The abscissa represents hours after hormone treatment, and the data are expressed as described for Fig. 1.

hormone, PRL, was also measured. In marked contrast to its effect on LH receptors, LH causes an increase in binding sites for PRL, whereas, in the absence of LH, granulosa cells show very little specific binding to PRL (Fig. 2). Thus, in association with luteinization, LH has dramatically opposite effects on different hormone receptors. Since in our system the effects of LH on available receptors for FSH resemble the effects of LH on its own receptor, the data on FSH receptors have been excluded from this report for the sake of brevity.

To determine if the doses of LH or of HCG used saturated binding sites *in vivo*, the uptake of a trace of iodinated HCG administered in the presence of 5 IU saline or 1,000 IU HCG was examined and compared to available receptor sites as measured *in vitro*. As seen in Fig. 3, uptake of the trace administered subcutaneously with the low dose of HCG or saline reached a maximum at 6 hr, remained constant for 6 to 12 hr, and

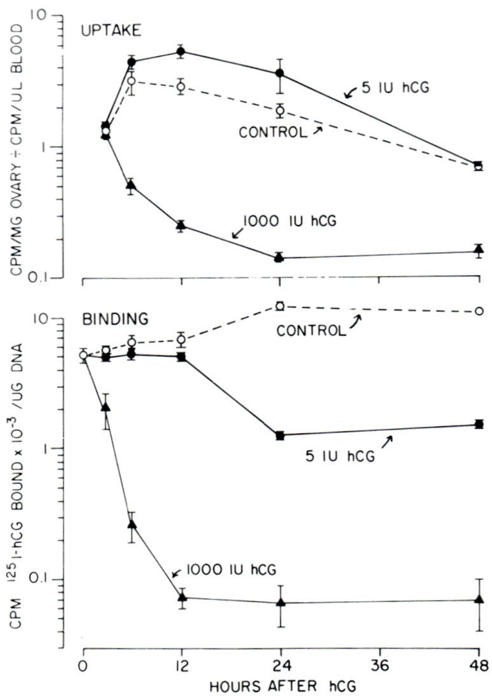

FIG. 3. *In vivo* uptake and *in vitro* binding of ^{125}I-HCG by granulosa cells. Top panel depicts uptake of a trace of ^{125}I-HCG administered *in vivo* as described in the text. Uptake is expressed as cpm in granulosa cells/mg ovary/cpm/µl blood. The bottom panel represents binding of ^{125}I-HCG *in vitro*. Solid lines with closed circles represent effects of 5 IU HCG and solid lines with closed triangles represent effects seen with 1,000 IU of HCG. Broken lines and open circles represent saline-treated controls. The ordinate of the lower panel and the abscissa are as described in Fig. 2.

declined by 24 and 48 hr. An excess (1,000 IU) of HCG blocked uptake of the trace completely. When binding was measured *in vitro*, no appreciable change in available sites was observed from 6 to 12 hr, but there was a marked decline between 12 and 24 hr which remained low till 48 hr. An excess of HCG saturated all available binding sites. If the maximal uptake between 6 and 12 hr observed with the low dose of HCG is due to saturation of sites, one would have anticipated a decline in available receptors over this time and an increase in available sites from 12 to 48 hr when uptake was declining. Conversely, if the persistent and relatively late decrease in binding caused by 5 IU of HCG

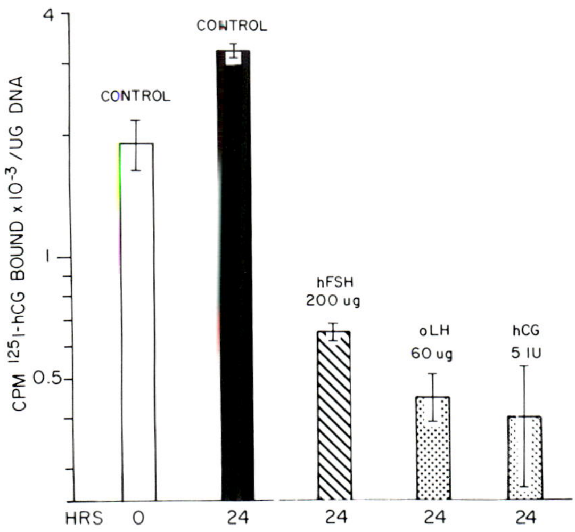

FIG. 4. Effect of HFSH, OLH, and HCG on LH receptor in granulosa cells. The ordinate and abscissa are as described for Fig. 2. The open bar and closed bar represent controls at 0 and 24 hr. The slashed bar depicts effects of HFSH. The stippled bar shows effects of OLH and the cross-hatched bar gives effects of HCG.

was due to occupancy of receptors, an increase or maintenance of uptake *in vivo* would be expected. However, as is evident, there was a simultaneous decrease in uptake and binding, suggesting again a mechanism other than the occupation of binding sites *in vivo*. This is further supported by evidence that LH in the serum declined more rapidly than the decrease in receptors. If the decrease in receptors is a phenomenon associated with the differentiation of granulosa cells into luteal cells, it could be predicted that agents other than LH or HCG capable of inducing luteinization should also cause a decrease in LH receptors.

Since FSH has been shown in a number of instances to induce luteinization, its ability to mimic the effects of LH in our model system was examined. FSH, which had been pretreated with chymotrypsin to reduce LH contamination, was further purified by adsorption in the presence of large amounts of LH receptor. The LH activity in 200 μg of this preparation was determined by radioreceptor assay to be 0.08% of the dose of LH normally administered in this study. As shown in Fig. 4, 200 μg of purified FSH caused a decline in available LH receptor comparable to the effects of LH and HCG. Furthermore, ovaries of FSH-treated rats appeared morphologically luteinized in 24 hr, and on

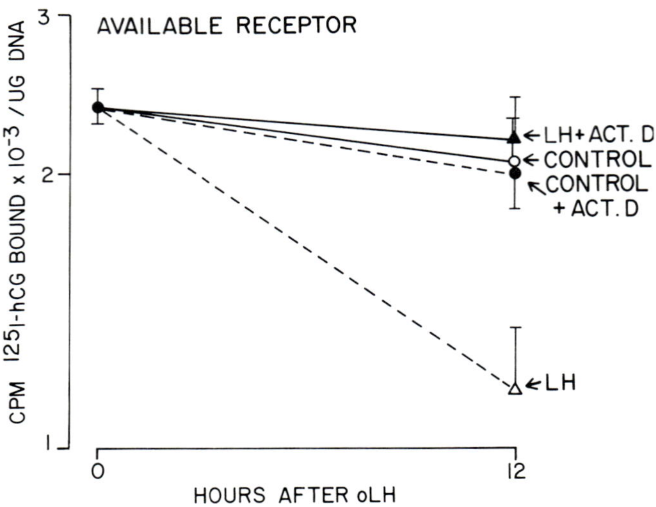

FIG. 5. Effect of actinomycin D on LH receptor regulation by OLH. Actinomycin D (2.5 mg/kg body weight) was administered intraperitoneally as described in the text. Abscissa and ordinate are as described for Fig. 2. Triangles depict LH-treated groups and circles depict controls. Solid symbols represent actinomycin D-treated groups; open symbols, saline-treated controls.

histological examination revealed cells in early luteinization. A high dose of FSH treatment also caused increased serum levels of progesterone. Since FSH and LH are known to bind to separate receptors specific for each hormone, these results indicate that the reduction of available binding sites for LH occurs by mechanisms other than occupancy.

To examine if this regulation of LH receptor by LH is dependent on intracellular processes like RNA synthesis, the effects of actinomycin D, a known inhibitor of RNA synthesis on the action of LH, was studied. When 2.5 mg/kg body weight of the inhibitor was administered intraperitoneally along with the subcutaneous injection of LH, actinomycin D prevented the ability of LH to decrease its own receptor (Fig. 5). These results support the view that the action of LH in inducing a decline in its own available receptor occurs by a mechanism other than occupancy, which may be mediated by a transcriptional-dependent process. The evidence provided indicates that during luteinization, LH is capable of inducing a decrease in its own available receptor, by mechanisms other than occupation of binding sites. The intriguing question of whether the desensitization of the target tissue to LH caused by ovulatory doses of the hormone could be a loss in receptors for LH will be the subject of future experimentation. This effect of LH has immense implications in the mechanism of LH action in numerous biological systems.

REFERENCES

1. Hunzicker-Dunn, M., and Birnbaumer, L. (1976): Adenylyl cyclase activities in ovarian tissues. III. Regulation of responsiveness to LH, FSH, and PGE_1 in prepubertal, cycling, pregnant and pseudopregnant rat. *Endocrinology,* 99:198-210.
2. Lamprecht, S. A., Zor, U., Tsafriri, A., and Lindner, H. R. (1973): Action of prostaglandin E_2 and of luteinizing hormone on ovarian adenylate cyclae, protein kinase and ornithine decarboxylase activity during postnatal development and maturity in the rat. *J. Endocrinol..,* 57:217-233.
3. Marsh, J. M., Mills, T. M., and LeMaire, W. J. (1973): Preovulatory changes in the synthesis of cyclic AMP by rabbit graafian follicles. *Biochimica Biophysica Acta,* 304:197-202.
4. Richards, J. S., and Midgley, A. R., Jr. (1976): Protein hormone action: A key to understanding ovarian follicular and luteal cell development. *Biol. Reprod.,* 14:82-94.
5. Richards, J. S., Ireland, J. J., Rao, C. M., Bernath, G. A., Midgley, A. R., Jr., and Reichert, L. E., Jr. (1976): Ovarian follicular development in the rat: Hormone receptor regulation by estradiol, follicle stimulating hormone and luteinizing hormone. *Endocrinology,* 99:1562-1570.

Ovarian Follicular Development and Function,
edited by A. R. Midgley and W. A. Sadler.
Raven Press, New York © 1979.

cAMP Accumulation Inhibitor in Follicular Fluid of Human Origin

Z. Kraiem and B. Lunenfeld

Institute of Endocrinology, Chaim Sheba Medical Center, Tel-Hashomer, Israel

Ovarian follicular fluid of rabbit (1,2) and porcine (3,4) origin has been shown to have an inhibitory influence upon oocyte maturation and granulosa cell luteinization. Channing and co-investigators (3) quantitated this inhibitory capacity by estimating the extent to which follicular fluid inhibited *in vitro* gonadotropic-stimulated cAMP accumulation in porcine granulosa cells. Also using an *in vitro* assay system, we obtained evidence indicating the presence of a cAMP accumulation inhibitor in follicular fluid derived from human origin. A study is being conducted in our laboratory correlating human ovarian morphology and hormonal profile, as part of a WHO-sponsored multicenter project (5), whose general objective is the selection of reliable parameters for predicting time of ovulation. Thus, a most valuable opportunity presented itself of analyzing follicular fluid inhibitor concentration in correlation with the above data on endocrine profile. This chapter is a progress report on this ongoing investigation.

MATERIALS AND METHODS

The study included normal, healthy subjects, aged 21 to 40, with documented evidence of regular ovulatory cycles and with medical diagnoses indicating laparotomic surgery. The experimental design is shown in Fig. 1. The operation was scheduled very close to the expected time of ovulation. The actual timing of surgery was adjusted according to the endocrine data obtained during the investigation. Day 0 was referred to as the day of surgery, blood samples were taken daily from days -8 to -4 and from days -3 to +3, three times daily, for plasma assays of FSH, LH, 17β-estradiol and

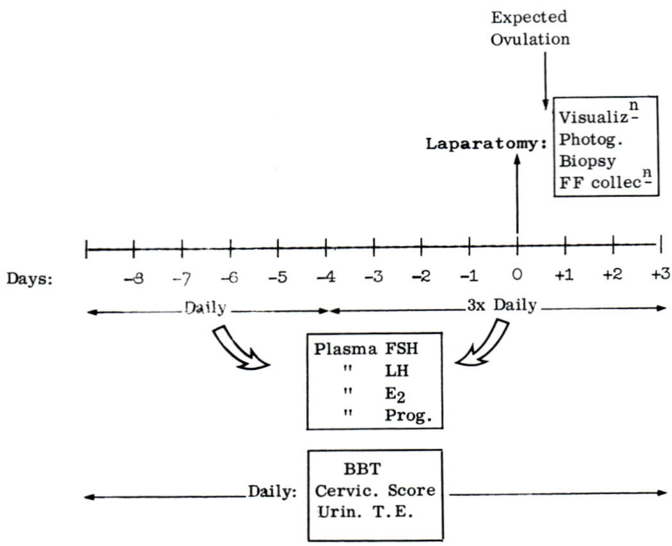

FIG. 1. Experimental design.

progesterone. BBT, cervical score, and urinary total estrogens were also monitored daily. During laparotomy, follicular fluid was aspirated from follicles of different sizes, and an ovarian biopsy was performed. Histological findings from the ovarian biopsy, together with morphological evidence based on visual observation and photography during laparotomy, provided characterization of the follicles from which fluid was collected. Moreover, the stage during the periovulatory period when laparotomy was performed was determined from the latter data in addition to the hormonal profile.

Assay of Follicular Fluid Inhibitory Capacity

Mouse ovarian tissue from 21-day-old mice was preincubated for 30 min under 95% O_2:5% CO_2 in a shaking water bath at 37°C. The incubation medium (1.3 ml) consisted of Krebs-Ringer bicarbonate buffer (with half the usual calcium concentration) containing glucose (1 mg/ml). Following preincubation, the incubation was continued for another 20 min in the presence and absence of HCG (1 IU) and follicular fluid (0.05 ml, free of any oocyte). cAMP was then determined in the tissue plus incubation medium by the protein-binding assay described by Gilman (6), modified--as suggested by Brown et al. (7)--by using charcoal for separating bound from free cAMP.

RESULTS AND DISCUSSION

Using, for simplicity's sake, data from an illustrative case (Table 1), the following deductions can be made. The difference between basal ovarian cAMP level in the presence (tube No. 2) and absence (tube No. 1)

TABLE 1. *Effect of human follicular fluid on cAMP accumulation*[a]

Ovarian tissue in the presence and absence of:				
Tube No.	Follicular fluid	HCG	cAMP accumulation (pmole/mg tissue)[b]	
1	-	-	1.12 ± 0.14	(6)
2	+	-	0.59 ± 0.01	(6)
3	-	+	4.54 ± 0.29	(6)
4	+	+	2.50 ± 0.03	(6)

[a]Experimental conditions and measurement of cAMP accumulation as described in "Materials and Methods."
[b]Means ± SEM with number of determinations indicated in parentheses.

of follicular fluid (1.12 - 0.59 = 0.53 pmole/mg tissue) represents the extent to which the follicular fluid degrades cAMP. In the absence of follicular fluid, the difference between basal ovarian cAMP level with (tube No. 3) and without (tube No. 1) HCG (4.54 - 1.12 = 3.42 pmole/mg tissue) constitutes the *de novo* net cAMP formation induced by HCG stimulation. Assuming that the follicular fluid inhibitor acts solely through degradation of cAMP, it follows that the cAMP level in HCG-stimulated ovarian tissue in the presence of follicular fluid (tube No. 4) should be 0.53 (i.e., the cAMP remaining following degradation by the inhibitor) + 3.42 (i.e., the *de novo* net cAMP formation induced by HCG) = 3.95 pmole/mg tissue. However, since the level obtained (2.50 pmole/mg tissue) is significantly lower ($p < 0.025$), this suggests that the inhibitor(s)--apart from stimulating cAMP degradation--might also exert an inhibitory influence on cAMP formation.

The data on the effect of follicular fluid on ovarian cAMP accumulation are shown in Table 2. In the first two cases shown (0 to 16 hr after the first significant rise in plasma LH and about 8 hr before the LH peak), an inhibition of close to 50% was observed for both basal and HCG-stimulated cAMP accumulation, with no apparent difference between the large, graafian

TABLE 2. *Effect of follicular fluid on ovarian cAMP accumulation*

Subject No.	Follicle	Stage during preovulatory period	% Inhibition of Basal cAMP[a]		% Inhibition of HCG-stimulated cAMP[b]	
1	Graafian	Early	48.2 ± 0.9	(4)	51.3 ± 3.4	(6)
2	Graafian	Early	51.2 ± 1.2	(4)	50.5 ± 5.1	(4)
	Medium	Early	49.7 ± 6.8	(4)	48.7 ± 0.3	(4)
3	Graafian	Early	33.0 ± 10.8	(6)	32.0 ± 4.4	(10)
	Medium	Early	50.0 ± 8.1	(8)	24.2 ± 5.8	(6)
4	Graafian	Late	4.2 ± 3.0	(4)	0.5 ± 13.8	(4)
5	Graafian	Late	-6.5 ± 10.9	(4)	-10.2 ± 18.3	(6)
	Medium	Late	4.5 ± 4.5	(4)		

[a]100 - [cAMP(ovarian tissue + follicular fluid)/cAMP(ovarian tissue)] × 100.

[b]100 - [cAMP(ovarian tissue + HCG + follicular fluid)/cAMP (ovarian tissue + HCG)] × 100.

follicle and medium-sized follicle. The third case (8 hr before the first significant rise in plasma LH and about 16 hr before the LH peak) again displayed around 50% inhibition of basal cAMP for the medium-sized follicle, whereas a lower inhibitory capacity--around 30%--was obtained with respect to HCG-stimulated cAMP in this follicle as well as the graafian follicle (both HCG-stimulated and basal cAMP). In contrast, in the last two cases--in which laparotomy was performed at a later stage (24 hr after the first significant rise in plasma LH and about the time of the LH peak)--the graafian and medium-sized follicles lacked the capacity to inhibit basal cAMP, and the graafian follicle also failed to restrain HCG-stimulated cAMP.

In another series of subjects, no inhibitory influence could be detected in fluid derived from follicular cysts.

Obviously, a number of interesting questions remain to be explored. For instance, is the cAMP accumulation inhibitor(s) identical, or related, to the inhibitors of oocyte maturation and luteinization, or perhaps to the female analog of the male "inhibit"? In any event, in view of the essential role cAMP plays as an intermediate in the processes governing ovarian function, the presence *in situ* of fluctuating levels of a cAMP inhibitor merits interest because of far-reaching implications regarding intraovarian regulatory mechanisms.

ACKNOWLEDGMENTS

This study was supported by the World Health Organization, Medical Research Foundation, and the Ford Foundation (Grant No. 67-470).

Follicular fluid was kindly supplied by Drs. N. Ben-Adereth and G. Potashnik from the Soroka Medical Center, Beer Sheba, Israel, and Drs. S. Grünstein and A. Shalit from the Hillel Yaffe Memorial Hospital, Hedera, Israel.

B. Lunenfeld is an Established Investigator of the Chief Scientist's Bureau, Ministry of Health, Israel.

REFERENCES

1. Chang, M. C. (1955): The maturation of rabbit oocytes in culture and their maturation, activation, fertilization, and subsequent development in the fallopian tubes. *J. Exp. Zool.*, 128:379.
2. El Fouly, M. S., Cook, B., Nekola, M., and Nalbandov, A. V. (1970): Role of the ovum in follicular luteinization. *Endocrinology*, 87:288.
3. Channing, C. P., and Tsafriri, A. (1974): Regulation of ovulatory processes: Ovum maturation, follicular rupture and luteinization. In: *Advances in Fertility Regulation Through Basic Research*, edited by W. A. Sadler and S. Segal. Plenum Press, New York.
4. Tsafriri, A., and Channing, C. P. (1975): An inhibitory influence of granulosa cells and follicular fluid upon porcine oocyte meiosis *in vitro*. *Endocrinology*, 96:922.
5. Ben-Aderet, N., Ben Michael, R., Grünstein, S., Kraiem, Z., Lunenfeld, B., Potashnik, G., and Shalit, A. (1976): Correlation of hormonal profile and ovarian morphological features during periovulatory period in humans. *Israel J. Med. Sci.* 12:1507.
6. Gilman, A. G. (1970): A protein binding assay for adenosine 3':5'-cyclic monophosphate. *Proc. Natl. Acad. Sci. USA*, 67:305.
7. Brown, B. L., Albano, J. D. M., Ekins, R. P., and Sgherzi, A. M. (1971): A simple and sensitive saturation assay method for the measurement of adenosine 3':5'-cyclic monophosphate. *Biochem. J.*, 121:561.

Ovarian Follicular Development and Function,
edited by A. R. Midgley and W. A. Sadler.
Raven Press, New York © 1979.

Steroidogenic Responsiveness of Ovulated Follicles from the Rabbit Ovary to LH and to Cyclic AMP

Thomas M. Mills

Medical College of Georgia, Augusta, Georgia 30902

Several published experiments have demonstrated that the ovarian follicle of the rabbit becomes refractory to stimulation with LH when the follicle is in the periovulatory interval (approximately 10 to 12 hr after mating or after injection of an ovulating dose of gonadotropin). Parameters such as steroidogenesis (1) and the synthesis of cyclic AMP (2) are included in this loss of responsiveness. Hunzicker-Dunn and co-workers (3) have recently put forward the proposal that in rabbit follicles LH stimulated cyclic AMP synthesis via an activation of adenylate cyclase, but, in addition, the LH also desensitized the enzyme to further LH stimulation. Studies from this laboratory prove that cyclic AMP is an intermediate in the mechanism of action of LH on follicular steroidogenesis (4). It was therefore considered possible that this loss of responsiveness of the ovulated follicles to LH stimulation was due to the desensitization of the cyclase activity following the coitally induced gonadotropin release. It would follow then that if the cyclase activity was inhibited in ovulated follicles, then these structures should remain sensitive to cyclic AMP. The following study was performed to test this possibility.

Ovarian follicles obtained from a rabbit 12 hr after the injection of an ovulatory dose (200 IU) of HCG were incubated for 3 hr in the presence of luteinizing hormone (5 μg LH/ml) or cyclic AMP (20 μmole cyclic AMP/ml). A third group of follicles from the same rabbit was incubated without further additions as control (C). The progesterone and estradiol content of the incubation mixture was determined by radioimmunoassay (4). The results of this study are presented in Fig. 1; each bar represents the mean of three to four separate rabbit experiments with brackets representing 1 SEM.

The results presented in Fig. 1 demonstrate that ovulated follicles are unresponsive to LH; the gonadotropin does not increase the follicular content of

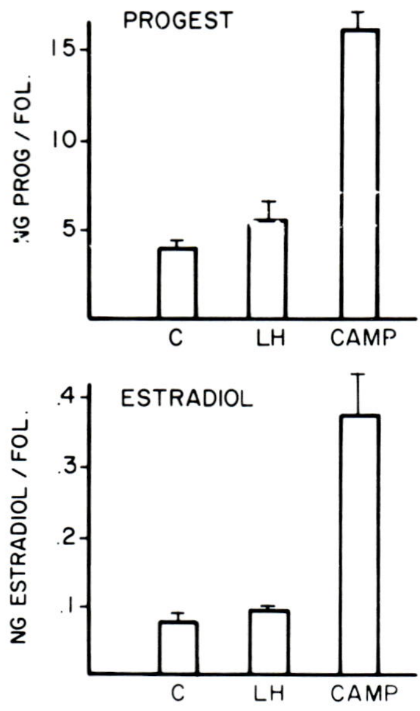

FIG. 1. The effect of incubation with LH (5 μg/ml) and cyclic AMP (CAMP-20 μmoles/ml) on the synthesis of progesterone (progest) and estradiol in ovulated follicles obtained 12 hours after HCG injection. No additions were made to control (C) incubations. Each bar represents the mean of three to four separate determinations with brackets representing one SEM.

either progesterone or estradiol. However, these follicles are highly responsive to cyclic AMP with significant increases in both steroids ($p < 0.05$ compared to both control and LH groups). These same concentrations of both LH and cyclic AMP have been previously shown to markedly stimulate the synthesis of several steroids but only in follicles from unmated rabbits (4).

This finding that ovulated follicles, which are insensitive to LH, do respond to cyclic AMP with increases in the synthesis of two steroids provides strong evidence that the cessation in response to LH at ovulation is due to a failure in the synthesis of cyclic AMP. The LH-induced desensitization of follicular adenylate cyclase, as proposed by Hunzicker-Dunn et al. (3) is, therefore, the most probable basis for the blockage of steroidogenesis in ovulated follicles of the rabbit.

REFERENCES

1. Mills, T., and Savard, K. (1973): Steroidogenesis in ovarian follicles isolated from rabbits before and after mating. *Endocrinology,* 92:788.
2. Marsh, J., Mills, T., and Le Maire, W. (1972): Cyclic AMP synthesis in rabbit Graafian follicles and the effect of luteinizing hormone. *Biochem. Biophys. Acta,* 273:389.
3. Hunzicker-Dunn, M., and Birnbaumer, L. (1976): Adenylyl cyclase activities in ovarian tissues. II. Regulations of responsiveness to LH, FSH, and PGE in the rabbit. *Endocrinology,* 99:185.
4. Mills, T. (1975): Effect of luteinizing hormone and cyclic adenosine 3'5'-monophosphate on the steroidogenesis in the ovarian follicle of the rabbit. *Endocrinology,* 96:440.

Ovarian Follicular Development and Function,
edited by A. R. Midgley and W. A. Sadler.
Raven Press, New York © 1979.

Influence of Follicular Maturation on the Ability of Porcine Granulosa Cells to Produce Cyclic AMP in Response to FSH and LH

Ada M. Lindsey and Cornelia P. Channing

Department of Physiology, University of Maryland
School of Medicine, Baltimore, Maryland 21201

Changes in granulosa cell (GC) responsiveness to luteinizing hormone (LH) and follicle-stimulating hormone (FSH) that occur with maturation of the porcine follicle were investigated. GCs were harvested from small (1 to 2 mm), medium (3 to 5 mm), and large (6 to 12 mm) porcine follicles according to the method of Channing and Ledwitz-Rigby (1). Subsequently, they were incubated in aliquots of approximately 2×10^7 cells in 1 ml of Eagles medium containing 1% BSA with either ovine FSH (NIH-FSH-S10) in concentrations of 0.01 to 100 μg/ml or ovine LH (NIH-LH-S18) ranging from 0.001 to 1.0 μg/ml for 1, 5, 15, 30, 60, or 120 min at 37°C in a Dubnoff metabolic shaking incubator. The ovine FSH was preincubated for 60 min at 37°C with antibovine LH to neutralize contaminating LH (6).

Following these short-term incubations, the intracellular cyclic adenosine 3',5'-monophosphate (cAMP) accumulation was measured by our modification of the protein binding assay (1) originally described by Gilman (2). After 15-min incubations, the stimulatory effect of 10 μg/ml FSH (pretreated with LH antiserum) on cAMP accumulation was on the average 16-fold greater in small follicle GCs when compared to large follicle GCs (Fig. 1). The greatest stimulatory effect of 1.0 μg/ml LH on cAMP accumulation occurred in the large follicle GCs.

Similar results have been obtained using a purified human FSH preparation (LER 1577) that has been treated with chymotrypsin to minimize the LH contamination and with a purified porcine FSH (3,5).

These data indicate that as the porcine follicle matures there is a decrease in GC responsiveness to FSH concurrent with an increase in responsiveness to LH in terms of cAMP accumulation. These findings extend our original finding that purified FSH and LH can

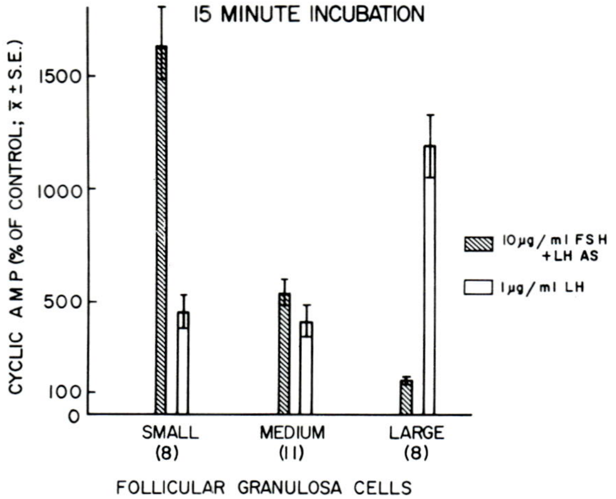

FIG. 1. Effects of ovine FSH (10 μg/ml) and LH (1.0 μg/ml) on cAMP accumulation by porcine granulosa cells harvested from small (1 to 2 mm), medium (3 to 5 mm), and large (6 to 12 mm) follicles when incubated for 15 min at 37°C. Ovine FSH was preincubated with antibovine LH for 60 min at 37°C. The data are represented as percents of control cAMP levels (mean ± SE) with the number of pooled observations indicated in parentheses at the bottom.

stimulate cAMP accumulation in GCs harvested from medium-sized porcine follicles (4).

ACKNOWLEDGMENTS

This research was supported by Grants HD-08834 and HD-00435 from the Institute of Child Health and Human Development (USPHS). We thank Dr. William Hansel for a generous supply of antibovine LH antiserum, and we thank the hormone distribution officer of the NIAMD for ovine LH and FSH. The gifts of purified human and porcine FSH from Drs. Leo Reichert and Maryce Jacobs, respectively, are gratefully acknowledged.

REFERENCES

1. Channing, C. P., and Ledwitz-Rigby, F. (1975): Methods for assessing hormone-mediated differentiation of ovarian cells in culture and in short-term incubations. In: *Methods in Enzymology*, Vol. 39, Part D, edited by J. G. Hardman and B. W. O'Malley, pp. 183-230. Academic Press, New York.

2. Gilman, A. G. (1970): A protein binding assay for adenosine 3',5'-cyclic monophosphate. *Proc. Natl. Acad. Sci. U.S.A.*, 67:305-312.
3. Jacobs, M. M., and Ward, D. N. (1976): Studies on the isolation and chemical properties of porcine follicle-stimulating hormone. *Proc. Soc. Exp. Biol. Med.*, 151:568-570.
4. Kolena, J., and Channing, C. P. (1972): Stimulatory effects of LH, FSH and prostaglandins upon cyclic 3',5'-AMP levels in porcine granulosa cells. *Endocrinology*, 90:1543-1550.
5. Reichert, L. E. (1967): Selective inactivation of the luteinizing hormone contaminant in human pituitary follicle stimulating hormone preparations. *J. Clin. Endocrinol. Metab.*, 27:1065-1067.
6. Snook, R. B. (1968): Immunological and biological properties of antiserum to bovine LH. In: *Protein and Polypeptide Hormones*, edited by A. Margoulies, pp. 398-400. Excerpta Medica Foundation, New York.

Ovarian Follicular Development and Function,
edited by A. R. Midgley and W. A. Sadler.
Raven Press, New York © 1979.

Discussion Summary: Session on Hormone Action in Follicles*

M. Hunzicker-Dunn

Recent research into the molecular mechanisms of protein hormone action has demonstrated that the response evoked by a stimulatory hormone depends not only on the concentration of circulating hormone that pertubates the target cell but also on a number of other variables, some of which include: (a) the number of specific, plasma-membrane hormone receptors, (b) the affinity of these receptors for the hormone, (c) the sensitivity of the receptor-coupled adenylyl cyclase system, and (d) the responsiveness of the receptor-coupled cyclase system to the stimulatory hormone.

The ovarian follicle offers an excellent model system in which to elucidate the molecular responses comprising hormone action. Depending on the specific stage of its life history, the follicle demonstrates varying degrees of hormonal responsiveness which, at least in part, can be attributed to the gonadotropin responsiveness of the receptor-coupled adenylyl cyclase system. Thus, when granulosa cells are proliferating, the small antral follicle contains large numbers of FSH receptors and a highly responsive, FSH-sensitive adenylyl cyclase system. Concurrently, few LH-receptors are present and the adenylyl cyclase system is much less response to LH stimulation. As the follicle matures to a preovulatory state, the highly responsive FSH receptor-coupled adenylyl cyclase system seems to be "replaced" by an LH receptor-coupled adenylyl cyclase system. Thus, at the time when the follicle is to receive the ovulatory surge of gonadotropins, it contains a greater proportion of LH receptors combined with a highly responsive LH-sensitive adenylyl cyclase system. These observations corroborate the general belief that FSH acts primarily on the

*References have been included only when they were referred to in the open discussion and were not already included in the printed texts.

growing follicle, stimulating cellular maturation, and that LH acts primarily on the preovulatory follicle, promoting ovulation and luteinization. It is clear, however, that this simple division of hormonal action is not sufficient, since it does not, for instance, account for the functionally synergistic effects characteristic of LH and FSH throughout follicular maturation. Steroidogenesis, for example, is such a response and consequently requires that antral follicles contain an adenylyl cyclase system that is functionally coupled *both* to LH and FSH receptors in all stages of follicular maturation. Results demonstrate this to be the case in all follicles thus far examined.

In summary, then, superimposed upon the highly responsive FSH-sensitive adenylyl cyclase system in small follicles and the highly responsive LH-sensitive adenylyl cyclase in large, preovulatory follicles is the less responsive, but probably equally significant, LH-sensitive adenylyl cyclase system in small follicles, the LH- and FSH-sensitive adenylyl cyclase systems in medium follicles, and the FSH-responsive adenylyl cyclase system in large follicles. The physiological significance of the varying degrees of hormonal responsiveness exhibited throughout follicular maturation by the follicular adenylyl cyclase system is unclear at this time and awaits further investigation.

During the final stages of follicular maturation, i.e., following the ovulatory surge of gonadotropins, the responsiveness of the follicle to gonadotropic stimulation is again altered. This alteration in hormonal responsiveness is characterized by an abrupt decline, and ultimately a cessation, in the ability of the follicle to respond to gonadotropic stimulation, in terms of steroidogenesis. The follicular refractoriness is, at least in part, due to an inability of the adenylyl cyclase system itself to respond to LH stimulation, i.e., desensitization of the adenylyl cyclase system to LH stimulation. Evidence presented suggests that follicular refractoriness does not extend enzymatically beyond the adenylyl cyclase system, since exposure of LH-refractory follicles to cAMP promotes steroidogenesis. These results imply that the cAMP-activated protein kinase enzyme system and distal steroidogenic enzymes are themselves not rendered insensitive to stimulation or activation by the ovulatory surge of gonadotropins.

Neither the mechanism nor the functional significance of LH-induced desensitization of the LH-sensitive adenylyl cyclase system is understood at this time. A number of possibilities may be considered to explain the phenomenon of desensitization.

(a) An increase in phosphodiesterase activity could account for the observed decrease in cAMP accumulation in some of those experiments that utilized an ATP-prelabeling technique; however, it need not be considered in experiments that directly accessed the activity of the adenylyl cyclase system. At any rate, phosphodiesterase activity in control follicles is not significantly different from that in refractory follicles. (b) Although the rise in PG levels in preovulatory follicles roughly correlates with the development of refractoriness of the adenylyl cyclase system, results from a number of laboratories have shown that the desensitization phenomenon is not mediated by PG. (c) Either a decrease in the number of available LH receptors or a significant decline in the affinity of the receptors toward LH could account for the observed refractoriness exhibited by the adenylyl cyclase system following exposure to gonadotropic stimulation. The injection of an ovulatory dose of LH or HCG into rats containing follicular- or luteal-predominated ovaries promotes a loss of available LH receptors by a mechanism other than receptor occupancy. This LH-induced loss of LH receptors correlates temporally with the LH-induced desensitization of the adenylyl cyclase system both in follicular and in luteal tissues of the rat, suggesting that LH-induced desensitization may indeed be associated with a specific decline in available LH receptors. On the other hand, rat follicles that were desensitized to LH in an *in vitro* culture demonstrated no decline in LH-binding capacity when compared to control, fully LH-responsive follicles. It is therefore impossible, at this time, to conclude whether LH-induced desensitization of the follicular adenylyl cyclase system is caused by a loss of available LH receptors. (d) Results presented demonstrate that the desensitization phenomenon induced in rat follicles exposed to LH or to PGE_2 in culture can be inhibited by actinomycin D or cycloheximide. Interestingly, the LH-induced loss of LH receptors in rat follicles, as demonstrated in *in vivo* experiments, is also inhibited by actinomycin D. These results indicate that a transcriptional-dependent event is involved in both phenomena. (e) Hormone-induced desensitization may be accomplished by the action(s) of a protein inhibitor, the synthesis of which is induced by the stimulatory hormone itself. Although the development of the refractory state may indeed require protein synthesis, as has been indicated, the available evidence suggests that the loss of hormonal responsiveness is not attributable to the accumulation of an inhibitor of cyclase activity (Franklin and Foster, 1973; Hunzicker-Dunn and Birnbaumer, 1976). (f) Although depletion of endogenous ATP could account for the adenylyl cyclase refractoriness, this possibility may be ruled out, since desensitization has been obtained in *in vitro* experiments where an ATP-regenerating system maintains constant ATP levels. (g) Desensitization of the adenylyl cyclase system does not seem to be due to direct "pertubation" of the catalytic subunit, since results presented demonstrate that the system retains its responsiveness to such modulators as GTP and NaF as well as to PG. (h) Evidence was alluded to which suggests that the desensitization phenomenon may be mediated by a phosphorylation reaction. In *in vitro* experiments, LH-induced desensitization of the adenylyl cyclase system occurs only when ATP and Mg are present in the incubation medium. Furthermore, substitution of ATP by AMP-P(NH)P, an analog of ATP which can serve as a substrate for adenylyl cyclase but which cannot donate its terminal phosphate, results in no desensitization of the adenylyl cyclase system (Bockaert et al., 1976). The inclusion of a phosphatase-containing preparation, moreover, promotes reversal of the LH-induced desensitization phenomenon. (i) The possibility exists that desensitization is due to an uncoupling of the LH receptor from the catalytic subunit, possibly accomplished by a reorganization of the adenylyl cyclase system or its subunits within the plasma membrane.

In summary, it is not yet possible to explain the mechanism by which the adenylyl cyclase system becomes unresponsive to its own stimulant. It seems, however, that desensitization of the adenylyl cyclase system is, at least in some cases, a transcriptionally mediated event that occurs coincident with a loss of hormone-specific receptors. Further research is clearly needed in order to elucidate the mechanism by which the adenylyl cyclase system becomes refractory to its own stimulant.

The biological significance of the phenomenon of desensitization of the adenylyl cyclase system in the follicle is unclear. It is not known, for instance, if desensitization is required for any specific follicular response or if it is simply a control mechanism to prevent the system from overshooting. There is evidence that desensitization is not "simply" the result of feedback inhibition by cAMP (Bockaert et al., 1976). A number of possible functions may be served by desensitization of the cyclase system, some of which were discussed in the text and will not be reiterated here. Desensitization may be required to permit ovulation. Although at least a partial refractoriness of the cyclase system always precedes ovulation, we have no way of knowing at the present time if it is required for ovulation. An interesting analogy was drawn between the involuting postpartum uterus and ovulation, both of which are reportedly induced by collagenase or similar proteolytic enzymes. The observation that estrogen can prevent collagenase-induced involution of postpartum rat uteri raises the possibility that ovulation itself may be impossible when estrogen levels are maintained at the levels reached immediately following the ovulatory surge of gonadotropins. Accordingly, desensitization of the cyclase system could accomplish this cessation in steroidogenesis, via a decline in cAMP synthesis, which accompanies the normal ovulatory response. It is, however, not even known whether ovulation can be accomplished in the presence of a fully responsive adenylyl cyclase system. Such a demonstration is necessary if one is to correlate desensitization of the adenylyl cyclase system with the ovulatory response.

In conclusion, the follicular adenylyl cyclase system demonstrates marked variations in its responsiveness to gonadotropic stimulation throughout follicular maturation. Although these alterations can temporally be related to specific responses evoked by follicles during different stages of maturation, the exact functional significance of these variations is not understood and clearly requires further investigation.

REFERENCES

Bockaert, J., Hunzicker-Dunn, M., and Birnbaumer, L. (1976): Hormone-stimulated desensitization of hormone-dependent adenylyl cyclase. *J. Biol. Chem.*, 251:2653-2663.

Hunzicker-Dunn, M., and Birnbaumer, L. (1976): Adenylyl cyclase activities in ovarian tissues. II. Regulation of responsiveness to LH, FSH and PGE_1 in the rabbit. *Endocrinology*, 99:185-197.

Franklin, T. J., and Foster, S. J. (1973): Hormone-induced desensitization of hormonal control of cyclic AMP levels in human diploid fibroblasts. *Nature (New Biol.)*, 246:146-148.

OOGENESIS

Ovarian Follicular Development and Function,
edited by A. R. Midgley and W. A. Sadler.
Raven Press, New York © 1979.

The Control of Oogenesis in Mammals

T. G. Baker

Hormone Laboratory, Department of Obstetrics and
Gynaecology, University of Edinburgh,
Edinburgh EH3 9EW, Scotland

The term "oogenesis" is often used but rarely defined. In its narrowest context, it has simply been used to denote the mitotic proliferation of oogonia (Franchi, et al., 1962). Many other investigators have used the term to denote the period of meiotic prophase, or the maturation of the ovum. However, the most widely used definition, which is used in this chapter, is that oogenesis is "...the formation, development and maturation of the female gamete or ovum." (Anonymous, 1974; Henderson and Henderson, 1960), since this most closely resembles the definition of comparable processes occurring in the testis (spermatogenesis).

The processes involved in oogenesis commence early in embryonic life and become arrested shortly after birth: the first part of this chapter will be concerned with these events. However, the principal events concerning us during this meeting occur shortly before ovulation, which can take place in women up to 50 years after the onset of oogenesis. Thus, the events of preovulatory maturation will form the major part of this review.

PROCESSES OCCURRING IN OVARIES OF MAMMALIAN EMBRYOS AND FETUSES

It is now well established that the germinal "stem" cells of the ovary (oogonia) are the direct progeny of primordial germ cells. The latter arise outside of the body (probably in the extraembryonic endoderm or mesoderm) and subsequently migrate to the presumptive gonadal ridges (see Moore, 1973). By contrast, the somatic components of the ovary arise from the celomic epithelium and mesenchyme of the dorsal body wall, and probably also from the mesonephros (primitive kidney:

see Baker and Wai Sum O, 1976). Once the gonad has been transformed into an ovary by the process known as sex differentiation, the female germ cells rapidly increase in number by mitosis before entering upon the prophase of meiosis to become oocytes (Baker, 1972; Franchi et al., 1962).

The process of meiosis (Fig. 1) is clearly of paramount importance during oogenesis and genetic errors occurring during this time may have serious chromosomal and phenotypic consequences. Until recently, nothing was known of the basic factors controlling meiosis, but some of the processes are less obscure in the light of recent experiments. Thus, when mouse or hamster ovaries, removed on Days 14 or 15 postconception (p.c.), respectively, are maintained in organ culture the majority of germ cells fail to enter meiosis and undergo premature degeneration after one or more mitotic divisions (e.g., Wai Sum O and Baker, 1976; Wolff and

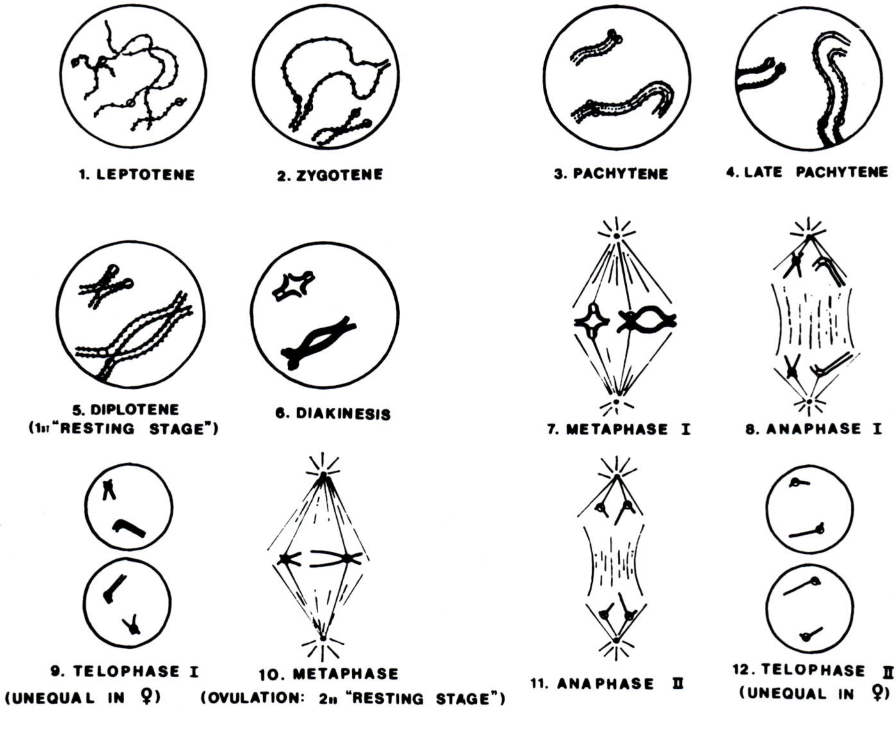

FIG. 1. Diagrammatic representation of the process of meiosis in an idealized situation depicting two pairs of chromosomes. Only the nuclei of the oocytes are shown: the cytoplasm has been disregarded. (From Baker, 1972.)

Haffen, 1965). By contrast, ovaries cultured after this critical period undergo normal differentiation and the germ cells proceed through meiotic prophase to the diplotene stage in a comparable period to that taken by germ cells *in vivo* (Baker and Neal, 1973; Challoner, 1975*a*). It would seem, therefore, that there is a gonadal inductor substance that "triggers" meiosis in the ovary (and seemingly prevents germ cells undergoing degeneration), which normally is not produced, or fails to act, in the fetal testis (but see below). The meiotic inductor is unlikely to be a hormone, since gonadotropins and maternal serum do not bring about the onset of meiosis in Day 15 hamster ovaries in organ culture (Challoner, 1975*a*). It seems more likely that the factor emanates from the rete ovarii, since meiosis does not occur in ovaries devoid of the rete and transplanted to suitable recipients (Byskov, 1974, 1975; Byskov and Lintern Moore, 1973). The rete secretes a PAS-positive substance at the time of meiotic induction, although attempts to isolate a chemical factor from the rete have so far been unsuccessful. An alternative explanation for the mode of action of the rete in the onset of meiosis is that it involves merely cellular contact between the rete cells and oogonia (Byskov, *personal communication;* see Saxén et al., 1976).

Further evidence that induction is brought about by cellular contact, or that the inductor is only effective over small distances, derives from experiments in which ovaries and testes from coeval hamsters are cultured together (Wai Sum O and Baker, 1976). When the ovary with rete ovarii is cultured in close apposition to a testis, meiosis is induced in some germ cells within the testis, as well as in the ovary. However, when the gonads are placed in the Petri dish 2 mm apart, meiosis occurs only in the ovary and testicular differentiation occurs as in noncultured controls (Wai Sum O and Baker, 1976).

Once meiosis has been initiated the process proceeds to the diplotene stage (Fig. 1) without further stimulus. Ovaries explanted after the critical age mentioned above, or those cultured with their rete ovarii, contain oocytes that reach the diplotene stage within a comparable time to those *in vivo*, irrespective of whether or not hormones or serum are added to the culture medium (Blandau, 1965; Baker and Neal, 1973; Challoner, 1975*a*; Wai Sum O and Baker, 1976). Indeed, when the ovary with rete ovarii is cultured with a coeval testis, even some of the male germ cells reach the diplotene stage (Wai Sum O and Baker, 1976). It should be emphasized, however, that follicular development in the first few days of life *is* dependent on a supply of gonadotropic hormones (Baker and Neal, 1973; Challoner, 1975*b*; Eshkol et al., 1971). For example, mouse ovaries explanted on Day 14 p.c. and cultured for 21 days (i.e., to a time equivalent to Day 14 postpartum *in vivo*) contain oocytes at the diplotene stage, a proportion of which complete their growth to a stage similar to those *in vivo* with four layers of granulosa cells. However, the follicles in culture are devoid of a follicular envelope unless FSH and LH are added to the nutrient medium (Baker and Neal, 1973).

ARREST OF MEIOSIS AT THE DIPLOTENE STAGE

In most mammalian species, meiosis is arrested at a diffuse diplotene or dictyate stage and does not resume until shortly before ovulation (see below). The period of meiotic arrest usually commences either before or shortly after birth, although in the pig and cat oocytes at prediplotene stages can often be detected at the hilus region of the prepubertal ovary (see Baker, 1972; Black and Erickson, 1968; Brambell, 1956; Mossman and Duke, 1973). A further exception to the general rule is seen in prosimian primates where "nests" of primordial germ cells, which enter meiosis during each reproductive cycle (Anand Kumar, 1968), persist in the adult ovary. It is unlikely, however, that any of these germ cells contribute to the definitive population of eggs, which are ovulated each cycle (David et al., 1974): They may merely represent an embryological curiosity and probably degenerate without undergoing follicular growth.

The mechanism causing the arrest of meiosis at the diplotene stage remains completely unknown at present, although germ cells that occasionally proceed beyond this stage in preantral follicles rapidly undergo spontaneous degeneration (so-called pseudomaturation division; see Franchi et al., 1962; Ingram, 1962). It is possible that meiotic arrest is brought about by the granulosa cells, since it has been suggested that oocytes at diplotene cannot survive unless they become invested with a layer of follicular cells (Ohno and Smith, 1964). A comparable period of meiotic arrest does not occur during spermatogenesis in the male where the homolog of the follicle cell is the Sertoli cell, but it is a feature of oogenesis in many groups of animals where it is characterized by the presence of lampbrush chromosomes within the oocyte nucleus (e.g., Callan, 1957; Davidson, 1968). These specialized chromosomes produce large quantities of ribonucleoprotein, which may be important in the maturation of the egg and its subsequent development (see Baker, 1971; Burkholder et al., 1971; Davidson, 1968). Further studies of these chromosomes, and of the general problem of meiotic arrest, are clearly called for, since they may shed light on the control of oogenesis, particularly the processes involved in preovulatory maturation (see below).

RESUMPTION OF MEIOSIS IN THE GRAAFIAN FOLLICLE

The period of meiotic arrest of the oocyte persists until a time (which is constant for each species) prior to ovulation, when germinal vesicle breakdown (GVB) occurs. For the mouse, this occurs about 13 hr before ovulation, while in man and pig the comparable times are 36 and 42 hr, respectively (see Edwards, 1965, 1966). The resumption of meiosis *in vivo* occurs only if the oocyte resides in a mature graafian follicle, and it is dependent on the high levels of gonadotropic hormones in serum and follicular fluid during proestrus (so-called "LH-surge"). It should be pointed out, however, that gonadotropins are only required if the oocyte is within its follicle. Oocytes removed from their follicles into chemically defined medium spontaneously undergo meiotic maturation in the absence of hormones, although the incidence of this change is dependent on the status of the follicle from which the egg is recovered. Thus, oocytes removed from mature graafian follicles progress *in vitro* from the diplotene (germinal vesicle) stage to metaphase II, while those from preantral follicles remain at the diplotene stage, and oocytes from small antral follicles are seemingly "blocked" at metaphase I (Erickson and Sorenson, 1974).

Essentially similar results are obtained when whole mouse ovaries, suitably pretreated with FSH or PMSG *in vivo*, are maintained in organ culture in medium containing LH or HCG (Baker and Neal, 1972). Only those oocytes within normal graafian follicles undergo meiotic maturation to metaphase II of meiosis: oocytes in preantral and abnormally large graafian follicles remain persistently at the diplotene stage, while those in small antral follicles are arrested at metaphase I rather than metaphase II. In the absence of added gonadotropins very few oocytes, irrespective of the size of the follicles in which they reside, undergo meiotic maturation unless they have been pretreated *in vivo* with superovulating doses of PMSG (Baker and Neal, 1972; Lindner et al., 1974).

The precise mechanism of action of gonadotropins in the induction of oocyte maturation prior to ovulation remains obscure, although the results of recent studies, employing the technique of organ culture of whole ovaries or isolated follicles, have shed some light on this problem (reviewed by Lindner et al., 1974). There can be no doubt that the induction of meiosis is *mediated* by LH, but it is unlikely that LH acts directly on the oocyte. The results of recent studies indicate that gonadotropins act by removing an inhibitor substance with a molecular weight of between 1,000 and 10,000 (Tsafriri and Channing, 1975). The precise nature of this substance remains to be established, but it is likely that it merely maintains, rather than initiates, meiotic arrest, since the latter process occurs at a time long before follicular growth begins (and this before "mature" granulosa cells can produce the substance).

It is now well documented that oocyte maturation beyond the diplotene stage can be induced by treating whole ovaries, or isolated follicles in organ culture, with LH or HCG (Baker and Neal, 1972; Lindner et al., 1974; Tsafriri et al., 1972). However, FSH is equally effective and its action is not due to contamination with LH (Lindner et al., 1974; Neal and Baker, 1975). The minimal effective dose of FSH required to induce resumption of meiosis in follicular oocytes is lower than that for LH and thus the small amount of LH in, for example, NIH-FSH-S8 (about 1.5%) is unlikely to have an effect (Neal and Baker, 1975). Furthermore, treatment of LH with antiserum to the β-subunit of LH abolishes its action on follicular oocytes, whereas treatment of FSH with the same antiserum has no effect (Lindner et al., 1974). Lindner and co-workers have also shown that FSH can induce meiotic maturation *in vivo*, although the levels of circulating FSH would normally be insufficient in the normal estrus cycle of rodents without the synergistic action of LH.

FSH and LH are also effective in inducing the resumption of meiosis in rabbit follicular oocytes in organ culture and these eggs can subsequently develop into viable young when transplanted into suitable recipients (Thibault, et al., 1975; cf. Foote and Thibault, 1969). By contrast, neither FSH nor LH seem able to induce meiotic maturation in isolated follicles derived from pigs (Baker et al., 1975), sheep (e.g., Hay and Moore, 1975), or women (Baker and Neal, 1974). Therefore, it seems that the control of oocyte maturation in these species may be slightly different to that occurring in rodents and lagomorphs. Before attempting to define these differences in the mechanism of preovulatory maturation it is first necessary to consider further data on the process in rodents.

It is well established that prostaglandin E_2 (and even $PGF_{2\alpha}$ at pharmacological doses), can mimic the action of gonadotropins when added to cultures of graafian follicles (Lindner et al., 1974; Neal et al., 1975, 1976; Tsafriri et al., 1972). PGE_2, like LH, stimulates the production of cyclic AMP, which in turn induces the synthesis of steroids. It is not surprising, therefore, the dibutyryl cyclic AMP, when instilled into the follicle, can also induce the resumption of meiosis (Lindner et al., 1974), although treatment of isolated eggs with this agent together with LH results in a block to *in vitro* maturation (Nekola and Smith, 1975). The results of these studies indicate that steroids are a necessary intermediate in LH-induced oocyte maturation in rodents, or that there is an intrafollicular inhibitor whose effects can be overcome by LH, PGE_2, and cyclic AMP within the follicle. The role of a potential inhibitor has been discussed above; and experiments involving the addition of inhibitors of steroid synthesis (aminoglutethimide or cyanoketone) at the time of LH treatment do not prevent the maturation of the oocyte (Lindner et al., 1974).

The precise mode of control of oocyte maturation in rodent follicles is thus obscure: LH appears to be essential, although FSH is also effective, and PGE_2 may play merely a permissive role (Lindner et al., 1974). It seems that oocyte maturation may require the *de novo* synthesis of a specific protein that may act by nullifying the effects of an intrafollicular inhibitor, seemingly produced by the granulosa cells (Tsafriri and Channing, 1975). It would appear that the inhibitor would be produced when the levels of gonadotropins in follicular fluid are low, while the other protein (antagonist) would be synthesized in response to the surge of LH at proestrus. This speculation would need to be tested by means of carefully controlled experiments.

There is no evidence at present for a completely different mechanism controlling oocyte maturation in the pig, sheep, or human ovary where LH is clearly required for meiosis and steroid synthesis *in vivo*. However, LH seemingly cannot "trigger" meiosis in the oocytes of these species within follicles in organ culture (see above). Provisional results for pig graafian follicles *in vitro* indicate that $PGF_{2\alpha}$ is capable of inducing GVB and progression of the oocyte to metaphase II, as is PGE_2, although with possibly slight damage to the metaphase spindle (Baker et al., 1977). The action of these prostaglandins might be to induce synthesis of cyclic AMP and thus mimic that of LH, in which case the mechanism of action would be similar but not identical in pigs and rodents.

In looking for an additional factor to potentiate or augment the action of LH on the follicle, it is necessary to examine the hormone content of follicular fluid, since this continuously bathes the oocyte within its cumulus. Surprisingly, there has been only one detailed report in the literature and this was concerned with the concentrations of pituitary and steroid hormones in human follicular fluid (McNatty et al., 1975). The concentrations of FSH and LH, apart from the midcycle peak of LH, remain more or less constant throughout the reproductive cycle, irrespective of the size of the follicle (4 to 12 mm) from which the fluid was aspirated. The levels of 17β-estradiol and progesterone are largely maximal when the values of LH in follicular fluid are also at a peak. McNatty et al. (1975) showed that the ability of granulosa cells to secrete steroids *in vitro* is dependent on a complex of hormones, including LH, FSH, and prolactin. The latter hormone also remains more or less constant in terms of concentration per milliliter of follicular fluid, except at the time of the LH surge when the level of prolactin drops considerably. The marked decline in prolactin levels in follicular fluid occurs at the time when the oocyte is about to resume meiosis (i.e., GVB): It would have seemed reasonable to speculate that LH acted on the follicle by changing blood flow or the membrane permeability of its basement membrane, and therefore removing prolactin from the follicular compartment. However, this hypothesis is rendered less likely, at least for sheep follicles, by recent experiments involving the radioactive microspere technique: Bruce and Moor (1976) could not detect changes in follicular blood flow after treatment with LH.

In an attempt to define a role for prolactin in oocyte maturation, intact graafian follicles from pig ovaries were maintained in culture with various hormone preparations (Baker et al., 1977). On the assumption tha the high levels of preovulatory prolactin in human follicular fluiIso occur in the pig and follow the same patter of LH-induced decline (see above), the follicles were treated with anitserum to prolactin. The results of thes experiments confirmed that LH, FSH, and estradiol, were effective in inducing meiosis, which also occurred in the presence of prostaglandins or prolactin antiserum: oocyt progressed to metaphase II and showed a seemingly normal appearance. These results indicate that prolactin may in self be an inhibitor of oocyte maturation in the pig foll and experiments are in progress to see if it plays a similar role in the human follicle. One might speculate that the intrafollicular concentration of LH might be sufficient to sustain meiosis irrespective of its concentration in serum or added to cultures, provided prolactin is removed from the follicular compartment.

Clearly, there are many unanswered questions that need to be solved, and the mechanism of action of hormones at the follicular level may vary even more widely between species. However, at least some of the cornerstones of the processes involved are becoming more clear and it would seem that the use of the techniques of organ culture may help to resolve this important and complex process.

ACKNOWLEDGMENTS

The financial support of the Population Council (New York), the Wellcome Trust (London), and Serono Laboratories (Boston) is gratefully acknowledged.

REFERENCES

Anand Kumar, T. C. (1968): Oogenesis in lorises: *Loris tardigradus lydekkerianus* and *Nycticebus coucang*. *Proc. R. Soc.* Lond. *[Biol.]*, 169:167.

Anonymous (1974): *Chambers Dictionary of Science and Technology*, W. and R. Chambers, Edinburgh.

Baker, T. G. (1971): Electron microscopy of the primary and secondary oocyte. In: *Advances in the Biosciences*, Vol. 6, edited by G. Raspé, p. 7. Pergamon, Oxford.

Baker, T. G. (1972): Oogenesis and ovarian development. In: *Reproductive Biology*, edited by H. Balin and S. R. Glasser, p. 398. Excerpta Medica, Amsterdam.

Baker, T. G., and Neal, P. (1972): Gonadotrophin-induced maturation of mouse graafian follicles in organ culture. In: *Oogenesis*, edited by J. D. Biggers and A. W. Schuetz, p. 377. University Park Press, Baltimore.

Baker, T. G., and Neal, P. (1973): Initiation and control of meiosis and follicular growth in ovaries of the mouse. *Ann. Biol. Anim. Biochim. Biophys.*, 13:137.

Baker, T. G., and Neal, P. (1974): Organ culture of cortical fragments and graafian follicles from human ovaries. *J. Anat.*, 117:361.

Baker, T. G., Hunter, R. H. F., and Neal, P. (1975): Studies on the maintenance of porcine graafian follicles in organ culture. *Experientia*, 31:133.

Baker, T. G., and O, Wai Sum (1976): Development of the ovary and oogenesis. In: *Clinics in Obstetrics and Gynaecology*, Vol. 3, edited by H. C. Macnaughton and J. D. Govan, p. 3.

Saunders, London.

Baker, T.G., Biggs, J.S.G., and Hunter, R.H.F. (1977): Control of oocyte maturation in mammals. In: Endocrinology, edited by V.H.T. James, Vol. 1, p.351. Excerpta Medica, Amsterdam.

Black, J. L., and Erickson, B. H. (1968): Oogenesis and ovarian development in the prenatal pig. *Anat. Rec.*, 161:45

Blandau, R. J. (1965): Observations on living oogonia and oocytes from human embryonic and fetal ovaries. *Am. J. Obstet. Gynecol.*, 104:310.

Brambell, F. W. R. (1956): Ovarian changes. In: *Marshall's Physiology of Reproduction* , Vol. 1, edited by A. S. Parkes, p. 397.

Bruce, N. W., and Moore, R. M. (1976): Capillary blood flow to ovarian follicles, stroma and corpora lutea of anaesthetized sheep. *J. Reprod. Fertil.*, 46:299.

Burkholder, G. D., Commins, D. E., and Okada, T. A. (1971): A storage form of ribosomes in mouse oocytes. *Exp. Cell Res.*, 69:361.

Byskov, A. G. S. (1974): Does the rete ovarii act as a trigger for the onset of meiosis? *Nature*, 252:396.

Byskov, A. G. S. (1975): The role of the rete ovarii in meiosis and follicle formation in different mammalian species. *J. Reprod. Fertil.*, 45:201.

Byskov, A. G. S., and Lintern Moore, S. (1973): Follicle formation in the immature mouse ovary: The role of the rete ovarii. *J. Anat.*, 116:207.

Callan, H. G. (1957): The lampbrush chromosomes of *Sepia officinalis* L., *Anilocra physodes* L. and *Scyllium catulus* Cuv. and their structural relationship to the lampbrush chromosomes of Amphibia. *Pubbl. Staz. Zool. Napoli*, 29:329.

Challoner, S. (1975*a*): Studies of oogenesis and follicular development in the golden hamster. 2. Initiation and control of meiosis *in vitro*. *J. Anat.*, 119:149.

Challoner, S. (1975*b*): Studies of oogenesis and follicular development in the golden hamster. 3. Initiation of follicular growth *in vitro*. *J. Anat.*, 119:157.

David, G. F. X., Anand Kumar, T. C., and Baker, T. G. (1974): Uptake of tritiated thymidine by primordial germinal cells in the ovaries of the adult slender loris. *J. Reprod. Fertil.*, 41:451.

Davidson, E. H. (1968): *Gene Activity in Early Development*. Academic Press, New York.

Edwards, R. G. (1965): Maturation *in vitro* of mouse, sheep, cow, pig, rhesus monkey and human oocytes. *Nature*, 208:349.

Edwards, R. G. (1966): Mammalian eggs in the laboratory. *Sci. Am.*, 215:72.

Erickson, G. F., and Sorenson, R. A. (1974): *In vitro* maturation of mouse oocytes isolated from late, middle and pre-antral graafian follicles. *J. Exp. Zool.*, 190:123.

Eshkol, A., Lunenfeld, B., and Peters, H. (1971): Ovarian development in infant mice. In: *Gonadotrophins and Ovarian Development*, edited by W. R. Butt, A. C. Crooke, and M. Ryle, p. 249. Livingstone, Edinburgh.

Foote, W. D., and Thibault, C. (1969): Recherches experimentales sur la maturation *in vitro* des ovocytes de truie et de veau. *Ann. Biol. Anim. Biochim. Biophys.*, 9:329.

Franchi, L. L., Mandl, A. M., and Zuckerman, S. (1962): The development of the ovary and the process of oogenesis. In: *The Ovary*, Vol. 1, edited by S. Zuckerman, p. 1. Academic Press, London.

Hay, M. F., and Moor, R. M. (1975): Functional and structural relationships in the graafian follicle population of the sheep ovary. *J. Reprod. Fertil.*, 45:583.

Henderson, I. F., and Henderson, J. H. (1960): *A Dictionary of Scientific Terms*. Oliver and Boyd, Edinburgh.

Ingram, D. L. (1962): Atresia. In: *The Ovary*, Vol. 1, edited by S. Zuckerman, p. 247. Academic Press, London.

Lindner, H. R., Tsafriri, A., Lieberman, M. E., Zor, U., Koch, Y., Bauminger, S., and Barnea, A. (1974): Gonadotropin action on cultured graafian follicles: Induction of maturation division of the mammalian oocyte and differentiation of the luteal cell. *Recent Prog. Hormone Res.*, 30:79.

McNatty, K. P., Hunter, W. M., McNeilly, A. S., and Sawers, R. S. (1975): Changes in the concentration of pituitary and steroid hormones in the follicular fluid of human graafian follicles throughout the menstrual cycle. *J. Endocrinol.*, 64:555.

Moore, K. L. (1973): *The Developing Human: Clinically Oriented Embryology*. Saunders, Philadelphia.

Mossman, H. W., and Duke, K. L. (1973): *Comparative Morphology of the Mammalian Ovary*. University of Wisconsin Press, Madison.

Neal, P., and Baker, T. G. (1975): Response of mouse graafian follicles in organ culture to varying doses of follicle stimulating hormone and luteinizing hormone. *J. Endocrinol.*, 65:27.

Neal, P., Baker, T. G., and McNatty, K. P. (1976): Effects of prostaglandin F-2α and E-2 on the production of progesterone by mouse ovaries *in vitro*. *J. Reprod. Fertil.*, 47:157.

Neal, P., Baker, T. G., McNatty, K. P., and Scaramuzzi, R. J. (1975): Influence of prostaglandins and human chorionic gonadotrophin on progesterone concentration and oocyte maturation in mouse ovarian follicles maintained in organ culture. *J. Endocrinol.*, 65:19.

Nekola, M. V., and Smith, D. M. (1975): Failure of gonadotropins to induce *in vitro* maturation of mouse oocytes treated with dibutyryl cyclic AMP. *J. Exp. Zool.*, 194:529.

Ohno, S., and Smith, J. B. (1964): Role of fetal follicular cells in meiosis of mammalian oocytes. *Cytogenetics*, 3:324.

Saxén, L., Lehtonen, E., Karkinen-Jääskeläinen, M., Nordling, S., and Wartiovaara, J. (1976): Are morphogenetic tissue interactions mediated by transmissible signal substances or through cell contacts? *Nature*, 259:662.

Thibault, C., Gérard, M., and Menezo, Y. (1975): Aquisition par l'ovocyte de lapine et de veau du facteur de decondensation du noyau du spermatozoide fecondant (MPGF). *Ann. Biol. Anim.*

Biochim. Biophys., 15:705.

Tsafriri, A., and Channing, C. P. (1975): An inhibitory influence of granulosa cells and follicular fluid upon porcine oocyte meiosis *in vitro*. *Endocrinology*, 96:922.

Tsafriri, A., Lindner, H. R., Zor, U., and Lamprecht, S. A. (1972): *In vitro* induction of meiotic division in follicle-enclosed rat oocytes by LH, cyclic AMP and prostaglandin E_2. *J. Reprod. Fertil.*, 31:39.

O, Wai Sum, and Baker, T. G. (1976): Initiation and control of meiosis in hamster gonads *in vitro*. *J. Reprod. Fertil.* 48:399.

Wolff, E., and Haffen, K. (1965): Germ cells and gonads. In: *Cells and Tissues in Culture, Methods, Biology and Physiology*, edited by E. N. Wilmer, p. 697. Academic Press, London.

Ovarian Follicular Development and Function,
edited by A. R. Midgley and W. A. Sadler.
Raven Press, New York © 1979.

Comments on the Control of Meiotic Maturation in Mammals

John D. Biggers and R. Douglas Powers

Laboratory of Human Reproduction and Reproductive Biology and Department of Physiology, Harvard Medical School, Boston, Massachusetts 02115

One of the most fascinating aspects of the behavior of chromosomes in the mammalian oocyte is its episodic nature (Fig. 1). Meiosis is first initiated during fetal life soon after both X chromosomes become functional (Gartler et al., 1975). About the time of birth, meiosis becomes arrested at the diplotene stage of the first meiotic division and remains in this stage until after puberty. Those oocytes destined to be ovulated resume meiosis only a few hours prior to ovulation; however, they do not complete the second meiotic division, since the process again becomes arrested--this time at the metaphase of the second meiotic division. Meiosis is only completed under natural conditions when a sperm penetrates the ovum. Thus, the entire life history of the mammalian oocyte in terms of the behavior of its chromosomes is characterized by an *on-off-on-off-on* sequence.

The mature ovum, however, acquires properties other than being in an arrested metaphase II condition, such as the ability to fuse with spermatozoa and a high energy charge (Biggers, 1976). The development of these capabilities may be initiated at the same time as meiotic maturation is resumed in response to the LH surge. The discussion concentrates on these final stages of oocyte maturation and emphasizes that several interacting factors may be involved in the *total maturation* of the mammalian oocyte as it emerges from arrest at the diplotene stage.

There is considerable electrophysiological evidence from studies on several nonmammalian systems that indicates that major changes in the oolemma take place in preparation for fertilization. Moreover, these changes are reversed in subsequent embryonic development. The results of these investigations can be combined to

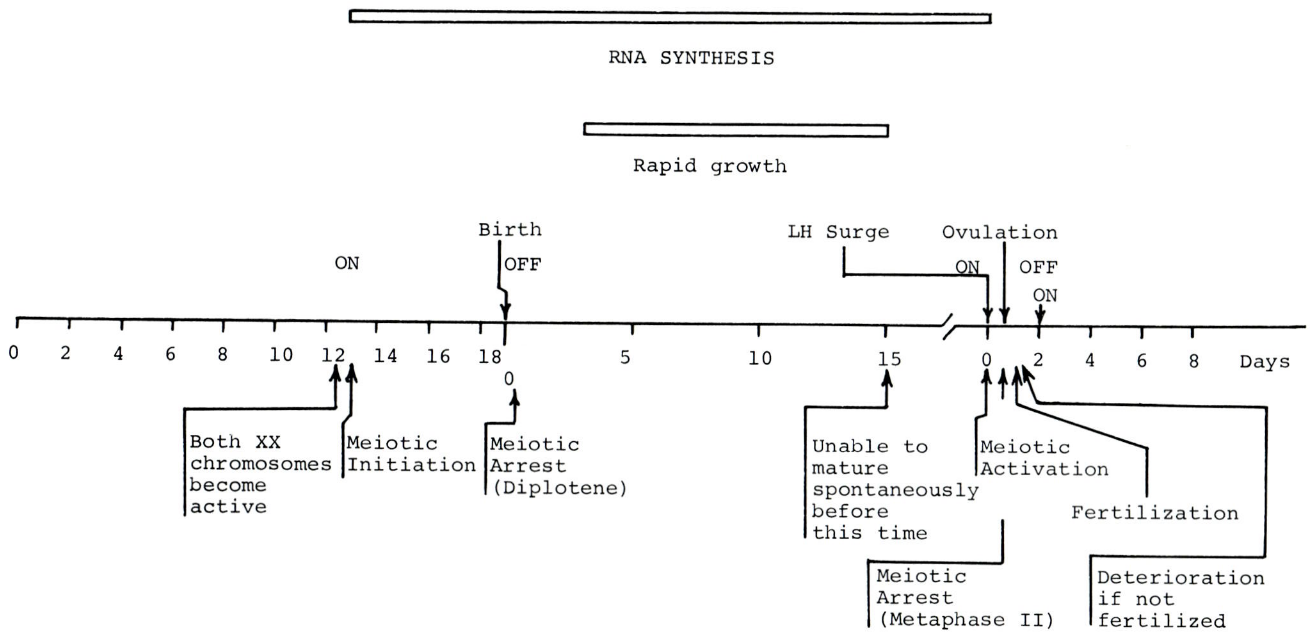

FIG. 1. The sequence of events in the life of a mammalian oocyte. The times given refer to the mouse.

suggest a general model. The membrane potential of the adult cells of most animals is relatively high (~75 mV, inside negative) and is maintained by the ion concentration gradients across the cell membrane and the relatively high permeability of the membrane to K^+ in comparison to Na^+ (Williams, 1970). This condition also exists in the immature oocyte of the echinoderm (Hagiwara and Takahashi, 1974) and amphibian (Maeno, 1959; Morill et al., 1966). However, by the end of meiotic maturation the membrane potential has been reduced (~15 mV) in the echinoderm (Tyler et al., 1956), amphibian (Ito and Hori, 1966; Maeno, 1959; Palmer and Slack, 1970), algae (Bentrup, 1970; Weisenseel and Jaffe, 1972), fish (Ito, 1962), and mammal (Powers and Tupper, 1974), usually as a result of a decrease in membrane K^+ permeability relative to Na permeability.

Fertilization is accompanied by a transient depolarization of the membrane (~2 mV, inside negative) (Steinhardt et al., 1971). Shortly thereafter, a slow hyperpolarization begins which reaches a plateau of at least 50 mV (inside negative) by the blastula stage (or formation of the cell wall in the algae). This increase in potential is usually caused by a reversal of the membrane events that occurred during oocyte maturation. Specifically, the ratio of membrane K^+ permeability to Na^+ permeability increases.

Through this process the electrical and permeability properties of the cells of the embryo become similar to those of the adult organism. Thus, the surface membrane of the mature oocyte is unique. Its membrane ionic permeability properties are unlike those found in any other cell in the body. These properties may well be a necessary prerequisite for the fusion of the egg and sperm at fertilization.

Some of the changes in membrane potential and permeability have also been demonstrated in the mouse embryo. The increase in membrane potential after fertilization has been described (Persianinov et al., 1973), and Powers and Tupper (*in preparation*) have shown that in the initial stages this rise is due to an increase in K permeability. Unfortunately, no studies have been made so far on the membrane properties of oocytes during maturation *in vivo*. Nevertheless, it is useful at this stage of our knowledge to assume that the behavior of the mammalian oocyte follows the general model derived from the studies on nonmammalian forms.

In 1935, Pincus and Enzmann showed that oocytes isolated with their surrounding follicle cells from unmated rabbits resumed meiosis spontaneously. This phenomenon has now been observed in several species (see Biggers,

1973, for a review) and, as a result, the isolated oocyte preparation is used frequently in studies of several aspects of oocyte physiology. The results obtained, however, may be limited, since the oocyte, and its supporting cells if present, have been removed from their natural environment, with the result that short-term responses are confounded with their adaptations to the new artificial environment. It is important to understand these limitations.

One of the first experimental studies with isolated oocytes was the demonstration that pyruvate is necessary for meiotic maturation of the isolated mouse oocytes, which have not been hormonally stimulated, and that glucose would only support the spontaneous maturation if cumulus cells were also present in the culture system (Biggers et al., 1967). The validity of this result is based on the fact that the cytological details of chromosome behavior in the spontaneously maturing oocyte follow details well known in other forms (Donahue, 1968), and by the fact that normal embryos have been obtained from spontaneously matured oocytes (Cross and Brinster, 1970).

The hormonally unstimulated, isolated oocyte preparation has also been used in a reverse manner to study conditions under which spontaneous meiotic maturation is prevented. An example is the action of dibutyryl cAMP and theophylline on the isolated mouse oocyte (Cho et al., 1974). Both drugs reversibly prevent germinal vesicle breakdown (GVB). This observation has been confirmed by several other investigators. The results indicate that cAMP arising from the action of LH is unlikely to be the direct initiator of meiotic resumption. The mode of action of cAMP at the molecular level is obscure, although it seems to interfere with GVB, and consequently chromatin condensation at the nuclear membrane (Wassarman et al., 1976).

Recently, Powers and Biggers (1976) have found two alternative methods of retarding GVB in the isolated mouse oocyte that may provide a clue as to the mechanism of spontaneous oocyte maturation in the mouse. Figure 2 shows the effect on GVB of mouse oocytes incubated in Na^+-free medium, or medium containing the K^+ ionophore valinomycin (0.1 μM). Both of these treatments cause a lag of about 2 hr in GVB, after which time they rapidly break down to the same extent as did the controls at an earlier time. Powers and Tupper (1974, 1975) have shown that the "constant field equation" accurately describes the membrane potential of the mouse oocyte. Using this equation, it can be predicted that removal of the external Na^+, or increasing permeability to K^+ with valinomycin, will result in

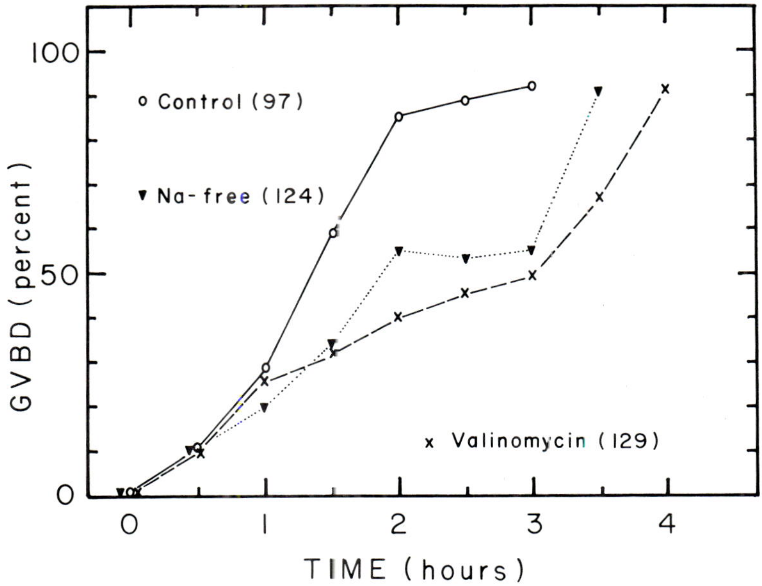

FIG. 2. The effect of Na^+-free medium and valinomycin (0.1 μM) on the time course of germinal vesicle breakdown (GVB) of the mouse in a simple chemically defined medium (Powers and Tupper, 1974). The numbers in parentheses indicate the number of oocytes exposed to each treatment.

hyperpolarization of the membrane. That hyperpolarization follows removal of external Na^+ has, in fact, been demonstrated experimentally by Powers and Tupper (1974). Eventually, both treatments will result in depolarization of the membrane that may be associated with spontaneous meiotic maturation. These preliminary experiments are of significance because they demonstrate the possibility of a connection between events in the oolemma and the behavior of the germinal vesicle. Both processes may be linked by being affected by a common mechanism or by mutually affecting each other.

Epstein et al. (1976) have recently demonstrated electrical coupling between the cumulus cells and oocyte of the rat. Moreover, the recent demonstration of gap junctions between the cumulus cells and oocyte provide a morphological basis for this electrical coupling (Albertini and Anderson, 1974; Amsterdam et al., 1976). From geometrical considerations it is clear that the total surface area of the coupled cumulus cells is far greater than that of the oocyte. As a result, the membrane potential of the oocyte may be controlled primarily by the permeability properties of the cell membranes of the cumulus. Any agent that modifies the permeability properties of the cumulus cells may automatically cause similar changes to be propagated to the oolemma. Alternatively, an agent that causes uncoupling of the oocyte and cumulus cells may allow the oocyte to undergo membrane changes reflected by depolarization.

It is important to remember that changes in the properties of the membranes of isolated oocytes could arise for a number of nonspecific reasons. Biggers (1963) divided the life of an organ culture into three phases on the basis of changes in the dry weight/wet weight ratio: (a) the initial "shock" phase, (b) the middle "stable" phase, and (c) the final phase of slow hydration. Possibly as an oocyte is freed from the graafian follicle, and is placed in foreign medium, it experiences "stresses" that could modify the potential of the oolemma. Different responses to various media may explain why some authors report follicular fluids or cumulus cells to be inhibitory (Chang, 1955; Edwards, 1962; Tsafriri and Channing, 1975*a*,*b*), while others report them to have a stimulatory effect (Cho et al., 1971; Chung et al., 1974; Hunter et al., 1972; Robertson and Baker, 1969).

Recently, Tsafriri and Channing (1975*a*,*b*) and Tsafriri et al. (1976) have utilized the isolated oocyte system of the pig as an assay for a hypothetical inhibitor of meiosis in follicular fluid. Should spontaneous maturation be an artifact of culture, as we have suggested, the physiological significance of these studies needs critical evaluation. It may turn out that the substances being isolated are merely agents of a nonspecific nature needed to stabilize the cultures. The fact that LH initiates meiosis in the presence of the "inhibitor" (Tsafriri et al., 1976) may merely mean that the hormone is being given an opportunity to act under normal conditions, perhaps by modifying the relationship between the oocyte and its cumulus cells. Pincus and Enzmann in 1935 prudently recognized the possibility of alternative hypotheses. In explanation of their observations that rabbit oocytes would spontaneously mature *in vitro*, they wrote: "...the associated follicle cells serve either to maintain the egg in a nutritional state wherein nuclear maturation is impossible, or that they actually supply to the ovum a substance or substances which directly inhibit nuclear maturation." Even if the substances being detected by the assay based on spontaneous oocyte maturation turn out to be more specific, it is important to remember that they may not be acting directly on the meiotic process but on some other physiologically coupled process involved in the total maturation of the oocyte. It may, therefore, be misleading to speak of a specific inhibitor of meiotic maturation.

At the Conference on the Regulation of Mammalian Reproduction held in Bethesda in 1970, Biggers compared the control of meiotic maturation in starfish, amphibia, and mammals. The paper concluded as follows: "It seems that a major area for research on oogenesis is to decide whether we need to postulate a pattern of control of meiosis in mammalian oocytes which is the converse of that which seems applicable in two lower forms" (Biggers, 1973). We suggest that this fundamental question is still unanswered.

REFERENCES

Albertini, D. F., and Anderson, E. (1974): The appearance and structure of the intercellular connections during the ontogeny of the rabbit ovarian follicle with particular reference to gap junctions. *J. Cell Biol.*, 63:234-250.

Amsterdam, A., Josephs, R., Lieberman, M. E., and Lindner, H. R. (1976): Organization of intramembrane particles in freezed-cleaved gap junctions of rat graafian follicles: Optical-diffraction analysis. *J. Cell Sci.*, 21:93-105.

Bentrup, F. W. (1970): Electrophysiologische Untersuchungen am Ei von *Fucus serratus:* Das Membranpotential. *Planta*, 94:319-322.

Biggers, J. D. (1963): Studies on the development of embryonic cartilaginous long-bone rudiments *in vitro*. In: *Symposium on Organ Culture*, edited by C. J. Dawe, pp. 1-18. National Cancer Institute Monograph No. 11.

Biggers, J. D. (1973): Oogenesis and ovum maturation. In: *The Regulation of Mammalian Development,* edited by S. J. Segal, R. Crozier, P. A. Corfman, and P. G. Condlifee, pp. 273-283. Thomas, Springfield, Ill.

Biggers, J. D. (1976): Bioenergetic aspects of fertilization and embryonic development in the mouse. Proc. 8th World Congr. Fertil. Steril. Excepta Medica, Amsterdam *(in press)*.

Biggers, J. D., Whittingham, D. G., and Donahue, R. P. (1967): The pattern of energy metabolism in the mouse oocyte and zygote. *Proc. Natl. Acad. Sci. U.S.A.*, 58:560-567.

Chang, M. C. (1955): The maturation of rabbit oocytes in culture and their maturation, activation, fertilization and subsequent development in the fallopian tube. *J. Exp. Zool.*, 128:378

Cho, W. K., Kim, M. K., and Chung, S. O. (1971): *in vitro* maturation of human follicular oocytes in the presence of follicular fluid. Proc. 7th World Congr. Fertil. Steril. Excerpta Medica, Amsterdam. Int. Congress Series No. 278, pp. 706-708.

Cho, W. K., Stern, S., and Biggers, J. D. (1974): Inhibitory effect of dibutyryl cAMP on mouse oocyte maturation *in vitro*. *J. Exp. Zool.*, 187:383-386.

Chung, S. O., Choi, Y. H., Kim, M. K., and Cho, W. K. (1974): The effects of follicular fluid on *in vitro* maturation of bovine follicular oocytes. *Yonsei Med. J.*, 15:147-155.

Cross, P. C., and Brinster, R. L. (1970): *In vitro* development of mouse oocytes. *Biol. Reprod.*, 3:298-307.

Donahue, R. P. (1968): Maturation of the mouse oocyte *in vitro*. I. Sequence and timing of nuclear progression. *J. Exp. Zool.*, 169:237-250.

Edwards, R. G. (1962): Meiosis in ovarian oocytes of adult mammals. *Nature*, 196:446-450.

Epstein, M. L., Beers, W. H., and Gilula, N. B. (1976): Cell communication between the rat cumulus oophorus and the oocyte. *J. Cell Biol.*, 70:302*a*. *(abstract)*

Gartler, S. M., Andina, R., and Gant, N. (1975): Ontogeny of X-chromosome inactivation in the female germ line. *Exp. Cell Res.*, 91:454-457.

Hagiwara, S., and Takahashi, K. (1974): The anomalous rectification and cation selectivity of the membrane of the starfish egg cell. *J. Membr. Biol.*, 18:61-80.

Hunter, R. H. F., Lawson, R. A. S., and Rowson, L. E. A. (1972): Maturation, transplantation and fertilization of ovarian oocytes in cattle. *J. Reprod. Fertil.*, 30:325-328.

Ito, S. (1962): Resting potential and activation potential of the *Orzyias* egg. II. Change of membrane potential and resistance after fertilization. *Embryologia*, 7:47-55.

Ito, S., and Hori, Y. (1966): Electrical properties of *Triturus* eggs during cleavage. *J. Gen. Physiol.*, 49:1019-1027.

Maeno, T. (1959): Electrical characteristics and activation potential of *Bufo* eggs. *J. Gen. Physiol.*, 43:139-157.

Morill, G. A., Rosenthal, J., and Watson, D. E. (1966): Membrane permeability changes in amphibian eggs at ovulation. *J. Cell Physiol.*, 67:375-382.

Palmer, J., and Slack, C. (1970): Some bio-electric parameters of early *Xenopus* embryos. *J. Embryol. Exp. Morphol.*, 24:535-553.

Persianinov, L. S., Limantsev, A. V., and Leonov, B. V. (1973): Dynamics of membrane potential of the cells of the mouse embryo during the preimplantation period of development. *Bull. Ekskop. Biol. Med.*, 9:10-11.

Pincus, G., and Enzmann, E. V. (1935): The comparative behaviour of mammalian eggs *in vivo* and *in vitro*. I. The activation of ovarian eggs. *J. Exp. Med.*, 62:665-675.

Powers, R. D., and Biggers, J. D. (1976): Inhibition of mouse oocyte maturation by cell membrane potential hyperpolarization. *J. Cell Biol. (abstract)*, 70:352*a*.

Powers, R. D., and Tupper, J. T. (1974): Some electrophysiological and permeability properties of the mouse egg. *Dev. Biol.*, 38:320-331.

Powers, R. D., and Tupper, J. T. (1975): Ion transport and permeability in the mouse egg. *Exp. Cell Res.*, 91:413-421.

Robertson, J. E., and Baker, R. D. (1969): Role of female sex steroids as possible regulators of oocyte maturation. *Proc. Soc. Study Reprod.*, Davis, Calif., *Abstract* 29.

Steinhardt, R. A., Lundin, L., and Mazia, D. (1971): Bioelectric responses of the echinoderm egg to fertilization. *Proc. Natl. Acad. Sci. U.S.A.*, 68:2426-2430.

Tsafriri, A., and Channing, C. P. (1975*a*): An inhibitory influence of granulosa cells and follicular fluid upon porcine oocyte meiosis *in vitro*. *Endocrinology*, 96:922-927.

Tsafriri, A., and Channing, C. P. (1975*b*): Influence of follicular maturation and culture conditions on the meiosis of pig oocytes *in vitro*. *J. Reprod. Fertil.*, 43:149-152.

Tsafriri, A., Pomerantz, S. H., and Channing, C. P. (1976): Inhibition of oocyte maturation by porcine follicular fluid: Partial characterization of the inhibitor. *Biol. Reprod.*, 14:511-516.

Tyler, A., Monroy, A., Kao, C., and Grundfest, H. (1956): Membrane potential and resistance of the starfish egg before and after fertilization. *Biol. Bull.*, 111:153-177.

Wassarman, P. M., Josefowicz, W. J., and Letourneau, G. E. (1976): Meiotic maturation of mouse oocytes *in vitro*: Inhibition of maturation at specific stages of nuclear progression. *J. Cell Sci.*, 22:531-545.

Weisenseel, M. H., and Jaffe, L. F. (1972): Membrane potential and impedance of developing fucoid eggs. *Dev. Biol.*, 27-555-574.

Williams, J. A. (1970): Origin of transmembrane potentials in non-excitable cells. *J. Theor. Biol.*,28:287-296.

Ovarian Follicular Development and Function,
edited by A. R. Midgley and W. A. Sadler.
Raven Press, New York © 1979.

Comments on the Control of Oogenesis in Mammals

Ralph A. Sorensen*

Department of Biological Sciences,
DePaul University, Chicago, Illinois 60614

The ovulation of a fully developed and meiotically mature oocyte is preceded by a period of oocyte growth and differentiation about which little information is available. In collaboration with Dr. Paul Wassarman at Harvard Medical School, I have investigated the growth and functional differentiation of oocytes recovered from juvenile mice aged 9 to 21 days *post-partum* (5). Our results confirm and extend the original observations of Szybec (1972). We feel that this is a potentially valuable system with which to explore both the biosynthetic correlates of each phase of this developmental process, and their possible regulation. In the spirit of a discussion of Dr. Baker's paper, I hope that these results will allow for a different perspective on the question of the control of meiotic maturation than that obtained from a graafian follicle.

A principal advantage to the use of juvenile mice is that a brief trypsin treatment of the ovaries, followed by physical disruption, releases a large number of oocytes of relatively uniform diameter. The mean diameter of oocytes recovered increases linearly with the age of the litter--50 μm for 9-day-old oocytes, 75 μm for 21-day-old oocytes. The increase in oocyte size with the age of the litter suggests that this procedure is releasing the lead population of oocytes in their progression through the growth phase of oocyte development. Even at 21 days, however, the oocytes are significantly smaller than those recovered from the larger follicles of an adult mouse.

As the mean diameter and the age of the mice increase, so does the percentage of "competent" oocytes. A "competent" oocyte is defined as an oocyte that undergoes germinal vesicle breakdown in a defined

*Present address: Gettysburg College, Gettysburg, Pennsylvania.

medium known to permit oocyte maturation (1). The earliest response was observed at 15 days. At 15 days of age the largest follicles present in the mouse ovary are composed of four to five follicle cell layers, while the first follicle with an incipient antrum is seen at 16 days (4). Clearly, then, the oocyte develops the capacity to resume meiosis *in vitro* long before it would normally (at ovulation) do so *in vivo*.

Two populations of "competent" oocytes are observed, those which complete maturation to metaphase II, and those that arrest prematurely at metaphase I. Interesting, the ratio of metaphase II to metaphase I arrest increases steadily from 15 to 21 days. The indication is that an oocyte first develops the capacity to *initiate* meiotic maturation, and subsequently develops the capacity to *complete* meiotic maturation. Again, the size data suggest that these two "states of competence" are acquired sequentially during the growth phase of oocyte development.

We have obtained similar results by isolating oocytes from follicles of various developmental stages dissected from adult mouse ovaries (2). Oocytes recovered from preantral follicles were incompetent to mature *in vitro*, whereas oocytes isolated from postantral follicles were competent to mature. We did not, however, observe oocytes arresting predominantly at metaphase I, presumably due to our failure to examine sufficient numbers of follicle stages.

Dr. Wassarman is utilizing this system to compare the electrophoretic spectra of proteins synthesized by oocytes at each stage of growth and meiotic maturation. The demonstration of unique patterns at each stage in the development of competence would suggest that these stages are controlled by sequential patterns of gene expression occurring during the growth of the oocyte.

These data also become a normal curve on which to base experimental studies of the regulation of oocyte growth and functional differentiation. We do not know if the acquisition of the competence to mature *in vitro* is a discrete inductive event. The role of a diffusible factor produced by the follicle cells of echinoderms (1-methyl adenine) and amphibians (presumably progesterone) in the induction of oocyte maturation in these species is at least suggestive of the possibility that maturation *competence* (rather than maturation at ovulation) may be similarly induced in mammals. The work of Eshkol and Lunenfeld (3) indicates that preantral folliculogenesis in juvenile mice is regulated by pituitary gonadotropins. The question of whether oocyte growth and differentiation are also regulated, directly or indirectly, by pituitary gonadotropins and/or gonadal steroids may be investigated using this system.

REFERENCES

1. Biggers, J. D. (1971): New observations on the nutrition of the mammalian oocyte and the preimplantation embryo. In *The Biology of the Blastocyst* (R. J. Blandau, ed.), pp. 319-327. University of Chicago Press, Chicago.
2. Erickson, G. F., and Sorensen, R. A. (1974): *In vitro* maturation of mouse oocytes isolated from late, middle, and pre-antral Graafian follicles. *J. Exper. Zool.*, 190:123-127.
3. Eshkol, A., and Lunenfeld, B. (1975): Use of antisera to gonadotrophins in reproduction research. In *Immunization with Hormones in Reproduction Research* (E. Nieschlag, ed.), pp. 55-81. North-Holland Publishing Co., Amsterdam.
4. Kent, J. (1972): Changes in the ovarian follicle population of mice aged 16 to 20 days. *J. Reprod. Fertil.*, 31:323-326.
5. Sorensen, R. A., and Wassarman, P. W. (1976): Relationship between growth and meiotic maturation of the mouse oocyte. *Devel. Biol.*, 50:531-536.
6. Szybec, K. (1972): *In vitro* maturation of oocytes from sexually immature mice. *J. Endocr.*, 54:527-528.

Ovarian Follicular Development and Function,
edited by A. R. Midgley and W. A. Sadler.
Raven Press, New York © 1979.

RNA Synthesis During *In Vitro* Meiotic Maturation of Mammalian Oocytes

Debra J. Wolgemuth-Jarashow and Georgiana M. Jagiello

Columbia University and College of Physicians and Surgeons, New York, New York 10032

RNA synthesis in the growing follicular oocyte is well documented (Moore and Lintern-Moore, 1974; Oakberg, 1968). Whether or not this synthesis occurs after the oocyte reaches its maximum size within the still growing follicle but before the oocyte is ovulated is not clearly resolved. Bloom and Mukherjee (1972) reported incorporation of radioactive uridine during meiotic maturation of mouse oocytes and further suggested that RNA synthesis is obligatory for meiotic maturation. A subsequent study by Moore et al. (1974) demonstrated no endogenous RNA polymerase activity in mouse oocytes in antral follicles and no label localized in antral oocytes following intraperitoneal injection of radioactive precursor to RNA. However, a recent report by Rodman and Bachvarova (1975) offered evidence that in the mouse there is synthesis of RNA by the oocyte during the last 7 days of dictyate preceding ovulation (an *in vivo* study). Further, Wasserman and Letourneau (1975) reported that RNA synthesis occurs during meiotic maturation of mouse oocytes within intact follicles and in an *in vitro* culture. We have initiated a series of experiments in an attempt to resolve these somewhat contradictory results by examining the *in vitro* synthesis of RNA during meiotic maturation of a variety of oocyte species.

MATERIALS AND METHODS

Oocyte Collection and Culture

Mature (40 to 60 days) Swiss Camm mice were killed by cervical dislocation and the oocytes recovered in Earles BSS. The oocytes were cultured in conical centrifuge tubes in 0.5 to 1.0 ml of McCoy's culture media containing 10 to 20 μCi/ml of either ^{14}C-uridine (sp. act., 50 mCi/mmole) or ^{3}H-uridine (sp. act., 50 Ci/mmole) for 2, 4, 6, or 14 hr at 37°C under CO_2 and air. Bovine oocytes were recovered from medium-sized follicles (2 to 8 mm) of ovaries obtained from young adult animals and cultures were carried out in microdishes containing McCoy's medium (with or without 15% FCS) and 20 μCi/ml of ^{3}H-uridine (sp. act., 39 Ci/mmole) for 2, 4, 20, and 24 hr at 37°C under CO_2 and air. Porcine oocytes were recovered from medium (3 to 5 mm) or large (>5 mm) follicles of ovaries from young adult animals. Collection and culture conditions were similar to those used in the cow oocyte experiments except that the culture period was extended to 40 hr.

Processing for Autoradiography

Oocytes from the different sources were processed similarly. They were removed from the culture media and washed by passage through five to seven changes of Earles BSS. Three different fixation procedures were utilized as follows: (a) 3:1 ethanol:glacial acetic acid as described by Bloom and Mukherjee (1972); (b) acetic-formalin vapors for whole mount examination (G. Jagiello, *unpublished observations*); (c) Bouin-Hollande for 2 hr followed by processing for paraffin embedding (M. B. Ducayen, *personal communication*). Five-micron sections were affixed onto albumin-coated slides and deparaffinized prior to autoradiography.

FIG. 1. Approximately 5-μm sections of paraffin-embedded oocytes. All oocytes cultured in the presence of 20 μCi/ml of ^{3}H-uridine. A: Cow oocyte, 2 hr in culture. Note labeling of germinal vesicle (arrow) and heavily labeled granulosa cells (double arrow). The zona and the oocyte cytoplasm are not labeled. ×560. B: Mouse oocyte, 3 to 4 hr in culture. Note more intense label over still intact germinal vesicle and lightly labeled cytoplasm. ×1,360. C: Pig oocytes, 24 hr in culture. Germinal vesicle breakdown has occurred. The label is distributed throughout the ooplasm. Only a single granulosa cell remained associated with oocytes. ×480. D: Higher magnification of pig oocyte, 20 hr in culture. Note generalized distribution of label. ×1,050. E: Cow oocyte, 2 hr in culture. Note labeling concentrated over germinal vesicle (similar to observation in A) and unlabeled zona. Inset: Same section but focused on a possible "process" between the intensely labeled granulosa cells and the oocyte cytoplasm. Both ×1,400. F: Pig oocyte, 20 hr in culture. Slide treated with RNase A as described in text prior to exposure to photographic emulsion. Note the removal of the majority of label. ×1,960.

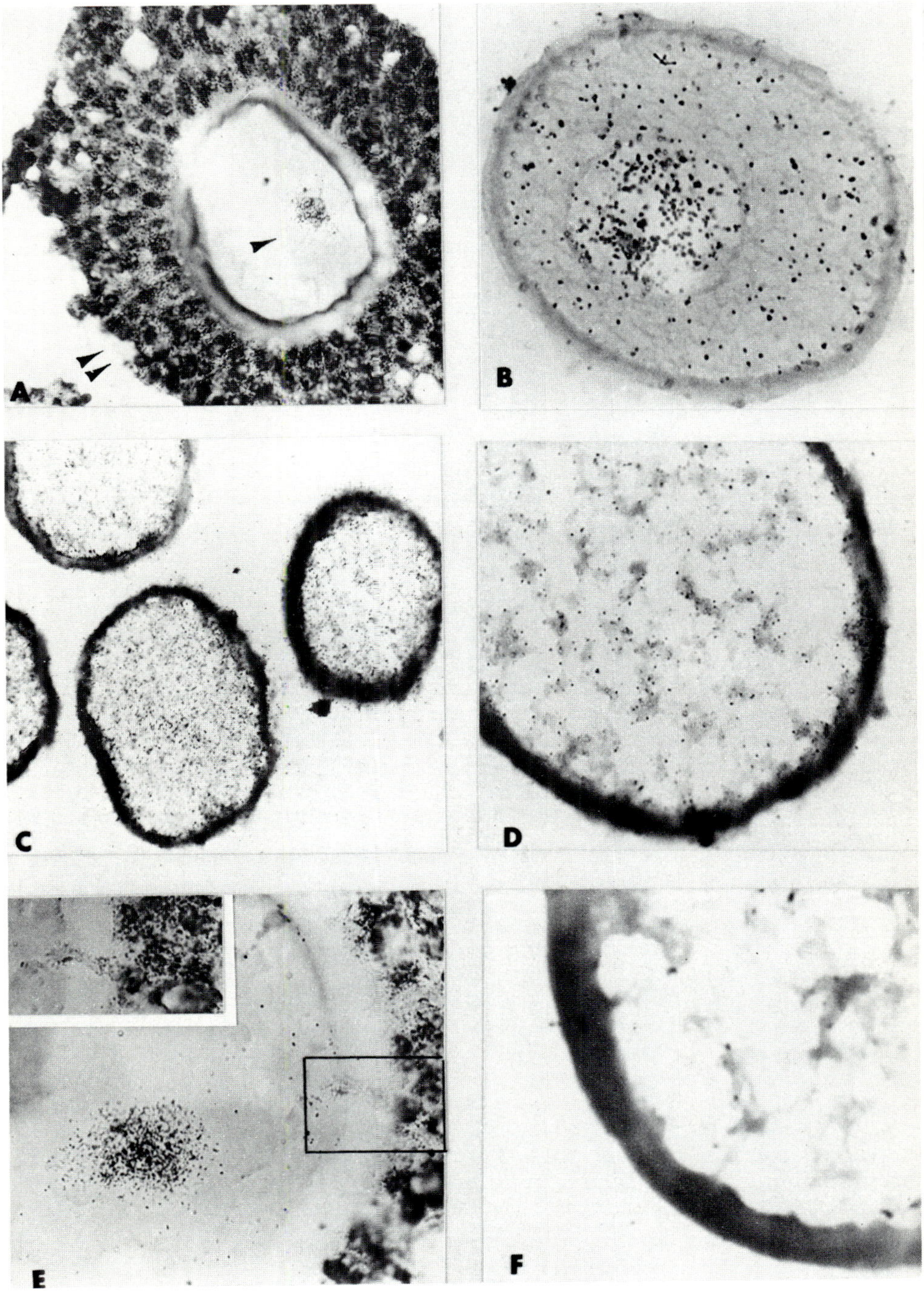
A
B
C
D
E
F

Slides were coated with Ilford K5 emulsion and exposed for 5 to 22 days. Control slides were subjected to RNase (Bovine Pancreatic, 100 μg to 1 mg/ml in 2 × SSC for 1 to 4 hr at 37°C) or DNase (100 μg/ml in 0.1 M Tris-HCl, pH 7.0 with 0.2 M $MgCl_2$ for 1 hr at 37°C) or 2 × SSC alone (1 to 4 hr at 37°C), rinsed in distilled water, treated with 5% TCA at 4°C for 20 min and rinsed in tap water.

RESULTS AND DISCUSSION

Of the three fixation procedures utilized, the paraffin-embedding technique was most informative in that it enabled light microscope analysis of localization of label with respect to time in culture. In short culture periods, the label is concentrated in the germinal vesicles of the oocytes (Fig. 1A). The label appears with time in the cytoplasm (Fig. 1B), and after germinal vesicle breakdown there is a generalized distribution of this RNase-sensitive label (Fig. 1F) throughout the ooplasm (Fig. 1C and D). The extent of this label appears to increase with the amount of time in the culture media but no attempt has yet been made to quantitate this accumulation of labeled product with respect to time.

Granulosa cells were very active in RNA synthesis during this period in which the oocyte is undergoing meiotic maturation (Fig. 1A). It was noted that almost all the granulosa cells were removed during the subsequent processing of the mouse and pig oocytes (Fig. 1B and C). However, bovine preparations usually included adhering granulosa cells and a possible interaction between the granulosa cells and the oocyte was noted (Fig. 1E). There appears to be a process from the granulosa cells traversing the zona into the oocyte cytoplasm which contained labeled material. This observation is particularly interesting in light of the report of Wasserman and Letourneau (1976) that mouse oocytes cultured in the presence of granulosa cells incorporated more radioactive uridine into RNA than did naked oocytes.

These preliminary observations support the notion that RNA synthesis occurs in the *in vitro* meiotic maturation of all the species investigated and further suggest a possible interaction between the granulosa cells and the oocyte in this process. We are continuing these investigations using inhibitors of specific polymerases and more precise localization techniques to determine if in fact this synthesis is obligatory, what species of RNA are synthesized, and which chromosomes are involved.

REFERENCES

Bloom, A. M., and Mukherjee, B. B. (1972): RNA Synthesis in Mature Mouse Oocytes. *Exptl. Cell Res.*, 74:577-582.

Moore, G. P. M., Lintern-Moore, S., Peters, H., and Faber, M. (1974): RNA Synthesis in the Mouse Oocyte. *J. Cell Biol.*, 60:416-422.

Moore, G. P. M., and Lintern-Moore, S. (1974): Correlation between Growth and RNA Synthesis in the Mouse Oocyte. *J. Reprod. Fert.*, 39:163-166.

Oakberg, E. F. (1968): Relationship between Stage of Follicular Development and RNA Synthesis in the Mouse Oocyte. *Mutation Res.*, 6:155-165.

Rodman, T. C., and Bachvarova, R. (1976): RNA Synthesis in Preovulatory Mouse Oocytes. *J. Cell Biol.*, 70:251-257.

Wasserman, P. M., and Letourneau (1976): RNA Synthesis in Fully-grown Mouse Oocytes. *Nature*, 261:73-74.

Ovarian Follicular Development and Function,
edited by A. R. Midgley and W. A. Sadler.
Raven Press, New York © 1979.

Characterization of Polyploidy in Ovarian Granulosa Cells

P. B. Coulson

Department of Zoology, Division of Life Sciences,
University of Tennessee, Knoxville, Tennessee 37916

Ovarian granulosa cells (GCs) are known to go through a sequence of developmental steps or differentiation stages culminating in matured ovarian granulosa cells in the graafian follicle capable of synthesizing large quantities of its steroid product, progesterone (Channing, 1970*a*, 1970*b*; Channing and Ledwitz-Rigby, 1975). The ploidy state of these differentiating cells has not, however, been described.

Abnormal systems exhibiting polyploidy include neoplastic tissues, ova exhibiting polyspermia, abnormal chromosome divisions in pig blastocysts (Moon et al., 1975), and plant polyploid wheat cells (Johnson, 1976), etc. Only a few examples of eukaryote cells exhibiting normally polyploid levels of DNA have been described. Trophoblast giant cells, apparently resulting from "endoreduplication" early in blastocyst development reach between 128N and 850N by the 11th day of pregnancy in the mouse (Barlow and Sherman, 1972; Herbert and Graham, 1974; Sherman, 1975; Sherman and Salomon, 1975). These should not be confused with syncitial trophoblastic cells, which are polynucleated. A second cell line, the megakaryocyte, has recently been described. It normally exhibits a dramatic increase in DNA per cell during early differentiation stages showing a mean ploidy in rats of 16 N and a maximum ploidy under hormonal stimulation with thrombopoietin of 32 N (Odell et al., 1970). The cell generation time using [^{3}H]-labeling studies averaged 9.3 hr with the "S" stage representing about 7.5 hr. Thus, the total differentiation for three to four cycles (going from 2 N to 32 N) requires 55 to 70 hr total for the megakaryocyte (O'dell, 1974; O'dell et al., 1976; O'dell et al., 1968).

Due to the dense granular nature of the immature ovarian follicle cells and the enlargement of the granulosa cell nucleus with increasing follicle size, the ploidy of these cells was questioned. Experiments were designed to evaluate the relationship between DNA/cell and the *in vivo* stages of granulosa cell differentiation.

MATERIALS AND METHODS

Porcine and bovine granulosa cells were selected for study because their large volumes of follicular fluids allowed easy sampling of individual follicles and yielded high granulosa cell counts per follicle. Ovaries were obtained fresh at the local slaughterhouse (East Tennessee Packing) and transported in media at 0 to 2°C. Ovaries were evaluated as to their approximate stage of the estrous cycle using criteria described previously (Akins and Morrissette, 1968, porcine; Ireland et al., 1976*a*, bovine). Follicles were aspirated with a finely drawn pipette, washed three times in media 199 plus Earles salts (GIBCO) plus 0.2 M Na citrate to inhibit clotting (which was most prevalent in large follicles) and with or without 1.0 μg/ml cytochalasin B to block further cytokinesis. Granulosa cells from 10 small follicles (1 to 3 mm diameter) were necessary from each ovary for analysis; however, medium-sized follicles (3 to 5 mm diameter) and large follicles (>5 mm diameter) were always analyzed individually.

Polyploidy

Normal autosomal cells have a double complement of DNA and are referred to as diploid (= 2N), gametes contain one set of chromosomes and are haploid (=1N), cells in division may contain DNA up to 4N while in S and G_2 postion of the cell cycle. Cells containing more than 4 times the haploid DNA value are referred to in this paper as polyploid.

Cell Counts

After washing, the granulosa cells were diluted to 0.5 ml with cold media and counted with a hemocytometer. Cover slips coated with 0.5% toluidine blue in ethanol were dried and used with the hemocytometer to allow better discrimination of nucleated granulosa cells from erythrocytes, platelets, or other contaminants.

DNA Assays

The Dische Assay for DNA using diphenylamine (Burton, 1956) was used and gave a sensitivity range of 10 to 100 µg DNA per determination. Granulosa cells were precipitated with 0.25 M perchloric acid and washed twice in 100% ethanol. The pellet was hydrolyzed in 1 M PCA for 3 min at 80°C. Samples were incubated with the Dische reagent 22 hr at 28°C and the OD was read at 600 mµ on a Beckman DB spectrophotometer.

For improved sensitivity a fluorometric assay for DNA was adopted (Kissane and Robbins, 1958; Setaro and Morely, 1976). The samples were precipitated with trichloroacetic acid, washed once with 0.1 N alcoholic potassium acetate and twice with 100% ethanol. The TCA pellet was resolubilized in 1 N NH_4OH. The fluorescent stain DABA (3',5'-diamino-benzoic acid; Aldrich) was added to the dried DNA sample at 6 mg/40 µl in distilled water. The reagent plus DNA was incubated for 30 min, 60°C, and diluted to 2 ml with 0.6 N PCA. The sample was then read for fluorescence on a Turner spectrofluorometer, model 430, at an activation wavelength of 405 nm and transmission wavelength of 520 nm. The reaction proved stable for at least 5 hr. Standards (purified Salmon DNA, Sigma) ranged from 0.01 to 5.0 µg per determination.

To test the reliability of the extraction procedure and the DNA analysis, two control tissues were prepared representing porcine liver (free cell preparation; Gospodarowicz D. and Gospodarowicz F., 1972), which has a reported DNA content of 6 to 8 pg/cell, and a nucleated erythrocyte preparation (*Bufo marinus*), which has a reported DNA content of 12 to 14 pg/cell (Sober, 1970). Table 1 shows that the extraction and assay values obtained were in good agreement with the values in the literature.

TABLE 1. *DNA extraction control*

Cell sample	pg DNA/cell ± SEM	Reported 2 N values (Sober, 1970)
Porcine (liver, free cell preparation	6.23 ± 1.42 (N = 8)	6-8 pg/cell
Bufo (nucleated RBC)	15.85 ± 2.11 (N = 4)	12-14 pg/cell

RESULTS

Bovine granulosa cells were characterized according to their stage of the estrous cycle and then analyzed for DNA content per cell. A consistent pattern emerged when multiple follicles and ovaries were pooled for small and large porcine follicles and analyzed via the Dische assay for average DNA per cell. The highest values occurred both in the smallest follicles during early days of the estrous cycle and in the immature ovaries. The lowest values appeared in the large follicles just prior to ovulation. The data with or without cytochalasin B gave identical results (data not shown).

The experiment was then repeated using a more sensitive fluorometric DNA assay (Kissane and Robbins, 1958). Porcine and bovine GCs from individual follicles were sampled (Tables 2 and 3). Cells from follicles taken from early estrous ovaries or from immature ovaries gave the highest average DNA (pg/cell) and also the highest maximum values per cell.

The wide range of values in each group may be due to the reported high portion of atretic follicles found on ovaries of single ovulating animals. However, the maximum levels of DNA/cell also gave a consistent pattern of decreasing DNA per granulosa cell as the follicle matured (Table 2).

Porcine granulosa cells were analyzed in a similar experiment in an attempt to show a lesser amount of variation per follicle (Table 3). The large variability seen in the porcine DNA per cell suggested a significant percent atresia and death in a large portion of follicles from multiple ovulating animals also.

When mean DNA/cell values were compared to the literature value for bovine diploid liver cells (= 7.0 pg/cell; Sober, 1970) the ploidy values ranged from 2.30 N to 5.45 N for bovine granulosa cells and 2.10 N to 11.24 N for porcine granulosa cells. These experiments have been repeated approximately seven times and, although absolute values for DNA/cell vary slightly between experiments, the pattern of relative ploidy changes is repeatable.

DISCUSSION

Polyploidy has been demonstrated to exist in porcine and bovine ovarian granulosa cells. It occurs with the highest values in granulosa cells from small follicles of immature ovaries, and decreased to 2 N

TABLE 2. *Polyploidy in bovine granulosa cells*

Diameter follicle	Stage ovary[a]	Mean pg DNA/cell ± SEM	Maximum pg DNA/cell	Ploidy relative to 2 N (= 7.0 pg/cell)[b]
Small ≦ 0.1 mm	Immature	16.76 ± 7.59 (N = 5)	45.769	4.78 N
Small ≦ 0.1 mm	Days 4-5 21-day cycle (MGS IV)	19.03 ± 5.66 (N = 5)	31.600	5.45 N
Medium 0.1 to 0.3 mm	Days 10-12 21-day cycle (MGS V)	12.93 ± 3.90 (N = 4)	20.250	3.70 N
Large > 0.3 mm	Days 17-18 21-day cycle (MGS VII)	6.88 ± 2.99 (N = 4)[b]	15.750	2.30 N

[a] MGS described elsewhere (Ireland et al., 1976*a*).
[b] Note: Bovine 2 N values range from 6 to 8 pg/cell for liver, kidney, etc. (Sober, 1970).

TABLE 3. *Polyploidy in porcine granulosa cells*

Diameter follicle	Stage ovary[b]	Mean pg DNA/cell ± SEM	Maximum pg DNA/cell	Mean ploidy relative to 2 N (= 7.0 pg/cell)[c]
Small, 1-3 mm	Immature	39.329 ± 2.13 (N = 6)[a]	140.00	11.24
Small 1-3 mm	2-4 days 18-day cycle	14.830 ± 8.03 (N = 6)	53.00	4.22
Medium 3-5 mm	10 days 18-day cycle	3.022 ± 1.31 (N = 6)	9.02	0.96
Large >5 mm	17-18 days 18-day cycle	7.378 ± 1.30 (N = 6)	11.12	2.10

[a]Experiment repeated six times in quadruplicate.
[b]Stage of estrous described for pig ovary (Akins and Morrissette, 1968).
[c]Note: Porcine 2 N value for diploid nucleus ranges from 6 to 8 pg/cell (Sober, 1970).

values in graafian follicles just prior to follicular rupture. Other known cell types that exhibit polyploidy are the megakaryocytes (differentiating toward mature cells capable of producing platelets), and the giant uterine trophoblast cells (differentiating toward mature cells capable of numerous functions including the production of pregnancy hormones such as chorionic gonadotropin and progesterone). Both the megakaryocytes and the trophoblast cells exhibit an increasing ploidy level during final differentiation (O'dell, 1976; Sherman, 1975).

Therefore, the newly described decreases in ploidy in granulosa cells represent a unique system. Efforts are currently under way to analyze the GC polyploidy by spectrophotometry to get absolute values of DNA/cell rather than average DNA extracted from 10^5 cells. Spectrophotometry will also allow the distinction between polyteny (multiple copies of selected segments of DNA and endoreduplication (even doubling of nuclear DNA).

Mechanisms controlling polyploidy are virtually unknown for the trophoblast cells except that they are apparently independent of continued maternal influences (Barlow and Sherman, 1972). The megakaryocyte stem cells appear to be influenced by the circulating platelet levels and by the relative levels of thrombopoietin (O'dell, 1976).

In granulosa cells, speculation concerning the control of ploidy can be focused at several different levels. Control may be exerted by the changing hormonal influences in the growing follicles. In bovine ovaries it has been shown that the progesterone/estrogen ratios in small follicles decrease as the follicle grows toward graafian size (Ireland et al., 1976*a* and 1976*b*). Many examples of hormonal control of cell division exist. These include the inhibitory effect of progesterone on uterine endometrium proliferation in the mouse (Martin et al., 1973), and the inhibition by progesterone of CHO cells in G_2 where they are prevented from entering mitosis or cytokinesis. These both result in an average ploidy equal to or less than 4 N (Remington, 1973; Koseki and Fujimoto, 1974).

Endoreduplication or inhibition of cytokinesis in granulosa cells may, on the other hand, be related to the hormonal-like influences controlling oocyte diakinesis in metaphase I (Schuetz, 1974) or the oocyte cytostatic factor causing inhibition in metaphase II of frog oocytes (Gallien, 1975).

The oocytes' ability to block differentiation of granulosa cells may be due to a mechanism involving an inhibitory substance. If ovoectomy will release the developing follicle and allow granulosa cells to luteinize (El-Fouly et al., 1970; Tsafriri and Channing, 1975) then a decrease in GC ploidy values may also follow ovoectomy.

CONCLUSIONS

High ploidy values were described for the first time in granulosa cells from bovine and porcine ovaries. This polyploidy appears highest in small immature follicles and decreases as the follicles mature. Average porcine ploidy values ranged from 2 N to 11 N in this study; however, maximum individual GC values may range higher. Ploidy changes in GCs may be an essential part of the differentiation process; however, control factors involved in the ploidy changes in ovarian granulosa cells are unknown and are currently under further investigation.

ACKNOWLEDGMENTS

The active support of Ms. Judy Stern, Jim Gillespie, David Rice, Ms. Jane Cohen, and Dr. Jim Ireland are gratefully acknowledged. The author would also like to thank Dr. Edward Howley for use of the Turner fluorometer. This work was supported in part by NSF Grant 37558 and NIH Biomedical Grant RR07088. Bovine and porcine tissue samples were obtained through the generous cooperation of the East Tennessee Packing Co., Knoxville, TN.

REFERENCES

Akins, E. L., and Morrissette, M. C. (1968): Gross ovarian changes during estrous cycle of swine. *Am. J. Vet. Res.*, 29:1953-1957.

Barlow, P. W., and Sherman, M. I. (1972): The biochemistry of differentiation of mouse trophoblast: Studies on polyploidy. *J. Embryol. Exp. Morphol.*, 27:447-465.

Burton, K. (1956): Assay for DNA. *Biochem. J.*, 62:315-323.

Channing, C. P. (1970*a*): Effect of stage of the estrous cycle and gonadotropin upon luteinization of porcine granulosa cells in culture. *Endocrinology*, 87:156-164.

Channing, C. P. (1970*b*): Influences of *in vivo* and *in vitro* hormonal environment upon luteinization of granulosa cells in culture. *Recent Prog. Horm. Res.*, 26:589-622.

Channing, C. P., and Ledwitz-Rigby, F. (1975): Methods for assessing hormone mediated differentiation of ovarian cells in culture and in short-term incubations. *Methods Enzymol.*, 39:183-230.

El-Fouly, M. A., Cook, B., Nekola, M., and Nalbandov, A. V. (1970): Role of the ovum in follicular luteinization. *Endocrinology,* 87:288-293.

Gallien, L. (1975): Sequential endocrine activity controlling oogenesis. *Am. Zool.,* 15:197-214.

Gospodarowicz, D., and Gospodarowicz, F. (1972): A technique for the isolation of bovine luteal cells and its application to metabolic studies of luteal cell *in vitro*. *Endocrinology,* 90:1427-1434.

Herbert, M. C., and Graham, C. F. (1974): Cell determination and biochemical differentiation of early mammalian embryo. In: *Current Topics in Developmental Biology,* edited by A. A. Moscona and H. A. Monroy. Vol. 8, pp. 151-171.

Ireland, J. J., Murphree, R. L., and Coulson, P. B. (1976*a*): Bovine estrous cycle identification via macroscopic changes in bovine corpora lutea. *(submitted)*

Ireland, J. J., Murphree, R. L., and Coulson, P. B. (1976*b*): Steroid content in ovarian follicular fluid compared to plasma in beef heifers. A comparative analysis of normal and erratic oestrous cycles. *(submitted)*

Johnson, B. L. (1976): Polyploid wheats and fraction 1 protein. *Science,* 192:1252.

Kissane, J. M., and Robbins, E. (1958): The fluorometric measurement of DNA in animal tissues with special reference to the central nervous system. *J. Biol. Chem.,* 233:184-188.

Koseki, Y., and Fujimoto, G. L. (1974): Progesterone effects contrasted with 17β-estradiol on DNA synthesis and Epith Nuc. Prot. in castrated rabbit uterus. *Biol. Reprod.,* 10:595-604.

Martin, L., Das, R. M., and Finn, C. A. (1973): The inhibition by progesterone of uterine epithelial proliferation in the mouse. *J. Endocrinology,* 57:549-554.

Moon, R. G., Rashad, M. N., and Mi, M. P. (1975): An example of polyploidy in pig blastocysts. *J. Reprod. Fertil.,* 45:147-149.

O'dell, T. T., Jackson, C. W., and Reiter, R. S. (1968): Generation cycle of rat megakaryocytes. *Exp. Cell Res.,* 53:321-328.

O'dell, T. T., Jackson, C. W., and Friday, T. J. (1970): Megakaryocytopoiesis in rats with special reference to polyploidy. *Blood,* 35:775-782.

O'dell, T. T. (1974): Megakaryocytopoiesis and its response to stimulation and suppression. In: *Platelets: Production Function Transfusion and Storage,* edited by M. G. Baldini and S. Ebbe, pp. 11-20. Grune & Stratton, New York.

O'dell, T. T., Murphy, J. R., and Jackson, C. W. (1976): Stimulation of megakaryocytopoiesis by acute thrombocytopenia in rats. *Blood,* 48:765-775.

Remington, J. A. (1973): Hormone-treated CHO cells exit cell cycle in G_2 phase. *Biochem. Biophys. Res. Commun.,* 50:140-146.

Schuetz, A. W. (1974): Role of hormones in oocyte maturation. *Biol. Reprod.*, 10:150-178.

Setaro, F., and Morley, C. G. D. (1976): A modified fluorometric method for the determination of microgram quantities of DNA from cell or tissue culture. *Anal. Biochem.*, 71:313-317.

Sherman, M. I. (1975): Long-term culture of cells derived from mouse blastocysts. *Differentiation*, 3(1-3):51-67.

Sherman, M. I., and Salomon, D. S. (1975): The relationship between the early mouse embryo and its environment. In: *33rd Symposium of Society of Developmental Biology*, Vol. 33, edited by C. L. Markert and J. Papaconstantinou, pp. 277-305. Academic Press, New York.

Sober, H. A. (1970): DNA content of eucaryotic cells. In: *Handbook of Biochemistry: Selected Data for Molecular Biology*, pp. 111-113. Chemical Rubber Company, Cleveland.

Tsafriri, A., and Channing, C. P. (1975): An inhibitory influence of granulosa cells and follicular fluid upon porcine oocyte meiosis *in vitro*. *Endocrinology*, 96:922-927.

Ovarian Follicular Development and Function,
edited by A. R. Midgley and W. A. Sadler.
Raven Press, New York © 1979.

A Hypothesis on the Factors that may Influence the Initiation of Meiosis

B. Lunenfeld, R. Weissenberg, and J. Blankstein

Institute of Endocrinology, Chaim Sheba Medical Center,
Tel Hashomer, and Department of Life Sciences,
Bar Ilan University, Ramat Gan, Israel

The factors that may trigger a germinal cell to stop its mitotic divisions and to enter into meiosis are still unknown. Recent observations on developmental features of male and female gonads from fetal life to puberty and their hormonal environment have clarified some of the events that may be involved in the initiation of the meiotic process.

If newborn mice are deprived of their endogenous gonadotropins, the first spermatogenic wave is not completed. Under such conditions mitotic division of spermatogonia continues to occur but the meiotic process is inhibited.

When only FSH was administered to the animals devoid of endogenous gonadotropins, some germinal cells completed meiosis and small numbers of spermatids were formed. However, HMG (FSH + LH) enabled initiation and completion of the first spermatogenic wave on a similar level as occurs under normal conditions (8).

Testosterone propionate administered from birth either to intact or to gonadotropin-deprived animals was incapable of causing initiation and completion of meiosis. Furthermore, administration of an antiandrogen (cyproterone acetate) to intact animals inhibited completion of meiosis (Table 1).

From experience by Steinberger et al. (10) it seems that culture conditions for germ cells that allow completion of the meiotic division or maturation of spermatids *in vitro* have not been defined. According to Steinberger and Steinberger (11), the limited spermatogenic activity and germ cell differentiation that occurred in organ culture, is dependent on close association of germ cell with the Sertoli cells. Addition of FSH to cultured neonatal testes induced morphological maturation of the Sertoli cells similar to that

TABLE 1. *The effect of hormonal deprivation and substitution in mice from birth until the age of 25 days*[a]

	Spermatogonia (type A)	Resting spermatocytes	Spermatids
Normal	3.1 ± 0.18	35	28.4 ± 3
aRG[b]	2.2 ± 0.15	23.8 ± 2	0
aRG + FSH	3.6 ± 0.24	41.8 ± 1.7	7.9 ± 0.8
aEG + HMG	3.8 ± 0.8	49.6 ± 3.9	37.8 ± 2.9
Testosterone propionate (60 μg/day)	2.3 ± 0.8	16.6 ± 2.3	0
aRG + testosterone propionate (60 μg/day)	2.3 ± 0.1	16 ± 2	0
Cyproterone acetate (6.5 μg/day)	2.3 ± 0.9	28.6 ± 2.6	0

[a]Each value represents the mean number of cells counted in 50 round tubular cross-sections from three to five mice.

[b]Antirat gonadotropin which neutralizes endogenous mouse gonadotropins and does not cross-react with gonadotropin of human origin.

observed *in vivo* (12-14). Furthermore, freshly isolated and cultured Sertoli cells specifically bound ^{125}I-FSH, responded to FSH (but not to LH) with increased levels of cAMP, and the secretion of androgen binding protein (ABP) was significantly stimulated (11).

The above observations support the concept that in the male meiosis can only be induced and completed if: (a) sufficient FSH is present for maturation and secretory activity of Sertoli cells (ABP); (b) sufficient LH activity is present to stimulate Leydig cells to produce testosterone. If either of these factors is missing, meiosis cannot be initiated.

We found (2) as did Giles et al. (6) that in the amniotic fluid of male fetuses testerone levels are significantly higher than in the presence of female fetuses (Fig. 1). However, FSH levels in amniotic fluid is significantly lower in the presence of male fetuses than in the presence of female fetuses (Fig. 2) (2,5). These findings are in accordance with the observations of Kaplan and Grumback (7) and Clements et al. (5) who reported that the mean LH and FSH contents of the pituitary was significantly lower in male than

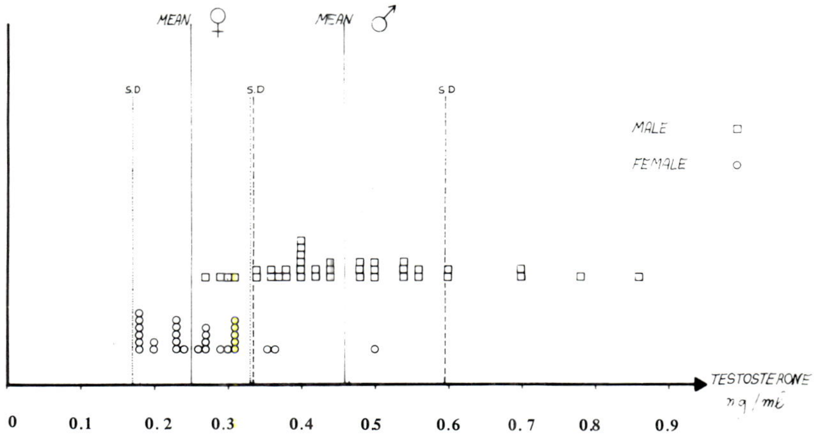

FIG. 1. Testosterone levels in amniotic fluid (16th to 19th week).

in female fetuses. However, high HCG levels were found in both male and female fetuses and were not sex dependent.

The finding of high testosterone levels in male fetuses is explained by the ability of the fetal Leydig cells, which appear at the 8th week, to secrete testosterone by the 8th to 10th week (9) and reach their highest peak activity by the 17th to 20th week.

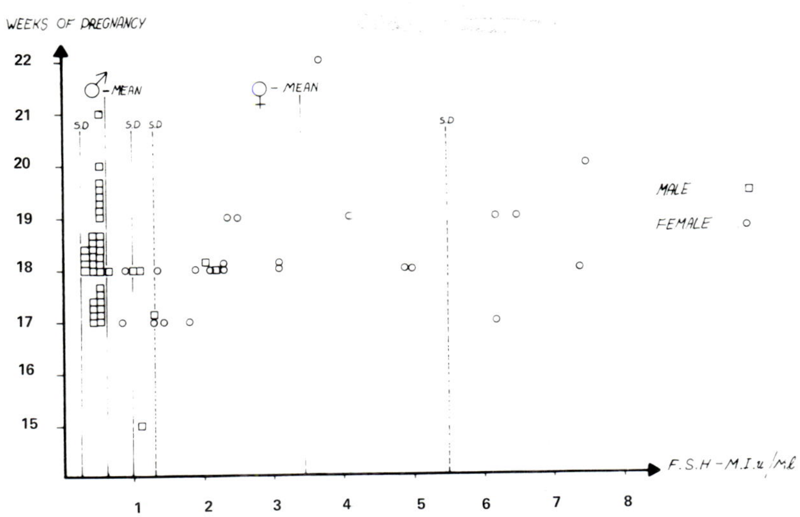

FIG. 2. FSH levels in amniotic fluid.

The onset and progression of sexual differentiation is thus temporally related to the sharp peak of serum and testicular testosterone. The changes in the synthetic and secretory activity of the fetal testis coincide with the high concentration of fetal serum HCG early in gestation (5). In the absence of sufficient FSH, however, the high testosterone level is incapable of inducing meiosis as we have explained above.

On the other hand, in female fetuses, in the absence of gonadal cells capable of responding to HCG with secretion of steroids, no testosterone is produced. The lack of testosterone brings about female sexual differentiation and is probably responsible for the significantly higher FSH production. Since gonadotropins are incapable of bringing about the onset of meiosis in organ cultures of fetal ovaries (4), and since oogonia do not enter meiosis in ovaries devoid of rete (3), it seems that to initiate meiosis in the female fetus the presence of both rete and FSH is essential. It could be speculated that FSH stimulates the rete to induce meiosis of fetal oogonia. The presence of increased amounts of FSH in pituitaries of female fetuses (7) and the fact that FSH levels are significantly higher in amniotic fluid (2,5) indicates increased secretory activity of FSH in the female fetus as opposed to in the male. This prompted us to propose the speculative hypothesis implicating FSH as one of the possible regulators of the initiation of meiosis.

ACKNOWLEDGMENTS

This study was supported in part by Ford Foundation Grant 67-470. B. Lunenfeld is an Established Investigator of the Chief Scientist's Bureau, Ministry of Health, Israel.

REFERENCES

1. Baker, T. G. (1976): The control of oogenesis in mammals. Workshop Meeting on Ovarian Follicular Development and Function, Santa Cruz.
2. Blankstein, J., Lunenfeld, B., and Serr, D. M. (1977): Antenatal sex determination by measuring FSH and testosterone levels in amniotic fluid during midgestation. *Submitted for publication*.
3. Byskov, A. G. S., and Lintern-Moore, S. (1973): Follicle formation in the immature mouse ovary: The role of the rete ovarii. *J. Anat.*, 116:207.

4. Challoner, S. (1975): Studies of oogenesis and follicular development in the golden hamster. 3. Initiation of follicular growth *in vitro*. *J. Anat.*, 119:157.
5. Clements, J. A., Reyes, F. I., Reyes, J. S. D., and Faiman, C. (1976): Studies on human sexual development. III. Fetal pituitary and serum and amniotic fluid concentrations of LH, CG, and FSH. *J. Clin. Endocrinol. Metab.*, 42:9.
6. Giles, H. R., Lox, C. D., Wayne-Heine, M., and Christian, C. D. (1974): Intrauterine foetal sex determination by radioimmunoassay of amniotic fluid testosterone. *Gynecol. Invest.*, 5:317.
7. Kaplan, S. L., and Grumbach, M. M. (1976): The antogenesis of human foetal hormones. III. Luteinizing hormone (LH) and follicle stimulating hormone (FSH). *Acta Endocrinol.*, 81:808.
8. Lunenfeld, B., and Weissenberg, R. (1972): The use of gonadotrophins in the induction of spermatogenesis. In: *Modern Trends in Endocrinology*, edited by E. T. G. Prunty and H. Gordiner-Hill. Butterworths, London.
9. Siiteri, P. K., and Wilson, J. D. (1974): Testosterone formation and metabolism during male secual differentiation in human embryo. *J. Clin. Endocrinol. Metab.*, 39:891.
10. Steinberger, E., Steinberger, A., and Sanborn, B. (1974): Endocrine control of spermatogenesis. In: *Physiology and Genetics of Reproduction*, edited by E. Countinho and F. Fuchs. Plenum Press, New York.
11. Steinberger, A., and Steinberger, E. (19): *Progress in Reproductive Biology*, edited by P. O. Hubinot and M. L'Hermite. Karger, Basel.
12. Steinberg, E., Steinberger, A., and Perloff, W. H. (1964*a*): Studies on growth in organ culture of testicular tissue from rats of various ages. *Anat. Rec.*, 148:581.
13. Steinberg, A., Steinberger, E., and Perloff, W. H. (1964*b*): Initiation of spermatogenesis *in vitro*. *Endocrinology*, 74:788.
14. Vilar, O., Steinberger, A., and Steinberger, E. (1967): An electron microscopic study of cultured rat testicular fragments. *Z. Zellorsch. Microsk. Anat.*, 78:221.

Ovarian Follicular Development and Function,
edited by A. R. Midgley and W. A. Sadler.
Raven Press, New York © 1979.

Discussion Summary: Session on Oogenesis

T. G. Baker

The discussion opened with a consideration of the life cycle of the female germ cell in terms of the initiation and cessation in sequence of the various processes of differentiation. Mitotic divisions commence toward the end of the period of germ cell migration, whereas meiosis appears to be associated with cell contact between oogonia and the rete ovarii. In most mammals, meiosis proceeds to the diplotene stage by about the time of birth: thereafter development of the oocyte is arrested until the follicle has completed its growth phase. It was pointed out that the resumption of meiosis beyond the germinal vesicle (diplotene) phase is related to the size of the follicle in which the oocyte resides: oocytes in preantral follicles do not resume meiosis, while those in small antral follicles are blocked at metaphase I. In response to precise endocrine signals (i.e., the LH surge) oocytes in mature graafian follicles reach metaphase II prior to ovulation, although their subsequent development is also blocked unless sperm penetration occurs. The control of these various switching mechanisms was discussed: some are clearly related to the levels of particular hormones but others may be induced by informational RNAs produced by the oocyte.

It was stressed that the follicle has to be considered as a composite structure (oocyte; cumulus; granulosa; theca; antral fluid) whose component parts may exert regulatory influences upon each other. Evidence was presented that some hormones are found at higher concentrations in follicular fluid than in blood.

Much of the discussion was concerned with potential inhibitors of oocyte maturation and as to whether such substances were produced by granulosa cells (especially those in small antral follicles). It was pointed out that "inhibition" could be due to a number of different factors without the need for a specific substance to be produced whereas an "inhibitor" was likely to be a specific substance (possibly a protein). "Inhibition" might be overcome by simply changing the permeability of the follicular basal lamina but a specific "inhibitor" would need to be eliminated or replaced by a stimulus.

There was some controversy as to whether the culture of isolated oocytes in chemically defined media is a satisfactory "model" for assessing the effects of potential inhibitors of oocyte maturation. The procedure was vigorously defended by some participants, while others were skeptical and suggested that the results obtained could arise from changes in the ability of the nutrient medium to support the cultures. It would seem that other systems will need to be used to assess the role of inhibitors of oocyte maturation.

A hypothesis was derived to explain the induction of meiosis during fetal life. The secretion of testosterone by the testis is enhanced by LH/HCG, and this in turn results in the suppression of fetal FSH. By contrast, when the levels of HCG (and consequently of testosterone) are low, FSH is produced in large quantities and this seemingly induces the initiation of meiosis in germ cells via an effect on the rete. This hypothesis would help to explain the various types of anomalous sexual development, including intersexuality.

The discussion concluded with a consideration of the binding sites available in the follicle for LH (and HCG). Using ^{125}I-labeled HCG it has been shown that the hormone is rapidly taken up by the theca and innermost (mural) layer of granulosa cells. Given sufficient time after treatment with the radioisotope, the label can be detected in the remaining layers of the membrana granulosa and the antral fluid. There was some doubt as to whether the labeling could be detected over the oocyte and cumulus cells, although this was found to be the case in one study using ^{3}H-HCG.

Subject Index